T0296577

A COURSE OF ANALYSIS

A COURSE OF ANALYSIS

BY

E. G. PHILLIPS, M.A., M.Sc.

Senior Lecturer in Mathematics in the University College of North Wales, Bangor; formerly Senior Scholar of Trinity College, Cambridge

CAMBRIDGE
AT THE UNIVERSITY PRESS
1962

CAMBRIDGE
UNIVERSITY PRESS

University Printing House, Cambridge CB2 8BS, United Kingdom

Cambridge University Press is part of the University of Cambridge.

It furthers the University's mission by disseminating knowledge in the pursuit of education, learning and research at the highest international levels of excellence.

www.cambridge.org
Information on this title: www.cambridge.org/9781316626139

First edition 1930
Second edition 1939
Reprinted 1944, 1946, 1948, 1950, 1956, 1960
Reprinted with additional examples 1962
First paperback edition 2016

A catalogue record for this publication is available from the British Library

ISBN 978-1-316-62613-9 Paperback

EXTRACT FROM THE PREFACE TO THE FIRST EDITION

The main purpose of the book is to give a logical connected account of the subject, by starting with the definition of "Number" and proceeding in what appears to me to be a natural sequence of steps.

Since modern Analysis requires great precision of statement, and demands from the student a very clear understanding of its fundamental principles, I have aimed at presenting the subject in such a way as to make every important concept clearly understood. The examples at the end of each chapter have been chosen mainly to illustrate the fundamental concepts, and most of them have been taken from a collection which I have made of questions suitable for examination and exercise work for my students.

It is extremely difficult to acknowledge indebtedness to all the different sources in a work of this kind, and I am fully aware that I have benefited largely from most of the existing text-books and standard works on the subject, as well as from the lectures of Prof. J. E. Littlewood, F.R.S., and Mr S. Pollard of Trinity College, Cambridge. Since it is so often difficult to discover the rightful originator of particular theorems or modes of demonstration, no systematic attempt has been made to cite authorities; but where I have definitely borrowed from any recent work which appears to possess originality, acknowledgement has been made either in footnotes or in the text itself.

Prof. W. E. H. Berwick very kindly read through part of the first draft of the manuscript, and made some helpful suggestions for which I am grateful. My sincere thanks are due to Prof. G. N. Watson, F.R.S., for valuable criticisms and suggestions which have helped very greatly to improve the form and presentation of the book; and to my colleague Mr W. M. Shepherd for his kindness in drawing all the diagrams and for his help with the proof-reading. I desire also to express my gratitude to the officials of the University Press both for their unfailing courtesy and for the excellence of their work.

E. G. PHILLIPS

UNIVERSITY COLLEGE OF NORTH WALES
BANGOR

January 1930

PREFACE TO THE EIGHTH IMPRESSION OF THE SECOND EDITION

The second edition of this book, based on a course of lectures on Analysis first prepared for the Honours students in the University College of North Wales, is largely a reproduction of the first edition. My experience of using the book with my own students has convinced me that it is already difficult enough for those who are using it as their first introduction to rigorous Analysis and so I have made no substantial changes in the subject-matter.

I have taken advantage of the opportunity presented by this reprint to correct one or two errors which had been overlooked previously.

I have included a set of Miscellaneous Examples at the end of the book. I wish to thank my colleague, Mr S. Moses, for his kindness in collecting these for me; and for providing Answers to the Examples.

E. G. P.

August 1962

CONTENTS

CHAPTER 1

NUMBER

1·1. Introduction.

The foundation upon which the whole structure of the subject of Mathematical Analysis rests is the *theory of real numbers.*

Accordingly an obvious starting-point for our study of this subject is the series of "natural numbers"

$$1, 2, 3, \ldots, n, \ldots.$$

These numbers are so familiar that it seems quite reasonable to assume that the concept of "number" is one of our primitive, or even intuitive notions, and that consequently it does not require *definition.* In fact most of the existing text-books do begin by accepting the natural numbers as "known," and therefore as not requiring definition. The reader who is satisfied with this point of view will be saved a good deal of preliminary difficulty by omitting much of the work with which the present chapter is concerned.

The problem with which we are faced is to decide what we are to accept as "*given.*" It would certainly be the *easiest* way out of the difficulty to accept the natural numbers as "given"; and without going outside what is usually understood to be "mathematics," this is the only reasonable starting-point which can be made.

There are two main reasons why we shall not accept the concept of number as a primitive concept which does not require definition. One reason is that by doing so we might be in danger of leading the reader to think that the foundations of the subject cannot be based upon anything more fundamental than "number" as a primitive undefinable concept. The second reason is that our preliminary investigation of a logical definition of number is the natural introduction to one of the best methods of defining an irrational number.

It is often stated that the foundations of Real Variable Theory have not yet reached an entirely satisfactory position, and to a certain extent this may be true; but no reader who wishes to make a systematic study of modern Analysis ought to remain in entire ignorance of the field of study which has been opened up by Frege, Bertrand Russell and others* in reducing to logic those arithmetical notions which had previously been shewn by Peano to be sufficient for mathematics. To consider this question fully would make this chapter unduly lengthy; accordingly it has been thought sufficient to indicate the main essential ideas, and leave the reader to consult other treatises for a more detailed explanation†.

No science is entirely self-contained; each borrows the strength of its ultimate foundations from something outside itself, such as experience, or logic or metaphysics. This is the case with Pure Mathematics, and the definition of number which is given in this chapter is a *logical* definition; hence some knowledge of that field of study which has come to be known as "mathematical philosophy" is unavoidable.

Although the natural numbers seem to represent what is easiest and most familiar in mathematics, very few people would be prepared with a *definition* of what is meant by "number" or "1" or "3," and a much greater difficulty arises when we consider how to define "0."

It may be remarked that 0 is a recent addition to the series of natural numbers, and the Greeks and Romans had no such digit.

All the essential ideas involved in the logical definition of number will be explained as simply and as untechnically as possible; but however carefully we attempt to avoid difficult and unfamiliar phraseology, this chapter will unavoidably appear difficult, and perhaps artificial to the beginner. As we have already remarked, however, we must deal with these unfamiliar ideas, unless we are content to shirk all the difficulties by the unjustifiable assumption that we already understand what is meant by "number."

Historically, the progress of mathematics has been *constructive* in the direction of rapidly increasing complexity. In the case of number, with which we are now concerned, the natural numbers

$$1, 2, 3, \ldots, n, \ldots$$

* Frege's *Grundlagen der Arithmetik* (1884) gave the first correct logical definition of number, but the book attracted little attention, and its contents remained practically unknown until they were rediscovered by Russell in 1901.

† The reader may profitably consult B. Russell's *Introduction to Mathematical Philosophy*, and J. E. Littlewood's *Elements of the Theory of Real Functions* (1926).

were the first to be considered, and their earliest use was in an ordinal sense, when they were employed for the purpose of *counting*. Our familiarity with the use of the natural numbers for counting is one of the chief obstacles to be removed before we are able to give a satisfactory definition of a cardinal number. The commonest every-day use of numbers (for the purpose of counting) is just the aspect of number which is least helpful for this purpose, and the definition of a *cardinal number* must not involve the use of counting. The importance of the distinction between cardinal and ordinal numbers will be emphasised as we proceed; and the reader will see later that counting, although so familiar, is logically a very complex operation. All that need be said at the moment is that counting employs the natural numbers in an ordinal sense, and the *logical* definition of "order" and "ordinal number" is by no means easy.

The impossibility of defining a cardinal number by the process of enumeration is very obvious when viewed psychologically. "Counting," it is said, "consists of successive acts of attention; the result of such a succession is a number." In other words, "the number *seven* is the result of seven acts of attention." This makes the vicious circle obvious.

The introduction of fractions into arithmetic was the next step, these arising naturally in connection with the problem of *measurement*; and their introduction was comparatively easy. On the contrary the negative numbers caused a great deal of trouble. For some time negative numbers were called absurd and fictitious, and the fact that the product of $-a$ and $-b$ could give a positive number ab was for a long time a difficulty to many minds*. The subsequent introduction of irrational numbers, such as $\sqrt{2}$, $\sqrt[3]{6}$, π and e, did not excite much comment. In actual calculations approximate rational values were used, and it seemed quite natural to subject them to the same laws as rational numbers. Irrational numbers arose first in connection with geometry, with the discovery by the early Greek geometers that there is no fraction of which the square is 2, a result which naturally emerges out of the problem of determining the length of the diagonal of a

* In the latter half of the eighteenth century, Maseres (1731–1824) and Frend (1757–1841) published works on Algebra and Trigonometry in which the use of negative numbers was disallowed, although Descartes had used them freely more than a hundred years before.

unit square. With the invention of algebra the same question arose in the solution of equations, but here it took a wider form involving also complex numbers, which will be discussed later.

Although irrational numbers were discovered as early as the time of Pythagoras, no real advance towards constructing a *rigorous* theory of irrational numbers was made until the time of Weierstrass (1815–97) and Dedekind (1831–1916). It may be remarked that if we agree to accept the natural numbers as fundamental, and thereby avoid the necessity of considering any mathematical philosophy, even then some rigorous theory of irrational numbers is necessary before Analysis can be founded on a satisfactory basis. The definition of an irrational number which will subsequently be given is due to B. Russell, and it is a slight modification of Dedekind's method.

The other method of pursuing the study of mathematics is the reverse of the historical order of progress. Instead of pursuing the constructive process towards increasing complexity, we proceed, by analysing, to greater abstractness and logical simplicity. Instead of considering what can be defined and deduced from our initial assumptions, we examine whether more general ideas and principles can be found in terms of which our original starting-point can be defined or deduced. This second method is what characterises the study which has come to be known as mathematical philosophy. Thus, if our foundations are to be based farther back than on the mere postulation of the existence of the natural numbers, it can only be done by considering some of the questions with which mathematical philosophy is concerned.

1·2. Fundamental notions.

We now state what concepts must be taken as fundamental in order to give a definition of number. The following remarks may perhaps clear the ideas of the beginner, and help him to appreciate the definition which will subsequently be given. A trio of men is an instance of the number 3, and the number 3 is an instance of number, but the trio itself is not an instance of number. The number 3 is not identical with any collection of terms having that number; it is something which all trios have in common, and which distinguishes them from other collections. It brings us a step nearer to the correct definition when we realise that number is to be regarded as something which characterises certain *collections*. The reference to a "collection" of objects introduces the first important concept, that of an aggregate.

An *aggregate* (or *collection*) of objects which is conceived of as containing more or fewer objects is a concept which will be taken as primitive, and no attempt is made to *define* it. A great deal can be known about an aggregate without our being able to enumerate its members. The elements composing an aggregate need not possess any parity as regards size or any other special quality. For example, an aggregate may be "all the living creatures in the city of London," "all the trees in a certain garden," or any other collection of entities of entirely diverse characteristics.

An aggregate, considered quite apart from the order of its members, is termed a CLASS or SET.

A great part of mathematical philosophy is concerned with RELATIONS, and although only a few important relations enter into the discussions in this book, a few remarks may be helpful to the beginner.

Amongst the most important kinds of relations is the class of "one-many," "many-one," and "one-one" relations*. If A and B denote two sets of entities, the relation between these two sets is "one-many," if more than one member of the set B bears the given relation to each member of the set A.

The following examples will make the ideas clearer. In countries where monogamy is practised the relation of *husband to wife* is one-one, in polygamous countries the relation is one-many, and in Tibet, where polyandry is practised, the relation is many-one. The relation of *father to son* is one-many, that of *son to father* is many-one, but that of *eldest son to father* is one-one.

The *domain* of a relation consists of all those terms which have the relation to something or other, and the *converse domain* consists of all those terms to which something or other has the relation.

The *field* of a relation consists of the domain and the converse domain together.

* The use of the word "one" in the description of these relations is justifiable, for a meaning can be assigned to the above relations which does not require any concept of the cardinal number "1." For an interesting remark on this point, see Littlewood's *Elements of the Theory of Real Functions* (1926), p. 2.

A further notion which is required is that of CORRESPONDENCE: this is the notion which underlies the process of tallying. The elements of one aggregate may be made to stand in some logical relation with those of another so that a definite element of one aggregate is regarded as correspondent to a definite element of another aggregate.

Two aggregates which are such that to each element of the first there corresponds one and only one element of the second, and to each element of the second there corresponds only one element of the first, are said to be in ONE-ONE CORRESPONDENCE.

1·21. Definition of number.

Two aggregates are said to be SIMILAR when there is a one-one correspondence which correlates their elements. Suppose now that all couples are in one bundle, all trios in another, and so on. In this way we obtain various bundles of collections. Each bundle is a set whose members are classes; thus each is a *set of classes.* To decide when two collections are to belong to the same bundle we use the notion of similarity defined above. Given any aggregate, we can define the bundle to which it must belong as being the set of all those aggregates which are similar to it.

We therefore give the following definition:

The NUMBER *of a class is the set of all those classes that are similar to it.*

According to this definition the set of *all* couples is the number 2; the set of *all* trios is the number 3, and so on.

Numbers in general have been defined as bundles into which similarity collects classes. A number is a set of classes such that any two of the classes are similar to each other, and none outside the set is similar to any inside the set.

On the same lines the number 0 can be defined. The number 0 is the number of terms in a class which has no members, and this class is called the *null class.* By the general definition of number, the number of terms in the null class is the set of all classes which are similar to the null class, and this is easily seen to be the set whose *only* member is the null class. The purely logical definition of the number 0 may therefore be given as follows:

The number 0 is the set whose only member is the null class.

1·3. Relations.

The important type of relation known as a one-one relation has already been mentioned. The concept of one-one correspondence is of fundamental importance, for upon it depends the definition of number given above. There are, however, many other kinds of relations, and one very important type, "serial relations," will be needed when we define "order."

The following examples of "one-one" relations may assist the reader to assimilate some of the essential ideas involved.

Take the first ten integers (excluding 0),

$$1, 2, 3, \ldots, 8, 9, 10 \quad \ldots\ldots\ldots\ldots\ldots\ldots(1).$$

This set can clearly be correlated with the set of integers

$$2, 3, 4, \ldots, 9, 10, 11 \quad \ldots\ldots\ldots\ldots\ldots\ldots(2);$$

and the relation which correlates these two classes can be described as *the relation of n to $n+1$*. The relation is clearly one-one; also the domain and converse domain overlap, for all the members of class (1), save the first, are repeated in class (2), and class (2) contains only one new member, 11.

If, instead of the relation of n to $n+1$, we take *the relation of a number to its square*, we obtain the set of integers

$$1, 4, 9, \ldots, 64, 81, 100 \quad \ldots\ldots\ldots\ldots\ldots\ldots(3),$$

which also bears a one-one relation to the set (1).

1·31. Serial relations.

An idea which obviously calls for attention is that of an aggregate whose members are arranged in a certain order. In defining number we considered aggregates quite apart from any question of order among their members. The numbers which we were able to define in this way are called CARDINAL NUMBERS.

An ordered aggregate, which appears to be a greater complexity than a class, is resolved in quite a simple way; but before we can deal with ordered aggregates, serial relations must be understood. It will be seen that an ordered aggregate leads naturally to the definition of an ordinal number.

In seeking a definition of order, the first thing to realise is that no set of terms has just *one* order to the exclusion of others; a set of terms has all the orders of which it is capable. It is true that the natural numbers (due to their employment for the purpose of counting) occur to us most readily in order of magnitude, but they are capable of an unlimited number of other arrangements. The definition of order is not therefore to be sought in the nature of the set of terms to be ordered, since the same set of terms has many different orders. The order lies, not in the *set* of terms, but in a relation among the members of the set, in respect of which some appear as earlier and some as later.

The essential characteristics of a relation which is to give rise to order may be discovered by considering that in respect of such a relation we must be able to say, of any two terms in the set which is to be ordered, that one "precedes" and the other "follows." We require the ordering relation to have the following three properties:

(1) ASYMMETRY. If x precedes y, y must not also precede x. We say that *any given relation is asymmetrical when, if it holds from x to y, then it does not hold from y to x.*

Relations which do not give rise to series often do not have the property of asymmetry. The relation "is the cousin of" is an example of this, for if x is the cousin of y, y is also the cousin of x.

(2) TRANSITIVENESS. If x precedes y and y precedes z, then x must also precede z. *A relation is transitive when, if it holds from x to y and from y to z, it also holds from x to z.*

The relation considered above is not transitive, for if x is the cousin of y, and y of z, x may not be the cousin of z, for x and z may be the same person. The relation "sameness of height" is transitive, but not asymmetrical. The relation "father" is asymmetrical but not transitive. The reader is advised to construct examples for himself.

(3) CONNECTIVITY. Given any two terms of the class which is to be ordered, there must be one which precedes and the other which follows.

A relation is connected, if given any two members x and y of its field, then either the relation holds from x to y or from y to x.

The relation "ancestor" has the first two properties, but not the third*. Its failure to possess the third property makes it an insufficient relation to arrange the human race in a series.

DEFINITION. *A relation is serial if it is asymmetrical, transitive and connected.*

A *series* is the same thing as a serial relation.

* We imply of course that the field of the relation considered is the human race.

1·32. Order.

The three properties required in order that any given relation may be serial have now been examined. It is not possible, without greatly increasing the number of new ideas and technical terms, to go very deeply into the question of *order*; but to be able to understand the fundamental relation "less than" the following definitions are required:

(1) A property is said to be *hereditary* in the natural number series, if whenever it belongs to a number n it also belongs to $n+1$, the successor of n.

(2) The *successor* of the number of terms in the class A is the number of terms in the class consisting of A together with x, where x is any term not belonging to A.

(3) A number m is said to be *less than* another number n when n possesses every hereditary property possessed by the successor of m.

It is not difficult to prove that the relation "less than" so defined is serial, and that it has the finite cardinal numbers for its field. By means of the relation "less than" the cardinal numbers acquire an *order*, and this order is the so-called "natural" order, or order of magnitude*.

The generation of series by means of relations more or less resembling that of n to $n+1$ is very common. The series of kings of England, for example, is generated by relations of each to his successor. This is probably the easiest way, where it is applicable, of conceiving the generation of a series.

1·33. Cardinal and ordinal numbers.

The distinction between a cardinal and an ordinal number is rendered difficult by the fact that each finite positive integer is made to serve two distinct purposes; it may be used to *count*, when it is acting in the ordinal sense, and it may be used to *number*, when it is acting as a cardinal number. Symbolically there is no distinction whatever between a cardinal and an ordinal number, but logically there is a fundamental difference between them. As we have already seen, the question of order is irrelevant to the definition of a cardinal number; and it is partly in order to secure this important distinction between cardinal and ordinal numbers that the discussion of "serial relations" and "order" has been given.

* The questions considered in this section are only briefly mentioned. For fuller reference the reader is referred to Russell's book, *loc. cit.* Chs. I–VI.

Briefly it may be said that cardinal numbers are obtained from the idea of *equivalence*, ordinal numbers from the idea of *likeness*. The precise significance to be attached to these two terms can be seen in the following definitions:

(1) *Two aggregates A and B are* EQUIVALENT (SIMILAR) *if there is a one-one correspondence between their members.*

(2) *Two series S and T are* LIKE (ORDINALLY SIMILAR) *if there is a one-one correspondence between their members, preserving the order**.

So long as our attention is confined to *finite* aggregates, rearrangement of the members of the aggregate cannot alter the ordinal number; in whatever way we count the members of a finite aggregate we always end up with the same number. Thus all the possible series which arrange the members of a finite class are *like* series.

Two finite series which are like, are said to have the same *ordinal number*, and by analogy with the definition of a cardinal number we can give the following definition:

The ORDINAL NUMBER *of a series is the set of all those ordered aggregates that are like (ordinally similar to) it.*

The important distinction between cardinal and ordinal numbers is best seen by considering *infinite aggregates.* A simple example will suffice.

Consider the two series

$$\text{(A)}\quad 1, 2, 3, \ldots, n, \ldots, 0,$$

$$\text{(B)}\quad 1, 2, 3, \ldots, n, \ldots.$$

The series (A) and (B) are not ordinally similar, for there is not a one-one correspondence which correlates each term of (A) with each term of (B) and which *preserves the order*, for there is no term in (B) with which the term 0 of (A) tallies. The ordinal number of the series (B) is usually denoted by ω, and the ordinal number of the series (A) is then denoted by $\omega+1$.

If these two series are considered as classes, no account being taken of order, there is a one-one correspondence between them.

$$\text{(A}'\text{)}\quad 0, 1, 2, 3, \ldots, n, \ldots,$$

$$\text{(B)}\quad 1, 2, 3, 4, \ldots, n+1, \ldots.$$

When (A) has been rewritten as (A′), the classes (A′) and (B) are correlated by the one-one relation of n to $n+1$. Thus both (A′) and (B) are classes which belong to the same set of classes, and the cardinal number which is characteristic of this particular set is denoted by $\aleph_0$ (Aleph zero).

* That is to say, a one-one correspondence in which, if x', y' correspond to any x, y, then x' precedes y' if x precedes y.

The question of *order* has not been very fully discussed, but the reader should realise that we have indicated the lines on which it is possible to give a purely logical definition of the relation "less than." The phrase "less than" is not used in its primitive sense as referring to magnitude. By virtue of the definitions in § 1·32, "less than" is a well-defined type of relation; it satisfies the three conditions which make it a serial relation (§ 1·31), so that when it is applied to the finite cardinal numbers it arranges them in a series. Thus whenever we have to deal with the ordered aggregate

$$0, 1, 2, \ldots, n, \ldots$$

we are, in reality, dealing only with the relation "less than." This relation, which is of the serial kind, is really what we mean by "the natural numbers arranged in ascending order."

1·4. Operations on classes.

The reader has doubtless realised that *classes* and *numbers* are entities of different kinds, although it has been shewn that the definition of number depends on the concept of a class. The signs $+$, $-$, $\times$ of elementary algebra are conveniently employed to denote those operations on classes which are the obvious analogues of the operations of addition, subtraction and multiplication in elementary algebra. If A and B denote two given classes, the meaning of $A+B$ depends upon the definition of the addition of two classes which follows.

When some or all of the formal laws of algebra are satisfied with new meanings assigned to the symbols $+$, $-$, $\times$, it is much more convenient to use these symbols than to try to invent new ones to replace them. The three concepts of class, cardinal number and ordinal number, involve three distinct algebras.

If A and B denote two given classes, a and b two given cardinals and α and β two given ordinals, the meaning of the sign $+$ is different in each of the expressions

$$A+B, \quad a+b, \quad \alpha+\beta.$$

Each new kind of number which is introduced involves a new algebra, for which it is necessary to shew, *from the definitions*, whether some or all of the fundamental laws of algebra hold.

We now give the definitions of addition, subtraction and multiplication of classes.

(1) *Addition of classes.* The sum class* ΣA of a set of classes A is the class consisting of every term which belongs to some class A of the set.

The sum class of A and B is denoted by $A+B$, that of A, B and C by $A+B+C$, and so on. The sum class is independent of the concept of order, and

$$A+B=B+A.$$

(2) *Subtraction of classes.* If the class B is contained in the class A, $A-B$ denotes the class of those terms which belong to A but not to B.

(3) *Product of classes.* The product ΠA of a set of classes is the class consisting of all those terms which belong to every class A of the set. For two given classes A and B, the product is written AB.

It can be shewn that the formal laws of algebra, *so far as they concern addition and multiplication*, are fulfilled, so that the notation used for addition and multiplication of classes is convenient by its analogy with ordinary algebra. Thus

$$(A+B)\,C=AC+BC=CB+CA.$$

1·41. Operations on cardinal numbers.

If a and b are two given cardinal numbers, $a+b$ is defined to be the cardinal number of the class $A+B$, where A and B are exclusive classes† having cardinal numbers a and b respectively.

The product ab is defined to be the cardinal number of the class of ordered couples (x, y), such that x is a member of the class A and y is a member of the class B.

It remains to define exponentiation. In order to render the definition of exponentiation suitable for extension to infinite classes, the best way to define it is as follows. Let P be the class of many-one correlations between the classes A and B; that is, correlations in which every member of B is partnered by some member of A, repetition of members of A being allowed. The cardinal number p of the class P (which is independent of the choice of the classes A

* Although we are concerned only with *finite* classes, and with sets of classes for which the number of classes in any given set is finite, the definition of the sum class does not *depend* upon the number of classes A being finite.

† Two classes are *exclusive* when they have no common member.

and B) depends only on a and b. The cardinal number p thus defined is a^b. We assume that $b \neq 0$ in the preceding definition. It is usual to define a^0 to be the cardinal number 1, and 0^0 is not defined.

The genesis of the above definition in common sense is "a kinds of things in b holes."

In the above definitions the classes A and B are required to be exclusive. Given any two classes A and B which are not necessarily exclusive, it is easy to construct two classes A' and B' which are exclusive, and such that A and A', B and B' are similar; for A' is the class of ordered couples (x, α), and B is the class of ordered couples (y, β), where α and β range respectively through the classes A and B*.

1·5. Real numbers.

Cardinal and ordinal numbers have both been defined, but neither of these is capable as it stands of the extensions of the concept of number to negative, fractional and irrational numbers. Logical definitions of these extensions will now be given.

At the outset it must be emphasised that if m denotes any finite cardinal number as defined in § 1·21, m is not the same as $+m$, nor is m the same as $\frac{m}{1}$. Further, it is not *necessary* that irrational numbers, such as the square root of 2, should find their place among rational fractions as being greater than some and less than others. In fact, distinctions of this kind must be made in order that precise definitions may be given.

(i) *Positive and negative integers.*

Both $+1$ and -1 are relations, and they are converses of each other. The obvious and sufficient definition of $+1$ is that it is the relation of $n+1$ to n, and -1 is the relation of n to $n+1$. If m be any cardinal number, $+m$ is the relation of $m+n$ to n (for any n), and $-m$ is the relation of n to $m+n$. The point to be emphasised is that $+m$ cannot be *identified* with m, which is not a relation, but a set of classes.

(ii) *Fractions.*

The fraction m/n is defined to be that relation which holds between the two cardinal numbers p and q when $pn = qm$. The

* For further details, the reader should refer to the books of Russell and Littlewood, *loc. cit.*

definition enables us to prove that m/n is a one-one relation provided neither m nor n is zero. n/m is the converse relation to m/n. Accordingly $m/1$ is the relation between the two cardinal numbers p and q which consists in the fact that $p = mq$. Hence again, since $m/1$ is a relation, it cannot be the same thing as m, which is a set of classes.

It can be seen that $0/n$ is always the same relation whatever number n may be: it is the relation of 0 to any other cardinal number. This is the zero of rational numbers, and it is not of course identical with the cardinal number 0. Conversely $m/0$ is always the same relation whatever m, but there is no *finite* cardinal number which corresponds to $m/0$: this may be called the "infinity of rationals." It is to be noted that this "infinity" does not require for its definition the use of any infinite classes or infinite integers.

Zero and infinity are the only relations among ratios which are not one-one: zero is one-many, and infinity is many-one.

Greater and *less* among fractions are easily defined: m/n is *less than* p/q if mq is less than pn. The relation *less than* so defined is serial, so that the fractions form a series in order of magnitude.

Positive and negative ratios can be defined in a similar way to that used for defining positive and negative integers. The addition of the two ratios m/n and p/q is defined to be the ratio $(mq + pn)/nq$, the product of the same two ratios is defined to be mp/nq. If x be any number, integral or fractional, the use of a ratio as an index is defined by the postulate

$$x^{p/q} . x^{m/n} = x^{p/q+m/n}.$$

The symbol $x^{p/q}$ is to be interpreted subject to this postulate provided that such interpretation is possible.

Positive and negative ratios can be defined just in the same way as positive and negative integers. Thus $+p/q$ is the relation of $m/n + p/q$ to m/n, where m/n is any ratio; $-p/q$ is of course the converse relation of $+p/q$.

The reader should note that the above method is not the only one which could be adopted for dealing with ratios, but it has the advantage of analogy with the method adopted for integers. It has been remarked by Hobson* that the above method is open to the objection that it is of an arbitrary character, and it is not easy to see why the particular laws of operation have

* *Theory of Functions of a Real Variable*, I (1921), §§ 12, 13.

been postulated, except as suggested by the traditional non-arithmetical concept of a fraction. To remedy this, Hobson gives another method of dealing with ratios, and the reader may refer, for an explanation of this, to the reference given in the footnote.

(iii) *Irrational numbers.*

The extensions of the domain of number by the introduction of fractional and negative numbers were suggested by the desirability of constructing a domain so complete that the operations of division and subtraction might always be possible. In the aggregate of rational numbers, the operations of addition, subtraction, multiplication and division are always possible operations, but it is easily shewn that the operation of determining a fractional power of a rational number is not, in general, a possible one.

A simple proof can be given* that *no positive integer m other than a square number has a square root within the aggregate of rational numbers.*

For, if possible, let m be the square of a rational fraction p/q in its lowest terms; thus $p^2 - mq^2 = 0$. There always exists a positive integer n such that $n^2 < m < (n+1)^2$; we then get $nq < p < (n+1)q$. Consider the identity

$$(mq - np)^2 - m(p - nq)^2 \equiv (n^2 - m)(p^2 - mq^2) \equiv 0:$$

from this identity it follows that m is the square of a rational number $(mq - np)/(p - nq)$ whose denominator is less than q. This contradicts the hypothesis that m is the square of the fraction p/q which is in its lowest terms. Hence there can be no rational number whose square is m.

As an illustration, consider the case in which $m = 2$. An ascending sequence of fractions, all of which have their squares less than 2, can be found; and, by taking enough terms of the sequence, a number will be reached whose square differs from 2 by less than any assigned amount.

Such a sequence, for example, is

$$\ldots,\ 1,\ 1{\cdot}1,\ 1{\cdot}2,\ 1{\cdot}4,\ 1{\cdot}41,\ 1{\cdot}411,\ 1{\cdot}412,\ 1{\cdot}413,\ \ldots \quad \text{(A)}.$$

Similarly a descending sequence can be found, such as

$$\ldots,\ 2,\ 1{\cdot}7,\ 1{\cdot}6,\ 1{\cdot}5,\ 1{\cdot}43,\ 1{\cdot}42,\ 1{\cdot}417,\ 1{\cdot}416,\ \ldots \quad \text{(B)},$$

in which all the terms have their squares greater than 2.

* The proof is due to Dedekind, *Stetigkeit und irrationale Zahlen* (Braunschweig, 1872).

This fact naturally suggests that between these two sequences of fractions the square root of 2, if it is to exist, must lie. On these lines attempts have been made to define $\sqrt{2}$. We shall now explain the method given by Dedekind*, and then, by a modification of this method, the definition of an irrational number will be obtained.

Suppose that all ratios are divided into two classes according as their squares are less than 2 or not. We then find that among those whose squares are *not* less than 2 (numbers such as appear in sequence (B) above) all have their squares greater than 2. There is no maximum to the ratios whose squares are less than 2, and no minimum to those whose squares are greater than 2. Clearly a similar argument applies to numbers other than 2.

Thus all ratios can be divided into two classes such that all the terms in one class are less than all the terms in the other, there being no maximum to the lower class and no minimum to the upper class.

This method of dividing all the terms of a series into two classes L and U, of which one wholly precedes the other, was adopted by Dedekind, and it is called a *Dedekind cut*†. There are four possibilities.

(1) There may be a maximum to the lower section L and a minimum to the upper section U; (2) there may be a maximum to L and no minimum to U; (3) there may be no maximum to L and a minimum to U; (4) there may be neither a maximum to L nor a minimum to U.

An example of case (1) is the series of integers. If all the integers not greater than 5 constitute class L, and all those greater than 5 constitute class U, L has the number 5 for maximum, and the number 6 is the minimum for U. The series of ratios illustrates cases (2) and (3). If the lower section L contains all ratios up to and including $\frac{1}{2}$, and U contains the remainder, then $\frac{1}{2}$ is the maximum for L while U has no minimum. If the ratio $\frac{1}{2}$ be

* Another method of defining an irrational number has been given by Cantor, but this theory depends upon the use of convergent infinite sequences, a concept which has not yet been considered. An account of Cantor's theory can be found in Hobson's *Functions of a Real Variable*, I (1921), § 23 *et seq*. A brief note on Cantor's theory is given in § 6·3.

† German *Schnitt*: French *coupure*. English writers usually speak of a *Dedekind section*.

included in U instead of in L, case (3) is similarly illustrated. An example of case (4) has already been considered, the sequences (A) and (B) above being composed of typical terms chosen respectively from the classes L and U. This fourth case is important, and we say that there is a "gap," or that we have an "irrational section," since sections of the ratios have gaps when they correspond to irrational numbers.

Of the four kinds of Dedekind section, the first three are similar in that each section has a boundary, upper or lower as the case may be. A series is "Dedekindian" if every section has a boundary. To draw the fourth case into line with the other three, Dedekind postulated the existence of an irrational limit to fill the gap: he set up his axiom that the gap must always be filled.

In order to give a precise logical definition of irrational number, the unjustifiable postulation of this irrational limit must be avoided. It is clear that an irrational Dedekind cut in some way "represents" an irrational number, and by a modification of the Dedekind definition, Russell has given one which avoids the logical difficulty involved in Dedekind's postulate.

The idea that an irrational number must be the limit of a set of ratios is rejected. Just as ratios whose denominator is 1 are not identical with integers, so those rational numbers which can be greater or less than irrationals, or can have irrational numbers as their limits, must not be identified with ratios.

We define a new kind of numbers called REAL NUMBERS, of which some will be rational, and some irrational. Those that are rational "correspond" to ratios in the same kind of way as the ratio $n/1$ corresponds to the cardinal number n; but they are not logically the same as ratios. To decide what they are to be, we observe that an irrational number is represented by an irrational cut, and that a cut will be represented by its lower section. If we confine ourselves to cuts in which the lower section has no maximum, the lower section is called a *segment**. Those segments which correspond to ratios are those which consist of all ratios less than the ratio to which they correspond, and this is their boundary;

* Since all the fractions a such that $a \leqslant \frac{1}{2}$, and all the fractions such that $a < \frac{1}{2}$ are both lower sections corresponding to the number $\frac{1}{2}$, ambiguity is avoided by defining the lower *segment* to be $a < \frac{1}{2}$.

while those which represent irrationals are those that have no boundary.

We now define a (signless) real number.

A (signless) REAL NUMBER *is a segment of the series of ratios in order of magnitude.*

If the segment has no boundary it is an IRRATIONAL NUMBER, *but if it has a boundary it is a* RATIONAL NUMBER.

In accordance with this definition the *real number* 3, for example, is the aggregate of all rational numbers which are less than the *rational number* 3 (more precisely 3/1). The irrational number $\sqrt{3}$ is the aggregate of all those rational numbers which are negative, and all those which are positive and which have their squares less than 3, the number 0 being included in the aggregate.

The reader will observe that the real number which corresponds to a rational number x, though logically distinct from x, has no properties which differ from those of x, and it is therefore usually denoted by the same symbol. This is analogous to the use of the same symbol n to denote both the cardinal number n and the ratio $n/1$.

There is no difficulty in defining addition and multiplication for real numbers as defined above. In fact, the extension of the term *number* to the real numbers is justified by the fact that it is possible to define the four fundamental operations for the real numbers in such a way that the formal laws of these operations are in agreement with those which hold for operations within the domain of the rational numbers.

Given two real numbers α and β, each being a class of ratios, take any member of α and any member of β and add them together by the law of addition of ratios. Form the class of all such sums obtainable by varying the selected members of α and β: this gives a new class of ratios which can easily be shewn to be a segment of the series of ratios. We define it to be the sum $\alpha + \beta$.

Similarly the product of two (signless) real numbers is the class of ratios generated by multiplying a member of the one class by a member of the other in all possible ways.

The definitions all extend to positive and negative real numbers, and to their addition and multiplication.

It will be seen that arithmetical operations between two real numbers are reduced to operations with rational numbers.

1·6. Complex numbers.

Although complex numbers are capable of a geometrical interpretation, they are not demanded by geometry in the same way as irrational numbers are demanded. A *complex* number means a number involving the square root of a negative number. Clearly since the square of a negative number is always positive, a number whose square is to be negative must be a new kind of number. It is customary to use the letter i to denote the square root of -1, and any number involving the square root of a negative number can be expressed in the form $x+iy$, where x and y are real.

The study of algebraic equations led to the introduction of complex numbers. We desire to be able to say that every quadratic equation has two roots, every cubic equation has three roots, and so on. If real numbers only are considered, the equation $x^2+1=0$ has no roots, and $x^3-1=0$ has only one. Every generalisation of number has first presented itself as needed for some simple problem, but the reader should realise that extensions of number are not created by the mere need of them: they are created by the definition, and our object is now to *define* complex numbers.

Complex numbers had been for some time used by mathematicians in spite of the absence of any precise definition. It was simply assumed that they would obey the formal rules of arithmetic, mainly because on this assumption their employment had been found profitable.

By choosing one of several possible definitions, we state that *a complex number is an ordered pair of real numbers.* Thus (2, 3), $(e, \sqrt{2})$, $(\frac{1}{2}, \pi)$ are all complex numbers. If we write $z \equiv (x, y)$, x is called the *real part* and y the *imaginary part** of the complex number z.

(i) *Equality.* Two complex numbers are *equal*, if, and only if, their real and imaginary parts are separately equal. The equation $z=z'$ implies that both $x=x'$ and $y=y'$.

(ii) *Modulus.* The modulus of z, written $|z|$, is defined to be $+\sqrt{(x^2+y^2)}$. It is an immediate consequence of the definition that $|z|=0$ if, and only if, $x=0$ and $y=0$.

* Although this terminology is now sanctioned by usage, it is very ill-chosen. The reader must realise that there is nothing imaginary about y; it is just as "real" as the real part x.

(iii) *Definition of the fundamental operations.*

If $z \equiv (x, y)$ and $z' \equiv (x', y')$ we have the following *definitions*:

(1) $z + z'$ is $(x + x', y + y')$.

(2) $-z$ is $(-x, -y)$.

(3) $z - z' = z + (-z')$ is $(x - x', y - y')$.

(4) $z . z'$ is $(xx' - yy', xy' + x'y)$.

From the definitions it is easy to deduce that the fundamental laws of algebra are satisfied.

(*a*) *The commutative and associative laws of addition hold, namely*

$$z_1 + z_2 = z_2 + z_1;$$
$$z_1 + (z_2 + z_3) = (z_1 + z_2) + z_3 = z_1 + z_2 + z_3.$$

(*b*) *The commutative and associative laws of multiplication hold,*

$$z_1 z_2 = z_2 z_1;$$
$$z_1 (z_2 z_3) = (z_1 z_2) z_3 = z_1 z_2 z_3.$$

(*c*) *The distributive law holds,*

$$(z_1 + z_2) z_3 = z_1 z_3 + z_2 z_3.$$

It will suffice, by way of illustration, to prove that the commutative law of multiplication holds. We have

$$z_1 z_2 = (x_1 x_2 - y_1 y_2, x_1 y_2 + x_2 y_1) = (x_2 x_1 - y_2 y_1, x_2 y_1 + x_1 y_2) = z_2 z_1.$$

Any of the others are proved similarly by direct appeal to the definitions.

We therefore see that complex numbers obey the same fundamental laws as real numbers: their algebra will accordingly be identical in form, though not of course in meaning, with the algebra of real numbers.

There is no order among the complex numbers; the algebra of complex numbers deals only with equalities; and the phrases "greater than" and "less than" have no meaning within the domain of complex numbers. Inequalities can only be introduced into relations between the *moduli* of complex numbers. The modulus of a complex number is of course a real positive number.

It remains to define division: this can be deduced from the definition of multiplication.

Consider the equation $z\zeta = z'$ ....(1),

where $z \equiv (x, y)$, $z' \equiv (x', y')$ and $\zeta \equiv (\xi, \eta)$.

The equation (1) is

$$(x\xi - y\eta,\ x\eta + y\xi) = (x', y'),$$

so that
$$x\xi - y\eta = x', \quad x\eta + y\xi = y',$$

and solving these equations for ξ and η, we get

$$\frac{\xi}{-(yy' + xx')} = \frac{\eta}{x'y - xy'} = \frac{-1}{x^2 + y^2}.$$

Hence provided that $|z| \neq 0$, there is a unique solution, and ζ is the quotient z'/z.

It will be seen that division by a complex number whose modulus is zero is meaningless; this conforms with the algebra of real numbers.

It cannot be emphasised too strongly that complex numbers and real numbers belong to entirely distinct domains. It is customary to denote complex numbers whose imaginary parts are zero by the real number symbol x. In order that no confusion of ideas may arise from this, it is essential to point out that the symbol x may have two meanings, (i) the real number x, (ii) the complex number $(x, 0)$; which of the two meanings is to be understood for it depends upon which domain is under consideration.

It is customary to define the complex number $(0, 1)$ to be i; then we have

$$i^2 = i \,.\, i = (0^2 - 1^2,\ 0\,.\,1 + 1\,.\,0) = (-1, 0).$$

In accordance with the above we see that i^2 is not the real number -1, but the complex number $(-1, 0)$. To say that i^2 is equal to -1 implies that we are interpreting the symbol -1 in the domain of complex numbers.

The abbreviated notation leads to no ambiguities, for if x denotes the complex number $(x, 0)$ and y denotes $(y, 0)$,

$$x + y = (x, 0) + (y, 0) = (x + y, 0),$$

$$x\,.\,y = (x, 0)\,.\,(y, 0) = (x\,.\,y - 0\,.\,0,\ x\,.\,0 + 0\,.\,y) = (x\,.\,y, 0).$$

Thus, so far as sums and products are concerned, complex numbers whose imaginary parts are zero can be treated as though they were real numbers. *With this convention for the abbreviated notation it is true that*

$$(x, y) \equiv x + iy.$$

For $$x+iy=(x, 0)+(0, 1).(y, 0)$$
$$=(x, 0)+(0.y-1.0, 0.0+1.y)$$
$$=(x, 0)+(0, y)=(x+0, 0+y)=(x, y).$$

In virtue of the relation just proved, we are allowed, in any operation involving sums and products, to treat x, y and i as though they were real numbers, replacing i^2 always by -1.

1·61. Complex numbers are only a special case of the theory of *vectors*, in which the n-dimensional vector can be denoted by $(x_1, x_2, \ldots, x_n)$, where the suffixes denote correlation with the integers used as suffixes, and the correlation is one-many, not necessarily one-one, because x_r and x_s may be equal when r is not equal to s. This definition, with a suitable rule of multiplication, serves all purposes for which n-dimensional vectors are needed.

In particular, in the three-dimensional case, we have the definitions

$$\text{(i)}\quad (x, y, z)+(x', y', z')=(x+x', y+y', z+z'),$$

$$\text{(ii)}\quad (x, y, z) \,.\, (x', y', z')=xx'+yy'+zz',$$

where the right-hand side is *not* a vector, and is called the "scalar product,"

$$\text{(iii)}\quad V.(x, y, z).(x', y', z')=(yz'-y'z, zx'-z'x, xy'-x'y),$$

where the symbol V denotes what is called the "vector product."

The components of the vector product are the co-factors of i, j, k in the determinant

$$\begin{vmatrix} i, & j, & k \\ x, & y, & z \\ x', & y', & z' \end{vmatrix}.$$

The reader can easily verify that

$$V.(x, y, z).(x', y', z')=-V.(x', y', z').(x, y, z),$$

which shews that the vector product is not commutative.

If $$i=(1, 0, 0),\quad j=(0, 1, 0),\quad k=(0, 0, 1),$$

then it is easily verified that

$$ij=0,\quad jk=0,\quad ki=0;\quad i^2=j^2=k^2=1;$$

the vector (x, y, z) then behaves like $xi+yj+zk$, and the fundamental laws of algebra hold when applied to the sum and to the scalar product.

Further information about vector algebra may be found by reference to any of the standard works which deal with the subject*. Vector algebra is of considerable importance in the modern treatment of certain branches of Applied Mathematics and Physics, but it does not come within the scope of this book.

* For example, E. B. Wilson's *Treatise on Vector Algebra* (Yale University Press, 1913).

1·62. Geometrical representation of a complex number.

If we choose a pair of rectangular axes Ox, Oy in a plane, then a point P whose coordinates referred to these axes are (x, y) may be regarded as representing the complex number $x+iy$. In this way, to every point of the plane there corresponds some one complex number; and conversely, to every possible complex number there corresponds one, and only one, point of the plane.

If we denote $(x^2+y^2)^{\frac{1}{2}}$ by r and choose θ so that $r\cos\theta = x$, $r\sin\theta = y$, then r and θ are clearly the radius vector and vectorial angle of the point P, referred to the origin O and axis Ox.

The representation of complex numbers thus afforded is often called the *Argand diagram*.

By the definition already given it is evident that r is the modulus of z, where $z \equiv (x, y)$; the angle θ is called the *argument*, or *amplitude* of z.

We write
$$\theta = \arg z.$$

The argument is not unique, for if θ be a value of the argument, so also is $2n\pi+\theta$ $(n=1, 2, 3, \ldots)$. The *principal* value of $\arg z$ is that which satisfies the inequality $-\pi < \arg z \leqslant \pi$.

The reader who wants further information about the geometrical representation of a complex number, or who is unfamiliar with de Moivre's theorem and its applications, must consult some other treatise*.

It may perhaps be worth while to remind the reader that theorems on complex numbers may be readily *illustrated* by means of the geometrical representation mentioned above, but these intuitions do not constitute an analytical proof.

The point of this remark will be illustrated in the next section.

1·63. THEOREM. *The modulus of the algebraic sum of any number of complex numbers is not greater than the sum of their moduli.*

To fix the ideas, consider first only two complex numbers z_1 and z_2: we shall prove that
$$|z_1 \pm z_2| \leqslant |z_1| + |z_2| \quad\ldots\ldots\ldots\ldots\ldots\ldots\ldots(1),$$

* See for example Hobson's *Trigonometry*, Ch. XIII, or Hardy's *Pure Mathematics*, Ch. III.

which is equivalent to proving that

$$\{(x_1 \pm x_2)^2 + (y_1 \pm y_2)^2\}^{\frac{1}{2}} \leqslant (x_1^2 + y_1^2)^{\frac{1}{2}} + (x_2^2 + y_2^2)^{\frac{1}{2}} \quad \ldots(2).$$

Squaring both sides, we see that (1) is true if

$$x_1^2 + x_2^2 \pm 2x_1x_2 + y_1^2 + y_2^2 \pm 2y_1y_2 \leqslant x_1^2 + y_1^2 + x_2^2 + y_2^2 + 2\{(x_1^2 + y_1^2)(x_2^2 + y_2^2)\}^{\frac{1}{2}},$$

i.e. if $\quad \pm(x_1x_2 + y_1y_2) \leqslant \{(x_1^2 + y_1^2)(x_2^2 + y_2^2)\}^{\frac{1}{2}},$

i.e. if $\quad (x_1x_2 + y_1y_2)^2 \leqslant (x_1^2 + y_1^2)(x_2^2 + y_2^2),$

i.e. if $\quad 0 \leqslant (x_1^2 + y_1^2)(x_2^2 + y_2^2) - (x_1x_2 + y_1y_2)^2,$

i.e. if $\quad 0 \leqslant (x_1y_2 - x_2y_1)^2,$

which is obvious, since the right-hand side is a perfect square.

In a similar way,

$$|z_1 \pm z_2 \pm z_3| \leqslant |z_1| + |z_2 \pm z_3|$$
$$\leqslant |z_1| + |z_2| + |z_3|.$$

Thus, in the case of n complex numbers we get

$$|z_1 \pm z_2 \pm \ldots \pm z_n| \leqslant |z_1| + |z_2| + \ldots + |z_n| \quad \ldots\ldots\ldots(3).$$

By adverting to the geometrical representation, we see that the inequality (1) expresses the well-known result that any two sides of a triangle are together greater than the third.

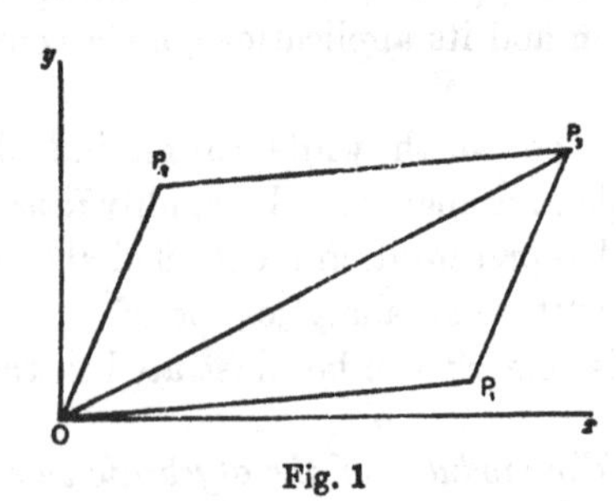

Fig. 1

If P_1 denote the point (x_1, y_1) and P_2 the point (x_2, y_2), then $|z_1| = OP_1$, $|z_2| = OP_2 = P_1P_3$ and $|z_1 + z_2| = OP_3$. Hence (1) is either equivalent to

$$OP_3 \leqslant OP_1 + P_1P_3$$

(equality holding only if P_1 lies in OP_3, that is when the complex numbers z_1 and z_2 have the same argument); or else (1) is equivalent to

$$P_2P_1 \leqslant OP_1 + OP_2,$$

since $|z_1 - z_2| = P_2P_1$.

1·7. Conjugate complex numbers.

If $z=x+iy$, the conjugate complex number $x-iy$ is denoted by $\bar{z}$. It is easy to see that the numbers conjugate to z_1+z_2 and z_1z_2 are $\bar{z}_1+\bar{z}_2$ and $\bar{z}_1\bar{z}_2$ respectively.

The following three formulae are easily proved:

$$|z|^2=z\bar{z},\quad 2Rz=z+\bar{z},\quad 2i\,Iz=z-\bar{z};$$

where Rz and Iz denote the real and imaginary parts of z.

Proofs of theorems concerning complex numbers may frequently be simplified by the use of conjugate complex numbers. The theorem of § 1·63 may be neatly proved in this way:

$$\begin{aligned}|z_1+z_2|^2&=(z_1+z_2)(\bar{z}_1+\bar{z}_2)\\ &=z_1\bar{z}_1+z_1\bar{z}_2+\bar{z}_1z_2+z_2\bar{z}_2\\ &=|z_1|^2+2R(z_1\bar{z}_2)+|z_2|^2\\ &\leqslant|z_1|^2+2|z_1\bar{z}_2|+|z_2|^2\\ &=(|z_1|+|z_2|)^2,\end{aligned}$$

and so

$$|z_1+z_2|\leqslant|z_1|+|z_2|.$$

Also we have

$$\begin{aligned}|z_1-z_2|^2&=|z_1|^2-2R(z_1\bar{z}_2)+|z_2|^2\\ &\geqslant|z_1|^2-2|z_1\bar{z}_2|+|z_2|^2\\ &=(|z_1|-|z_2|)^2,\end{aligned}$$

leading to the important and useful inequality

$$|z_1-z_2|\geqslant\big||z_1|-|z_2|\big|.$$

EXAMPLES I.

1. If α and β are two real numbers for which $A_1, B_1, \ldots$ are members of the respective L classes and $A_2, B_2, \ldots$ are members of the corresponding U classes of a Dedekindian section of the rational numbers, prove that the class determined by $A_1+B_1, \ldots$ and the class determined by $A_2+B_2, \ldots$ form a definite section of the rational numbers.

[One member of each class plainly exists; it remains to prove that there is at most *one* rational number which is greater than every A_1+B_1 and less than every A_2+B_2. Assume that there are two such, and shew that this leads to a contradiction.]

2. Shew that the preceding example gives a unique meaning to the number $\alpha+\beta$. Discuss similarly the definition of the real number $\alpha\beta$.

3. By taking Russell's definition of a real number, shew that Example 1 can be modified to give the definition of $\alpha+\beta$.

[Dedekind defines the real number (when irrational) to be the "cut" which then separates the L class from the U class. Russell defines it as the lower segment of the series of ratios. Example 1 establishes the existence of a definite unique lower segment which is the real number $\alpha+\beta$.]

4. Prove that between any two real numbers there lie both rational and irrational numbers.

5. Shew that a parabola can be drawn to pass through the representative points of the complex numbers

$$2+i,\quad 4+4i,\quad 6+9i,\quad 8+16i,\quad 10+25i.$$

6. Prove that, if z_1 and z_2 are two complex numbers,

$$\text{(i)}\quad |z_1 z_2| = |z_1| . |z_2| \text{ and } \left|\frac{z_1}{z_2}\right| = \frac{|z_1|}{|z_2|},$$

$$\text{(ii)}\quad \arg(z_1 z_2) = \arg z_1 + \arg z_2;\ \arg\left(\frac{z_1}{z_2}\right) = \arg z_1 - \arg z_2,$$

$$\text{(iii)}\quad |z_1 - z_2|^2 + |z_1 + z_2|^2 = 2|z_1|^2 + 2|z_2|^2.$$

7. Discuss the loci on which

$$\left|\frac{z-\alpha}{z-\beta}\right| \text{ and } \arg\left(\frac{z-\alpha}{z-\beta}\right)$$

respectively are constant, where z is a variable complex number, while α and β are fixed complex numbers.

Prove that these loci cut orthogonally.

8. If t be a complex number such that $|t|=1$, prove that the point z given by

$$z = \frac{\alpha t + \beta}{t - \gamma}$$

describes a circle as t varies, unless $|\gamma| = 1$.

What is the locus of z when $|\gamma|=1$?

9. Prove that $z^{p/q}$ has exactly q different values, when p and q are positive integers which are prime to each other.

Factorise completely $\qquad z^5 - (a+ib),$

and shew that the five values of z represent the vertices of a regular pentagon on the Argand diagram.

10. Shew that

$$|c_0 + c_1 z + \ldots + c_n z^n| \geqslant |c_n| \, |z|^n (1 - np/|z|),$$

provided that $|z|$ exceeds the greater of 1 and np, and p is the greatest of $|c_0/c_n|, \ldots, |c_{n-1}/c_n|$.

1. Shew that the area of the triangle whose vertices are the points z_1, z_2, z_3 on the Argand diagram is

$$\Sigma \{(z_2 - z_3) \mid z_1 \mid^2 / 4iz_1\}.$$

Shew also that the triangle is equilateral if

$$z_1^2 + z_2^2 + z_3^2 - z_2 z_3 - z_3 z_1 - z_1 z_2 = 0.$$

12. If $0 < a_0 < a_1 < \ldots < a_n$, prove that all the roots of

$$f(z) \equiv a_0 z^n + a_1 z^{n-1} + \ldots + a_n = 0$$

are outside the circle of unit radius $|z| = 1$.

13. If α, β are real and a is a complex constant, shew that if

$$\alpha z_1 \bar{z}_1 + \bar{a} z_1 \bar{z}_2 + a \bar{z}_1 z_2 + \beta z_2 \bar{z}_2 = 0$$

and $\alpha\beta - a\bar{a} < 0$, then the values of z_1/z_2 lie on a real circle, or on a straight line in a special case.

14. Shew that, if c is real, the equation

$$(a\bar{z} + \bar{a}z)^2 = 2(b\bar{z} + \bar{b}z) + c$$

represents a parabola.

15. Determine the regions of the z-plane specified by

$$\left| \frac{z - a}{1 - \bar{a}z} \right| < 1, = 1 \text{ or } > 1,$$

where a is a complex number such that $|a| < 1$.

16. If z_1 and z_2 are two complex numbers such that $|z_1| < 1$, $|z_2| < 1$, shew that there exists a positive constant k (depending only on z_1 and z_2) such that, for every point z (other than $z = 1$) which lies inside the triangle whose vertices are the points $1, z_1, z_2$,

$$|1 - z| \leqslant k \{1 - |z|\}.$$

17. If p, q, r, s are all rational numbers, and

$$(ps - qr)^2 + 4(p - r)(q - s) = 0,$$

prove that either (i) $p = r$, $q = s$ or else (ii) $1 - pq$ and $1 - rs$ are squares of rational numbers or are zero.

18. If all the solutions of the equations

$$ax^2 + 2bxy + cy^2 = 1, \quad lx^2 + 2mxy + ny^2 = 1$$

are rational solutions, prove that both

$$\sqrt{\{(b - m)^2 - (a - l)(c - n)\}} \text{ and } \sqrt{\{(an - cl)^2 + 4(am - bl)(cm - nb)\}}$$

are rational, when a, b, c, l, m, n are rational numbers.

CHAPTER II

BOUNDS AND LIMITS OF SEQUENCES

2·1. Introduction.

We now introduce two concepts which are fundamental in Analysis; they are (i) the concept of a *bound*, and (ii) the concept of a *limit*.

The definition of a bound follows immediately from the definition of a real number, so that it is naturally the next subject for consideration in a logical order of development of Analysis. Before the actual definition can be given we must examine an important property of the *aggregate of real numbers*, or as it is sometimes termed, the *continuum*.

If an irrational number is defined by Dedekind's method, a theorem, known as *Dedekind's theorem*, must be proved. The theorem can be stated as follows:

If the aggregate of REAL NUMBERS *is divided into two classes L and U in such a way that* (i) *every number belongs to one or other of the two classes,* (ii) *each class contains at least one member, and* (iii) *any member of L is less than any member of U, then there is a number α which has the property that all numbers less than it belong to L, and all numbers greater than it belong to U. The number α itself may belong to either class*.*

Dedekind's theorem expresses the fact that the *continuum is closed.* In other words, the real numbers do not produce any further extension in the field of numbers. Real numbers are obtained as definite collections of rational numbers (see § 1·5), but collections of real numbers only yield real numbers, not some further extension to some still more general species of number.

2·11. The Principle of Continuity.

The reader will recall that the definition of a real number given in the previous chapter differs slightly from the one given by Dedekind†. The concept of a segment of the series of ratios, which was used to formulate the definition of a real number in Chapter I, will now be applied to the continuum.

* The reader should notice that the aggregate of *rational numbers* does not possess this property.

† See Exs. I, note on Question 3; also § 1·5.

Imagine that all the real numbers are laid out before us, and that we select one of these numbers, say α; it is clear that α determines two segments of the aggregate of real numbers, a *lower segment* and an *upper segment*: the number α itself is the dividing number, or the number which determines the segments.

The characteristic property of a lower segment is that if β belongs to the segment, then so do all the numbers less than β.

Similarly the characteristic property of an upper segment is that if β belongs to the segment, then so do all the numbers greater than β.

The assertion that every segment of the aggregate of real numbers determines a definite real number constitutes the Principle of Continuity.

When allowance is made for the difference between Dedekind's method and Russell's method of defining real numbers, the reader will see that the Principle of Continuity is equivalent to Dedekind's theorem. For this reason no proof of the latter theorem was given in § 2·1.

In the case of a *lower segment*, the boundary number α has the following properties:

If $x < \alpha$, then x belongs to the segment.

If $x > \alpha$, then x definitely does not belong to the segment.

The number α itself may or may not belong to it.

Similarly in the case of an *upper segment*, the number α has the properties:

If $x > \alpha$, then x belongs to the segment.

If $x < \alpha$, then x definitely does not belong to the segment.

The number α itself may or may not belong to it.

2·2. Upper and lower bounds of a set of numbers*.

We now define two kinds of bounds, rough bounds and exact bounds. Let S denote any linear set of numbers x; for example, S might be the set of numbers defined by the relation $0 < x < 1$.

If there exists a number H which is less than every number x of the set, then S is said to be *bounded below*, and H is called a *rough lower bound.*

* We shall hereafter use the word "number" to mean "real number," as defined in § 1·5.

Similarly if there exists a number K which is greater than every number x of the set S, then S is said to be *bounded above*, and K is a *rough upper bound*.

The Principle of Continuity is needed in order to give the definition of an "exact bound." If rough lower bounds exist at all, they form a lower segment, for if H' is less than H, then H' is another rough lower bound. The aggregate of rough lower bounds thus determines a definite real number. This real number, which will be denoted by m, is the *(exact) lower bound** of the set of numbers S.

THEOREM 1. *The lower bound m of a set of numbers S has the properties:*

(i) *$m \leqslant$ every number x of the set,* (ii) *$m + \epsilon >$ some one number of the set at least, where ϵ is an arbitrary positive number as small as we please.*

To prove (i), suppose that $m > x_1$ (say), where x_1 is a member of the set S, then

$$m = x_1 + \delta,$$

where δ is positive; hence

$$m - \tfrac{1}{2}\delta > x_1;$$

but $m - \frac{1}{2}\delta < m$ and it is a rough lower bound.

Hence the supposition that $m > x_1$ leads to a contradiction; and this establishes (i).

To prove (ii), if possible suppose that

$$m + \epsilon \leqslant \text{every one of the numbers } x \text{ of the set } S$$

then $\quad m + \frac{1}{2}\epsilon <$ every one of the numbers x;

so that $m + \frac{1}{2}\epsilon$ is a rough lower bound. This is impossible, for $m + \frac{1}{2}\epsilon > m$. This contradiction establishes (ii), and the theorem is proved†.

By a similar argument it may be shewn that there is an *(exact) upper bound M* determined by the upper segment which consists of the aggregate of rough upper bounds.

* The word "exact" has only been used to distinguish between the two kinds of bound which have now been defined. In what follows, m will be called the "lower bound" of the set S, and the word "exact" will not be used unless it is desired to *emphasise* that the bound in question is not a "rough" bound.

† The reader should be familiar with this mode of proof: it is frequently used in Analysis.

THEOREM 2. *The upper bound M of a set of numbers S has the properties:*

(i) $M \geqslant$ *every number x of the set,* (ii) $M - \epsilon <$ *some one number of the set at least.*

The proof, which follows the same lines as that of Theorem 1, is left as an exercise for the reader.

It follows directly from the properties of M and m that

$$M \geqslant m.$$

2·3. Limit points of a set of numbers.

Before a formal definition is given, it may be said by way of general description that a "limit point" of an infinite set is a number in whose immediate neighbourhood is concentrated an indefinite number of members of the set. Consider, for example, the aggregate of points representing the numbers

$$1, \frac{1}{2}, \frac{1}{3}, \ldots, \frac{1}{n}, \ldots.$$

In any interval, however small, extending from the origin to the right, there is an indefinite number of points of this aggregate. The origin is therefore a limit point, for an indefinite number of points of the aggregate are clustered about the origin. At a limit point the concentration of the members of the set has the property of endlessly great density.

We now give a formal definition.

The point α is a limit point of an infinite set E if, however small ϵ may be, there is a point of the set E, other than α, whose distance from α is less than ϵ.*

If there is one such point within the interval $(\alpha - \epsilon, \alpha + \epsilon)$ there must be an indefinite number; for if there were only n of them, and α_n were the nearest to α, there would not be in E a point other than α whose distance from α was less than $|\alpha - \alpha_n|$†.

* The letter E, the initial letter of the French word *ensemble*, is often used to denote an arbitrary set of points. The German word corresponding to aggregate or set is *Menge*.

† When z is a complex number, we have already defined the meaning of $|z|$. If x be real, the symbol $|x|$ means simply the positive (or absolute) value of x.

2·31. WEIERSTRASS'S THEOREM. *Every bounded infinite aggregate has at least one limit point.*

To fix the ideas, consider only a linear infinite set of points included in the interval (a, b). Take any point c, such that $a<c<b$; then since (a, b) contains an infinite set, there must be an infinite set either in (a, c) or in (c, b) or in both. Suppose, for definiteness, that there is an infinite set in (a, c). The set of points x in (a, c), such that none or only a finite number of members of this infinite set is less than x, is bounded above by c, so that c, and every number greater than c, is a rough upper bound. The aggregate of rough upper bounds determines a unique number ξ, the upper bound of the set. The number ξ has the properties,

$$\xi \geqslant \text{every } x,$$
$$\xi - \epsilon < \text{at least one } x;$$

hence in any interval $(\xi - \epsilon, \xi + \epsilon)$ there are points of the infinite set in (a, c) other than ξ. The number ξ is therefore a limit point.

2·4. The concept of a function.

A *function* is merely a relation between real numbers. If x denote any real number, since "the square of x" is the relation between x^2 and x, the phrase "the square of x" implies the existence of a function of x. The concept of a function of the real variable x will be more fully discussed in the next chapter, but for our present needs the concept of a function of the positive integral variable n must be understood.

Let us consider the sequence of numbers

$$s_1, \; s_2, \; s_3, \; \ldots, \; s_n, \; \ldots,$$

it being understood that some law exists according to which the general term s_n can be written down*. For example, if $s_n = 2n + \frac{1}{n}$, we see that

$$s_1 = 3, \; s_2 = 4\tfrac{1}{2}, \; s_3 = 6\tfrac{1}{3}, \; s_4 = 8\tfrac{1}{4}, \; \ldots.$$

Since, when n assumes in turn the integral values 1, 2, 3, ... there is a corresponding set of numbers $s_1, s_2, s_3, \ldots, s_n$ may be

* The law of formation need not be embodied in any *explicit formula* which enables us to obtain s_n for a given n by direct calculation. For example, a sequence $\{s_n\}$ may be defined by the statement that "s_n = the decimal fraction for $\sqrt{2}$ terminated at the nth digit."

described as a *function of the integral variable n*. Such a function may be denoted by $\phi(n)$, so that all that is implied by the equation

$$y = \phi(n)$$

is that as n assumes in turn the values 1, 2, 3, ... there is a corresponding set of values of y, say $y_1, y_2, y_3, \ldots$, where

$$y_1 = \phi(1), \quad y_2 = \phi(2), \quad y_3 = \phi(3), \ldots.$$

The notation $\{s_n\}$ will be used to denote the infinite sequence of numbers

$$s_1, s_2, s_3, \ldots, s_n, \ldots.$$

2·5. Limits of sequences.

The limit of a sequence $\{s_n\}$ as n increases indefinitely is an extremely important concept which needs to be discussed in some detail. Two phrases are of constant use, and they must be understood. They are,

(1) "*For all sufficiently large values of n*,"

(2) "*For values of n as large as we please.*"

We now assign a definite meaning to each of these phrases.

(1) Suppose that it is possible to find a fixed integer ν such that whenever $n \geqslant \nu$ the number s_n possesses some definite property. It need not possess this property for *every* value of n; all that we need to know is that it possesses this property for every value of n as soon as n exceeds a fixed integer ν. If this is the case, we say that the property holds *for all sufficiently large values of n*.

To illustrate this, consider an example of the kind in which this phrase occurs quite frequently in the theory of limits of sequences. If ϵ denote an arbitrary positive number which may be as small as we please*, then we can say that, if $s_n = 1/n$,

"$s_n < \epsilon$, for all sufficiently large values of n."

This is easily seen, for $1/n < \epsilon$ whenever $n \geqslant \nu$, if ν is the greatest integer in $1 + \dfrac{1}{\epsilon}$.

The reader to whom this idea is novel, is advised to consider the above example in greater detail by assigning different small numerical values to ϵ, and calculating the corresponding values of ν.

* The symbol ϵ will be used in this sense throughout the book. In future, therefore, ϵ will be used without the above description being repeated each time.

Two facts should be noticed, (1) that ν is always an integer, and (2) that the value of ν *depends upon the chosen value of* ϵ. The latter statement is especially important, and it must be clearly understood by every reader who wishes to acquire a proper grasp of the theory of limits.

To take numerical examples, if

$$\epsilon = 0{\cdot}1, \quad \nu = 11;$$
$$\epsilon = 0{\cdot}01, \quad \nu = 101;$$
$$\epsilon = 0{\cdot}0001, \quad \nu = 10{,}001;$$

and so on.

We shall often write $\nu(\epsilon)$ to indicate the dependence of ν upon the arbitrarily chosen value of ϵ. It is to be emphasised that ϵ is quite arbitrary, and to each chosen value of ϵ there will be a *corresponding* value of ν.

(2) A property is said to hold "for values of n as large as we please," if it holds for at least one value of n greater than any assigned N, however large. The idea is perhaps best comprehended if we think of an attempt to prevent some property holding for $\{s_n\}$. We endeavour to assign some large value N of n, such that when $n \geqslant N$ a certain property of $\{s_n\}$ fails to hold. If no such value N of n can be found, we say that the property holds "*for values of n as large as we please.*"

For example, it is true that $(-1)^n$ is greater than zero for values of n as large as we please, for no matter how large N may be, so long as n is an *even* integer $(-1)^n > 0$. Thus no integer N can be found so large that $(-1)^n$ is not greater than zero when $n \geqslant N$. The reader should observe that we could not say that $(-1)^n > 0$ for all sufficiently large values of n, for whatever value is selected for ν, every odd integer which exceeds ν renders the inequality $(-1)^n > 0$ untrue.

2·51. Upper and lower limits of a bounded sequence.

Let $\{s_n\}$ be any bounded sequence, H and K its lower and upper bounds. Any real number A which is such that

$$A \leqslant s_n$$

for all sufficiently large values of n, is called an *inferior number* for $\{s_n\}$. Similarly a number B which is such that

$$B \geqslant s_n$$

for all sufficiently large values of n, is called a *superior number* for $\{s_n\}$.

A number C which is such that $C < s_n$ for an infinity of values of n, and also $C > s_n$ for an infinity of values of n, is called an *intermediate number*.

Consider the aggregate of all superior numbers. It is bounded below, for none of its members is less than H: this aggregate therefore has a lower bound which is denoted by Λ.

Similarly the aggregate of inferior numbers has an upper bound which is denoted by λ.

The number Λ is called the *upper limit* of the sequence $\{s_n\}$ as $n \to \infty$, and the number λ is called the *lower limit* of $\{s_n\}$ as $n \to \infty$. It is customary to write

$$\Lambda = \overline{\lim_{n \to \infty}}\, s_n, \quad \lambda = \underline{\lim_{n \to \infty}}\, s_n,$$

or more simply, by omitting "$n \to \infty$,"

$$\overline{\lim}\, s_n \text{ and } \underline{\lim}\, s_n.$$

It is clear from the mode of definition of the numbers λ and Λ that

$$H \leqslant \lambda \leqslant \Lambda \leqslant K.$$

We have also the following elementary properties, stated for convenience in the form of theorems.

THEOREM 1. *The numbers Λ and λ are the upper and lower bounds of the aggregate of intermediate numbers, if any such exist.*

Since $\Lambda \geqslant \lambda$ there are two cases to consider.

(1) If $\Lambda = \lambda = l$, there can be at most one intermediate number l and there is nothing to prove.

(2) If $\Lambda > \lambda$, since, by definition, any intermediate number C is less than any superior and greater than any inferior number, we have

$$\lambda \leqslant C \leqslant \Lambda.$$

If $\lambda < C < \Lambda$, then C must be an intermediate number, since it is clearly neither superior nor inferior. Hence there are intermediate numbers as near as we please to λ or to Λ.

THEOREM 2. *Given ϵ, then*

(i) $s_n < \Lambda + \epsilon$ *for all sufficiently large values of n, and*

(ii) $s_n > \Lambda - \epsilon$ *for an infinity of values of n.*

For, the number $\Lambda+\epsilon$ is a superior number, and $\Lambda-\epsilon$ is either intermediate or inferior. The result is then an immediate consequence of the definition.

Similarly we have

THEOREM 3. (i) $s_n > \lambda - \epsilon$ *for all sufficiently large values of* n,

(ii) $s_n < \lambda + \epsilon$ *for an infinity of values of* n.

We now give two illustrative examples.

Example 1. Let
$$s_n = (-)^n \left\{\frac{1}{2} - \frac{4}{n}\right\}.$$

By forming a table of values

n	1	2	3	4	5	6	8	16	48
s_n	$3\frac{1}{2}$	$-1\frac{1}{2}$	$\frac{5}{6}$	$-\frac{1}{2}$	$\frac{3}{10}$	$-\frac{1}{6}$	0	$\frac{1}{4}$	$\frac{5}{12}$

we see that, as soon as n exceeds 4 every value of s_n lies between $-\frac{1}{2}$ and $\frac{1}{2}$. As n increases indefinitely, $4/n$ tends to zero, and the limit of s_n as $n \to \infty$ is either $-\frac{1}{2}$ or $\frac{1}{2}$ according as n ranges through odd or even values. Hence
$$\lambda = -\tfrac{1}{2} \text{ and } \Lambda = \tfrac{1}{2}.$$

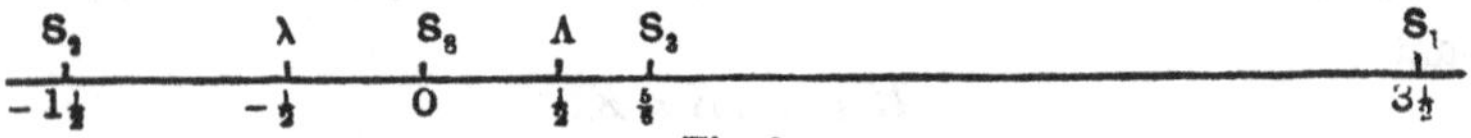

S_2 λ S_8 Λ S_3 S_1

$-1\frac{1}{2}$ $-\frac{1}{2}$ 0 $\frac{1}{2}$ $\frac{5}{6}$ $3\frac{1}{2}$

Fig. 2

As n increases through odd values, s_n tends to the value $-\frac{1}{2}$, and we see that the points representing values of s_n when n is large and odd cluster about the point $-\frac{1}{2}$ on its right. Similarly the points representing values of s_n when n is large and even cluster about the point $\frac{1}{2}$ on its left. The upper and lower limits of a bounded sequence therefore both have the character of limit points of a set as described in § 2·3.

It is evident in this example, that with few exceptions, the points representing values of s_n lie within the interval (λ, Λ). In fact, for this sequence, the only points lying outside this interval are those representing s_1, s_2, s_3, and s_4 coincides with λ.

It is however possible that an *infinite* number of terms of the sequence may lie to the left of λ and to the right of Λ. This is illustrated by the following example.

Example 2. Let
$$s_n = (-)^n\left(1+\frac{1}{n}\right).$$

The first few terms of this sequence are
$$-2,\ \tfrac{3}{2},\ -\tfrac{4}{3},\ \tfrac{5}{4},\ -\tfrac{6}{5},\ \ldots,$$
and evidently $\lambda = -1$ and $\Lambda = 1$. In this case an infinite number of terms lie to the left of λ and to the right of Λ and *no* term of the sequence lies *between*

λ and Λ. It is not therefore necessary that there should be only a finite number of terms of the sequence outside the interval (λ, Λ).

In virtue of Theorems 2 and 3 above, $s_n < \Lambda + \epsilon$ and $s_n > \lambda - \epsilon$ for all sufficiently large values of n, so that at most a finite number of terms of the sequence lie to the left of $\lambda - \epsilon$ and to the right of $\Lambda + \epsilon$.

2·52. The unique limit of a sequence.

If the upper limit coincides with the lower limit, so that $\Lambda = \lambda$, their common value l is defined to be the *unique limit* of the sequence $\{s_n\}$ as $n \to \infty$. The unique limit is often called simply *the limit*. When the sequence $\{s_n\}$ has a unique limit it is said to *converge* to that limit. Symbolically this may be expressed in either of two ways,

$$(1) \quad \lim_{n \to \infty} s_n = l,$$

or

$$(2) \quad s_n \to l \text{ as } n \to \infty.$$

Since $\Lambda - \lambda \geqslant 0$, it is a direct consequence of the above definition that *the necessary and sufficient condition for the existence of a unique finite limit (convergence of the sequence) is that*

$$\Lambda - \lambda < \epsilon,$$

for every arbitrary $\epsilon > 0$.

Thus, when a unique limit exists, the lower bound of the superior numbers is equal to the upper bound of the inferior numbers $(\Lambda = \lambda = l)$, and so *every* number greater than l is superior, and *every* number less than l is inferior.

2·53. Theorems on unique limits.

A different definition of a unique limit is frequently given. *The sequence $\{s_n\}$ is said to tend to a limit l, if, given ϵ, there exists a number $\nu(\epsilon)$, such that*

$$|s_n - l| < \epsilon$$

whenever $n \geqslant \nu(\epsilon)$.

It is immaterial which of the two possible definitions is adopted, but whichever one is selected as the initial *definition*, the other must be *proved* to be a consequence of it. Since we have taken as our initial definition the one given in § 2·52, it is necessary to prove that the definition just given is equivalent to the one which we have adopted. This is done in Theorem 1 which follows.

The reader is advised also to adopt the second definition and then prove as an exercise that the first definition is equivalent to it.

THEOREM 1. *The necessary and sufficient condition that the sequence* $\{s_n\}$ *converges to a limit* l *is that, given* ϵ, *there exists a* $\nu(\epsilon)$ *such that*

$$|s_n - l| < \epsilon$$

whenever $n \geqslant \nu(\epsilon)$.

(*a*) *The condition is necessary*, for if $s_n \to l$, then every number greater than l is superior for $\{s_n\}$, and every number less than l is inferior for $\{s_n\}$. Hence

$$l - \epsilon < s_n, \quad \text{for } n \geqslant \nu_1;$$

$$l + \epsilon > s_n, \quad \text{for } n \geqslant \nu_2.$$

If* $$\nu = \max(\nu_1, \nu_2),$$

$$l - \epsilon < s_n < l + \epsilon, \quad \text{for } n \geqslant \nu,$$

that is to say $$|s_n - l| < \epsilon, \quad \text{for } n \geqslant \nu.$$

(*b*) *The condition is sufficient*, for suppose that $|s_n - l| < \epsilon$ for $n \geqslant \nu$, then $l - \epsilon$ is an inferior number for $\{s_n\}$, but since ϵ is arbitrary *every* number less than l is an inferior number for $\{s_n\}$. Similarly every number greater than l is a superior number for the sequence. Hence a unique limit exists.

THEOREM 2. CAUCHY'S GENERAL PRINCIPLE OF CONVERGENCE. *The necessary and sufficient condition for the convergence of any sequence* $\{s_n\}$, *is that corresponding to every arbitrary* ϵ *there exists an integer* ν, *such that*

$$|s_\nu - s_{\nu+p}| < \epsilon$$

for all positive integral values of p.

(*a*) *The condition is necessary*, for by Theorem 1, if s_n tends to a limit l, whenever $n \geqslant \nu$,

$$|s_n - l| < \tfrac{1}{2}\epsilon.$$

In particular we have

$$|s_\nu - l| < \tfrac{1}{2}\epsilon, \qquad |s_{\nu+p} - l| < \tfrac{1}{2}\epsilon.$$

Now $$|s_\nu - l + l - s_{\nu+p}| \leqslant |s_\nu - l| + |l - s_{\nu+p}|$$

$$< \tfrac{1}{2}\epsilon + \tfrac{1}{2}\epsilon = \epsilon;$$

that is, $$|s_\nu - s_{\nu+p}| < \epsilon.$$

* $\max(\nu_1, \nu_2)$ means the greater of the two numbers ν_1 and ν_2. Similarly we use $\min(\nu_1, \nu_2)$ to denote the lesser of the two numbers concerned. The symbolism applies equally well to more than two numbers; thus $\max(a, b, c, d, \ldots, k)$ means the greatest of the numbers $a, b, c, d, \ldots, k$.

(b) *The condition is sufficient*, for, if $|s_\nu - s_{\nu+p}| < \epsilon$ for all positive integral values of p, we have

$$s_\nu - \epsilon < s_{\nu+p} < s_\nu + \epsilon,$$

that is $s_\nu - \epsilon$ is an inferior number and $s_\nu + \epsilon$ is a superior number for the sequence $\{s_{\nu+p}\}$. Hence

$$\Lambda - \lambda \leqslant (s_\nu + \epsilon) - (s_\nu - \epsilon) = 2\epsilon;$$

and so, by remembering that $\Lambda - \lambda \geqslant 0$ and that ϵ is arbitrary, it must follow that

$$\Lambda - \lambda = 0.$$

Since this is the condition that a unique limit exists, the sufficiency of the condition is established.

2·54. Unbounded sequences.

In the preceding sections only *bounded* sequences have been considered, and the upper and lower limits of such sequences must be finite. It is sometimes convenient, though not indispensable, to extend the concept of upper and lower limits to unbounded sequences. In order to do this it is necessary to say that in certain cases $\Lambda = +\infty$ and $\lambda = -\infty$. So far we have assigned no meaning to the symbol "∞" standing alone, and up to the present the symbol has been regarded as meaningless except when it has occurred in the notation "$n \to \infty$," which is merely a convenient abbreviation for the phrase "n is indefinitely increased." A precise *definition* of what is meant by the statements $\Lambda = \infty$, $\lambda = -\infty$ is given below, but the reader should be able to see the need for their employment in the following examples.

The sequence defined by the relations

$$s_{2n-1} = 1/n, \quad s_{2n} = n \qquad (n = 1, 2, 3, \ldots)$$

is clearly one for which λ exists and has the value 0, but unless we say that $\Lambda = +\infty$ we cannot assign a value to Λ.

Again, if the sequence $\qquad s_n = (-)^n n$

is to be considered as possessing upper and lower limits at all, we must say that

$$\lambda = -\infty, \quad \Lambda = +\infty.$$

The reader must clearly understand that the symbol "∞" and the terms *infinite*, *infinity* and *tends to infinity* have purely conventional meanings. Phrases in which these terms are employed have a meaning only when *by a definition* some suitable meaning has been assigned to them.

When we say that a *number n tends to infinity* $(n \to \infty)$, we are using a short and convenient phrase (or a still shorter symbolical expression) to express the fact that n assumes an endless sequence of values which eventually become and remain greater than any arbitrary positive number, however large.

For this reason the reader is advised not to use the concept of λ and Λ being supposed infinite until he fully realises what that concept really means. The precise meaning to be attached to the symbols $\Lambda=\infty$, $\lambda=-\infty$ is contained in the following definitions:

(i) A sequence $\{s_n\}$ may be such that, when n exceeds ν, inferior numbers exist and have a finite upper bound λ, but superior numbers do not exist because an infinity of values of s_n can be found which exceed any arbitrary positive number G, however large. *In this case λ exists and is finite, and we say that* $\Lambda=\infty$.

(ii) Similarly $\{s_n\}$ may be such that, when n exceeds ν, superior numbers exist and have a finite lower bound Λ, but inferior numbers do not exist because an infinity of values of s_n can be found which are less than any arbitrary negative number $-G$, however large G may be. *In this case Λ exists and is finite, and we say that* $\lambda=-\infty$.

(iii) Sequences $\{s_n\}$ exist for which there are neither inferior nor superior numbers, so that λ and Λ do not exist finitely, but there is always an infinity of values of s_n which exceed any arbitrary large number G, and an infinity of values of s_n which are less than any arbitrary number $-G$. *It is this case which is represented symbolically by the statements* $\lambda=-\infty$, $\Lambda=\infty$.

2·55. Divergent and oscillatory sequences.

When the sequence $\{s_n\}$ does not converge to a unique finite limit, there are several possibilities which will now be considered.

(1) *Divergence.* If the terms of the sequence have the property that, however large the positive number G may be, an integer ν can be found such that

$$s_n > G$$

whenever n exceeds ν, then $\{s_n\}$ is said to *diverge.* This may be expressed symbolically by writing

$$s_n \to \infty \quad \text{as} \quad n \to \infty.$$

Similarly, if an integer ν can be found such that

$$s_n < -G$$

whenever $n \geqslant \nu$, then $\{s_n\}$ diverges. This may be expressed symbolically by writing

$$s_n \to -\infty \quad \text{as} \quad n \to \infty.$$

In both cases $\{s_n\}$ is a *divergent sequence,* and $\{s_n\}$ may be said to *diverge to* ∞, or to *diverge to* $-\infty$.

(2) *Oscillation.* When the sequence $\{s_n\}$, while not convergent, possesses a finite upper limit and a finite lower limit, then it is said to *oscillate finitely.*

For example, the sequence $s_n = (-)^n \left\{1 + \frac{1}{n}\right\}$ oscillates finitely, for $\lambda = -1$, $\Lambda = 1$.

Now that the meaning of the expressions $\Lambda = \infty$, $\lambda = -\infty$ has been precisely defined, a sequence $\{s_n\}$ for which Λ and λ have these values may be said to *oscillate infinitely*. Infinite oscillation does not differ much from divergence, and some writers use the latter term to include infinite oscillation. The difference is clearly seen by considering the following examples:

If $s_n = n^2$, $\{s_n\}$ diverges, and $s_n \to \infty$ as $n \to \infty$.

If $s_n = -n^2$, $\{s_n\}$ diverges, and $s_n \to -\infty$ as $n \to \infty$.

If $s_n = (-)^n n^2$, $\{s_n\}$ oscillates infinitely, the even terms increasing and the odd terms decreasing without limit. In other words $s_{2m} \to \infty$ as $m \to \infty$, and $s_{2m+1} \to -\infty$ as $m \to \infty$.

2·56. Theorems on limits of sequences.

For the sake of brevity the following notation is adopted. If $\{s_n\}$ converges to a unique limit, say s, as n is indefinitely increased, we shall write simply $s_n \to s$, leaving the phrase "as n tends to infinity" to be implicitly understood. Also the word "limit" will be taken always to imply "unique limit," unless the contrary is stated.

In § 2·53 we proved two fundamental theorems, the second of which, *the general principle of convergence*, is of immense theoretical importance. It gives a means of deciding whether a sequence tends to a limit or not, without any previous information concerning the value of the limit.

Although in elementary work the limit theorems which will now be proved may be employed more frequently than the "general principle," the reader must not underestimate the importance of the general principle on the grounds that it seems to be rarely used. Most of the abstract theorems which involve the concept of a limit depend for their proof upon the employment, in some form or other, of this important result.

Proofs of the fundamental limit theorems will now be given. We begin by proving two lemmas.

LEMMA 1. *If* $|s_n - s| < k\epsilon$ *for* $n \geq \nu$, *where* k *is a positive constant, then* $s_n \to s$.

Replace ϵ by ϵ/k, as is clearly permissible, since ϵ is arbitrary and ϵ/k is positive. Thus if $\epsilon' = \epsilon/k$ we can find an integer ν, depending on ϵ', such that

$$|s_n - s| < k\epsilon' \quad \text{for } n \geq \nu,$$

that is

$$|s_n - s| < \epsilon \quad \text{for } n \geq \nu;$$

and, since k is a constant, ν is a function of ϵ, so that $s_n \to s$ by the ordinary criterion.

The reader may be inclined to regard the above lemma as trivial, but it has been proved for a special reason. In the proof of Theorem 2, § 2·53, the first inequality was taken to be $|s_n - l| < \frac{1}{2}\epsilon$. The reason for the choice of $\frac{1}{2}\epsilon$ was to ensure that the final result should read $|s_\nu - s_{\nu+p}| < \epsilon$. Although such a procedure is frequently adopted for the sake of artistic elegance, the lemma shews that it is quite unnecessary. If ϵ had been chosen at the outset, the final result would have read $|s_\nu - s_{\nu+p}| < 2\epsilon$, which would do just as well.

The reader should observe that it is rarely possible to decide beforehand whether $\frac{1}{2}\epsilon$, or $\frac{1}{3}\epsilon$, or $\frac{1}{4}\epsilon$ must be chosen first, so as to be left with ϵ at the final step of the proof. The lemma shews that this artificial choice of some fraction of ϵ is unnecessary.

LEMMA 2. *Every convergent sequence is bounded.*

If $s_n \to s$ and s is finite, then when $n \geqslant \nu$,

$$s - \epsilon < s_n < s + \epsilon.$$

Let g and G be the least and greatest respectively of the $\nu - 1$ numbers

$$s_1, s_2, \ldots, s_{\nu-1}.$$

If $h = \min(s - \epsilon, g)$ and $H = \max(s + \epsilon, G)$, then

$$h \leqslant s_n \leqslant H \quad \ldots\ldots\ldots\ldots\ldots\ldots\ldots\ldots(1)$$

for all values of n; in other words $\{s_n\}$ is bounded.

Corollary. An equivalent inequality to (1) above is

$$|s_n| \leqslant M,$$

where M is a constant. Plainly $M = \max(|h|, |H|)$.

2·561. The fundamental limit theorems.

If $s_n \to s$ and $t_n \to t$, where s and t are finite, then

(1) $s_n + t_n \to s + t$,

(2) $s_n t_n \to st$,

(3) $\dfrac{1}{s_n} \to \dfrac{1}{s}$, *so long as* $s \neq 0$.

Proof of (1). We have

$$|s_n - s| < \epsilon \text{ for } n \geqslant \nu_1,$$

and

$$|t_n - t| < \epsilon \text{ for } n \geqslant \nu_2;$$

now $$(s_n + t_n) - (s + t) = (s_n - s) + (t_n - t),$$

hence $$|(s_n + t_n) - (s + t)| \leqslant |s_n - s| + |t_n - t| \leqslant 2\epsilon,$$

for $n \geqslant \nu \equiv \max(\nu_1, \nu_2)$. This proves that $s_n + t_n \to s + t$, by Lemma 1.

First proof of (2).

By Lemma 2, $\{s_n\}$ and $\{t_n\}$ are bounded; hence there is a positive constant M such that, for *all* values of n,

$$|s_n| \leqslant M, \quad |t_n| \leqslant M.$$

If we write S for $|s|$ and T for $|t|$, then

$$s_n t_n - st = s_n(t_n - t) + t(s_n - s),$$

hence $$|s_n t_n - st| \leqslant M|t_n - t| + T|s_n - s| < (M + T)\epsilon,$$

for $n \geqslant \nu$. Thus, by Lemma 1, $s_n t_n \to st$.

Second proof of (2).

We can write

$$s_n t_n - st = (s_n - s)(t_n - t) + t(s_n - s) + s(t_n - t),$$

so that, for $n \geqslant \nu$,

$$|s_n t_n - st| < \epsilon . \epsilon + T\epsilon + S\epsilon;$$

and, since we may suppose that $\epsilon < 1$, when $n \geqslant \nu$ we have

$$|s_n t_n - st| < (1 + T + S)\epsilon.$$

Hence, by Lemma 1, the theorem follows.

Proof of (3).

Since we must assume that* $s \neq 0$, we may write $|s| = 2\delta$, where $\delta > 0$. Since $s_n \to s$, a number ν_1 can be chosen so that, when $n \geqslant \nu_1$,

$$|s_n - s| < \delta,$$

and accordingly, when $n \geqslant \nu_1$,

$$|s_n| > \delta.$$

Now $$\frac{1}{s_n} - \frac{1}{s} = \frac{s - s_n}{s . s_n},$$

and $$\left|\frac{1}{s_n} - \frac{1}{s}\right| = \frac{|s - s_n|}{|s||s_n|} < \frac{|s - s_n|}{2\delta . \delta}$$

* The reader should observe that the condition that $s \neq 0$ is necessary, for $1/s$ has no meaning when $s = 0$. If $s = 0$, $1/s_n$ may not even possess a unique limit; for example, $s_n = (-)^n \frac{1}{n} \to 0$, but $\frac{1}{s_n} = (-)^n n$, and this oscillates infinitely.

for values of n which exceed ν_1. A number ν_2 can be found such that, when $n \geqslant \nu_2$,

$$|s_n - s| < \epsilon;$$

so that if $\nu = \max(\nu_1, \nu_2)$, as soon as $n \geqslant \nu$,

$$\left|\frac{1}{s_n} - \frac{1}{s}\right| < \frac{\epsilon}{2\delta^2},$$

hence by Lemma 1, since $\dfrac{1}{2\delta^2}$ is a constant, the theorem follows.

2·57. Theorems on upper and lower limits.

Let $\{s_n\}$ and $\{t_n\}$ be two bounded sequences, and let their upper and lower limits be respectively denoted by Λ, λ; Λ', λ'. A set of general limit theorems applicable to upper and lower limits can be obtained, but the proofs of some of these are rather difficult, and there is a great variety of results.

The reader should observe that the relations which hold between upper and lower limits are not the same as the corresponding ones for unique limits.

It is not proposed to deal exhaustively with the relations between upper and lower limits*; it will suffice, by way of illustration, to prove two of the simpler results. Several others have been set as examples for solution at the end of this chapter.

THEOREM. *To prove that*

$$(a)\quad \overline{\lim}\, s_n + \overline{\lim}\, t_n \geqslant \overline{\lim}\, (s_n + t_n),$$

$$(b)\quad \underline{\lim}\, s_n + \underline{\lim}\, t_n \leqslant \underline{\lim}\, (s_n + t_n).$$

Proof of (*a*).

Since $\overline{\lim}\, s_n = \Lambda$ and $\overline{\lim}\, t_n = \Lambda'$, then, in virtue of Theorem 2, § 2·51, there is at most a *finite* number of values of n for which

$$s_n > \Lambda + \tfrac{1}{2}\epsilon \quad\text{.........................(1)},$$

and at most a *finite* number of values of n for which

$$t_n > \Lambda' + \tfrac{1}{2}\epsilon \quad\text{.........................(2)};$$

hence it follows that

$$s_n + t_n > \Lambda + \Lambda' + \epsilon \quad\text{.....................(3)},$$

* For fuller information the reader should consult Carathéodory, *Vorlesungen über Reelle Funktionen* (Berlin, 1918), §§ 85–95. Unfortunately the author is unable to refer the reader to any English text-book which deals fully with this topic.

for at most that *finite* number of values of n for which the inequalities (1) and (2) hold.

Since $\Lambda + \frac{1}{2}\epsilon$ is a superior number for $\{s_n\}$ we have

$$s_n \leqslant \Lambda + \tfrac{1}{2}\epsilon,$$

for $n \geqslant \nu_1$, and since $\Lambda' + \frac{1}{2}\epsilon$ is a superior number for $\{t_n\}$,

$$t_n \leqslant \Lambda' + \tfrac{1}{2}\epsilon,$$

for $n \geqslant \nu_2$. Hence for $n \geqslant \max(\nu_1, \nu_2)$

$$s_n + t_n \leqslant \Lambda + \Lambda' + \epsilon \quad \text{.....................(4).}$$

It follows that

$$\overline{\lim}\,(s_n + t_n) \leqslant \Lambda + \Lambda' + \epsilon;$$

and since ϵ is *arbitrary*, the inequality (a) is established.

The proof of (b) follows the same lines, and is left as an exercise for the reader.

If a unique limit exists for both $\{s_n\}$ and $\{t_n\}$, the above theorem reduces to the fundamental theorem (1) of § 2·561.

2·58. Monotonic sequences.

A very important type of sequences will now be considered. Let $\{s_n\}$ be any given sequence, then if either

$$s_1 \leqslant s_2 \leqslant \ldots \leqslant s_n \leqslant \ldots \quad \text{.....................(1),}$$

or

$$s_1 \geqslant s_2 \geqslant \ldots \geqslant s_n \geqslant \ldots \quad \text{.....................(2),}$$

$\{s_n\}$ is said to be *monotonic* (*monotone*). Sequences of type (1) are said to be *monotonic increasing*, those of type (2) are said to be *monotonic decreasing*.

A sequence is said to be *strictly monotonic*, if the equality signs in (1) and (2) are not allowed. The sequence (1) may be said to be *increasing in the wide sense*, but if the equality signs are disallowed the sequence is said to be *increasing in the strict sense*.

THEOREM. *A monotonic increasing sequence tends to its upper bound.*

Let the upper bound of the sequence $\{s_n\}$ be M; then either (i) M is finite or (ii) M is infinite.

In case (i) we have

$$s_n \leqslant M \text{ for all values of } n \quad \text{.........................(3)},$$

$$s_n > M - \epsilon \text{ for at least one value of } n;$$

suppose that
$$s_\nu > M - \epsilon.$$

Since the sequence increases, for $n \geqslant \nu$ we have

$$s_n \geqslant s_\nu,$$

so that
$$s_n > M - \epsilon \quad \text{.........................(4)}.$$

By comparing (3) and (4) we have, for $n \geqslant \nu$,

$$M - \epsilon < s_n < M + \epsilon,$$

which implies that, as $n \to \infty$,

$$s_n \to M.$$

In case (ii) M is infinite. This means that, given any number G

$$s_n > G,$$

for at least one value of n, say $n = \nu$, so that

$$s_\nu > G.$$

Since $\{s_n\}$ increases we must have, for $n \geqslant \nu$,

$$s_n > G,$$

in other words
$$s_n \to \infty.$$

Similarly it may be proved that *a monotonic decreasing sequence tends to its lower bound.*

The proof is left as an exercise for the reader.

2·6. An important limit theorem.

To prove that if $s_n \to l$, then $\dfrac{s_1 + s_2 + \ldots + s_n}{n} \to l.$

Write $s_n = l + t_n$, and we have to prove that if $t_n \to 0$, then so does $(t_1 + t_2 + \ldots + t_n)/n$.

Divide the numbers $t_1, t_2, \ldots, t_n$ into two groups:

$$(1)\ t_1, t_2, \ldots, t_k, \quad (2)\ t_{k+1}, t_{k+2}, \ldots, t_n.$$

The number k is a function of n which tends to infinity *more slowly than n**, in other words we suppose that $k \to \infty$ but $k/n \to 0$.

* The number $\sqrt{n}$ is a number which tends to infinity more slowly than n, for $\sqrt{n} \to \infty$, but $(\sqrt{n})/n \to 0$ as $n \to \infty$.

However small ϵ may be we can find a number ν_1 such that every term in group (2) is numerically less than ϵ when $n \geqslant \nu_1$,

hence $$\left|\frac{t_{k+1}+t_{k+2}+\ldots+t_n}{n}\right| < \epsilon \cdot \frac{n-k}{n} < \epsilon,$$

there being $n-k$ terms on the left-hand side.

Let T denote the greatest of $|t_1|, |t_2|, \ldots$, then we have

$$\left|\frac{t_1+t_2+\ldots+t_k}{n}\right| < \frac{kT}{n};$$

and since $k/n \to 0$ as $n \to \infty$ this number kT/n can be made less than ϵ for $n \geqslant \nu_2$. Thus

$$\left|\frac{t_1+t_2+\ldots+t_n}{n}\right| \leqslant \left|\frac{t_1+t_2+\ldots+t_k}{n}\right| + \left|\frac{t_{k+1}+\ldots+t_n}{n}\right| \quad \ldots(1)$$
$$< 2\epsilon$$

for $n \geqslant \nu$, where $\nu = \max(\nu_1, \nu_2)$. The theorem is therefore proved.

Instead of making k a function of n, we can prove the theorem by choosing a k such that $|t_\nu| < \epsilon$ for all $\nu > k$. Now k is a *fixed* number and we shew in a similar way that each of the terms on the right-hand side of (1) is less than ϵ. The details of this alternative method of proof are left as an exercise for the reader.

2·7. Complex sequences.

For complex sequences inferior and superior numbers do not exist, nor do upper and lower limits. In fact there is no general theory of complex sequences, and they can only be treated by considering their real and imaginary parts separately. A systematic theory of *convergent* complex sequences may however be developed, for which the following inequalities are fundamental:

$$\left.\begin{matrix}|x|\\|y|\end{matrix}\right\} \leqslant |x+iy| \leqslant |x|+|y|.$$

Since $|x+iy| = \sqrt{(x^2+y^2)}$, the first inequality is obvious; and since $|i|=1$, the second depends on the theorem in § 1·63.

Definition. *The sequence* $\{s_n\}$, *where* $s_n \equiv \sigma_n + i\tau_n$, *is said to converge to the limit* $s \equiv \sigma + i\tau$ *if* $\{\sigma_n\}$ *converges to* σ *and* $\{\tau_n\}$ *converges to* τ.

We write $\sigma_n + i\tau_n \to \sigma + i\tau$, which implies that both $\sigma_n \to \sigma$ and $\tau_n \to \tau$.

THEOREM. *The necessary and sufficient condition that the sequence* $\{s_n\}$, *where* $s_n = \sigma_n + i\tau_n$, *should converge to a limit* s, *where* $s = \sigma + i\tau$, *is that for* $n \geqslant \nu$, $|s_n - s| < \epsilon$.

(1) *It is necessary*; for, if $s_n \to s$, we have

$$|\sigma_n - \sigma| < \epsilon, \text{ for } n \geqslant \nu_1,$$

and

$$|\tau_n - \tau| < \epsilon, \text{ for } n \geqslant \nu_2,$$

so that, for $n \geqslant \max(\nu_1, \nu_2)$,

$$|s_n - s| = |\sigma_n + i\tau_n - (\sigma + i\tau)| \leqslant |\sigma_n - \sigma| + |\tau_n - \tau| < 2\epsilon.$$

(2) *It is sufficient*, for, if, when $n \geqslant \nu$,

$$|s_n - s| < \epsilon,$$

it follows that, for $n \geqslant \nu$,

$$|\sigma_n - \sigma| \leqslant |s_n - s| < \epsilon,$$

and

$$|\tau_n - \tau| \leqslant |s_n - s| < \epsilon.$$

Hence $\sigma_n \to \sigma$ and $\tau_n \to \tau$; that is to say the sequence $\{s_n\}$ converges to a limit s.

2·71. The general principle of convergence holds for complex sequences.

When $\{s_n\}$ is complex the statement of the general principle assumes exactly the same form as in the enunciation of it for real sequences in § 2·53.

The proof follows at once in virtue of the inequalities

$$\left.\begin{array}{l}|\sigma_\nu - \sigma_{\nu+p}| \\ |\tau_\nu - \tau_{\nu+p}|\end{array}\right\} \leqslant |s_\nu - s_{\nu+p}| \leqslant |\sigma_\nu - \sigma_{\nu+p}| + |\tau_\nu - \tau_{\nu+p}|.$$

The details of a formal proof are left to the reader.

The reader should observe that the criterion for convergence and the general principle of convergence for complex sequences are precisely the same as for real sequences. This is because in the statement of these theorems only the *moduli* of the terms appear, and the modulus of any complex number is real and positive.

No inequality can hold between complex numbers, so that the preceding theorems exist only because they involve the moduli of the complex numbers concerned.

2·8. An important limit.

It will now be proved, on the hypothesis that, when x is real and positive, the series

$$s = 1 + x + \frac{x^2}{2!} + \frac{x^3}{3!} + \dots \qquad \text{..................(1)}$$

is convergent, that the limit of the sequence

$$s_n = \left(1 + \frac{x}{n}\right)^n$$

is the sum-function s of the series (1).

By the Binomial Theorem for a positive integral index,

$$s_n = \left(1+\frac{x}{n}\right)^n = 1 + n\frac{x}{n} + \frac{n(n-1)}{2!}\frac{x^2}{n^2} + \frac{n(n-1)(n-2)}{3!}\frac{x^3}{n^3} + \dots$$

$$= 1 + x + \frac{\left(1-\frac{1}{n}\right)}{2!}x^2 + \frac{\left(1-\frac{1}{n}\right)\left(1-\frac{2}{n}\right)}{3!}x^3 + \dots,$$

the series terminating since n is a positive integer.

We observe that (i) each term increases as n increases, and (ii) as n increases new positive terms are added. Hence $\{s_n\}$ is monotonic increasing, and therefore tends to its upper bound. Thus if we shew that $\{s_n\}$ is bounded above, the limit sought must be finite.

Clearly $$s_n < 1 + x + \frac{x^2}{2!} + \dots + \frac{x^n}{n!},$$

and, since every term is positive, this is less than the sum of the infinite series* s. On the hypothesis that the series (1) converges, s must be finite, so that

$$s_n < s,$$

from which we deduce that

$$\lim_{n\to\infty} s_n \leqslant s \qquad \text{........................(2).}$$

If p be fixed, and $n > p$, we have

$$s_n > 1 + x + \frac{\left(1-\frac{1}{n}\right)}{2!}x^2 + \dots + \frac{\left(1-\frac{1}{n}\right)\dots\left(1-\frac{p-1}{n}\right)}{p!}x^p;$$

* Although the proof is simple, the reader should realise that the above statement *requires* proof. For a series of positive terms the sum to n terms, s_n, *increases* to the limit s; hence s must exceed the sum of any number of terms, taken arbitrarily, in the series, for n can be chosen large enough to ensure that s_n includes all these terms. See §§ 2·58, 5·11, 5·12.

so that

$$\lim_{n\to\infty} s_n \geqslant \lim_{n\to\infty}\left\{1+x+\frac{\left(1-\frac{1}{n}\right)}{2!}x^2+\ldots+\frac{\left(1-\frac{1}{n}\right)\ldots\left(1-\frac{p-1}{n}\right)}{p!}x^p\right\}.$$

The only restriction on p is that it must be less than n, so that p can be as large as we please, hence

$$\lim_{n\to\infty} s_n \geqslant 1+x+\frac{x^2}{2!}+\ldots+\frac{x^p}{p!},$$

however large p may be. It follows that

$$\lim_{n\to\infty} s_n \geqslant s \qquad\qquad (3).$$

By comparing (2) and (3) we have

$$\lim_{n\to\infty} s_n = s.$$

Note 1. The sequence s_n is a function of n only, for throughout the preceding proof x is regarded as having a *fixed* real and positive value. The reader may recognise that the series (1) above is the *exponential series*, which converges absolutely for all real values of x, and the sum-function s is usually denoted by e^x (or $\exp x$). Further reference to the exponential function will be found in §§ 4·33, 10·12.

Note 2. The number e may be defined to be the sum of infinite series

$$1+1+\frac{1}{2!}+\frac{1}{3!}+\ldots \qquad\qquad (4),$$

which we assume to be convergent*; and from what we have just proved, e is also the value of

$$\lim_{n\to\infty}\left(1+\frac{1}{n}\right)^n.$$

Now

$$\left(1+\frac{1}{n}\right)^n = 1+1+\frac{1-\frac{1}{n}}{2!}+\frac{\left(1-\frac{1}{n}\right)\left(1-\frac{2}{n}\right)}{3!}+\ldots<1+1+\frac{1}{2}+\frac{1}{2^2}+\ldots+\frac{1}{2^{n-1}},$$

and, by summing the geometrical progression,

$$\left(1+\frac{1}{n}\right)^n < 1+2-\frac{1}{2^{n-1}} < 3.$$

Hence e is not greater than 3, and from the series (4) it is clear that $e>2$.

* The series is easily proved to be convergent by Test III, § 5·2.

EXAMPLES II.

1. For the following sequences examine whether or not they are bounded and whether or not they are monotonic. Determine the bounds, the upper and lower limits or the unique limits, whichever exist.

$$(a)\ \frac{1}{2},\ \frac{2}{3},\ \frac{3}{4},\ \ldots,\ \frac{n}{n+1},\ \ldots.$$

$$(b)\ 2,\ 1,\ 1\tfrac{3}{4},\ 1\tfrac{1}{4},\ 1\tfrac{5}{8},\ 1\tfrac{3}{8},\ \ldots.$$

$$(c)\ 1,\ 2,\ \tfrac{1}{2},\ 2\tfrac{1}{2},\ \tfrac{1}{4},\ 2\tfrac{3}{4},\ \tfrac{1}{8},\ 2\tfrac{7}{8},\ \ldots.$$

$$(d)\ \sqrt{2},\ \sqrt[3]{3},\ \sqrt[4]{4},\ \ldots,\ \sqrt[n]{n},\ \ldots.$$

$$(e)\ 7,\ -2,\ 2,\ 3,\ 1\tfrac{1}{2},\ 3\tfrac{1}{2},\ 1\tfrac{1}{4},\ 3\tfrac{3}{4},\ 1\tfrac{1}{8},\ 3\tfrac{7}{8},\ \ldots.$$

2. Shew that, if $a > 0$ and $0 < s_1 < b$, the sequence

$$s_1 = a,\quad s_2 = \sqrt{\left(\frac{ab^2 + s_1^2}{a+1}\right)},\ \ldots,\ s_{n+1} = \sqrt{\left(\frac{ab^2 + s_n^2}{a+1}\right)},\ \ldots$$

is an always increasing bounded sequence, and find its limit as $n \to \infty$.

3. If $a > 0$ and

$$s_1 = \frac{a^a}{(a+1)^{a+1}},\quad s_2 = \frac{s_1}{(1-s_1)^a},\ \ldots,\ s_n = \frac{s_1}{(1-s_{n-1})^a},\ \ldots,$$

prove that $\{s_n\}$ is monotonic, and has the unique limit $1/(a+1)$ as $n \to \infty$.

4. Prove that, if $a > 1$, the sequence $\{x_n\}$ defined by

$$x_1 = a,\quad x_2 = a^{x_1},\quad x_3 = a^{x_2},\ \ldots,\ x_n = a^{x_{n-1}},\ \ldots$$

converges if $$a \leqslant e^{1/e}.$$

5. If s_1 and s_2 are positive and $s_{n+1} = \frac{1}{2}(s_n + s_{n-1})$, prove that the sequences $s_1, s_3, s_5, \ldots;\ s_2, s_4, s_6, \ldots$ are the one increasing and the other decreasing; and shew that their common limit is $\frac{1}{3}(s_1 + 2s_2)$. Discuss similarly the sequence $s_{n+2} = \sqrt{(s_{n+1} s_n)}$, shewing that the common limit in this case is $(s_1 s_2^2)^{\frac{1}{3}}$.

6. If $\{s_n\}$ is any bounded sequence, and U_n, L_n are respectively the upper and lower bounds of the sequence

$$s_n,\ s_{n+1},\ s_{n+2},\ \ldots,$$

shew that the sequences $U_1, U_2, U_3, \ldots, L_1, L_2, L_3, \ldots$ are each monotonic and bounded.

Prove that the sequence $\{s_n\}$ is convergent, if, and only if, the unique limits U and L of the sequences $\{U_n\}$ and $\{L_n\}$ are equal, and in this case

$$U = \lim s_n = L.$$

Shew also that if $U \neq L$ the numbers U and L are respectively the upper and lower limits of $\{s_n\}$.

[This last result is sometimes taken as a *definition* of upper and lower limits for a bounded sequence $\{s_n\}$.]

7. If $\{s_n\}$ and $\{s_n'\}$ are two bounded sequences having lower and upper limits respectively λ, Λ; λ', Λ'; prove that

$$\text{(i)}\quad \overline{\lim}\,(-s_n) = -\underline{\lim}\, s_n,$$

$$\text{(ii)}\quad \underline{\lim}\,(s_n + s_n') \leqslant \begin{matrix}\lambda + \Lambda' \\ \lambda' + \Lambda\end{matrix} \leqslant \overline{\lim}\,(s_n + s_n').$$

8. With the notation of the last example, prove that, if the sequences $\{s_n\}$ and $\{s_n'\}$ are of *positive terms*,

$$\text{(i)}\quad \lambda\lambda' \leqslant \underline{\lim}\,(s_n s_n'),$$

$$\text{(ii)}\quad \overline{\lim}\,(s_n s_n') \leqslant \Lambda\Lambda'.$$

9. If $\phi(n) - \phi(n-1) \to l$ as $n \to \infty$, prove that $\phi(n)/n \to l$.

10. If s_n denote the sum to n terms of the series

$$\sin\theta + \tfrac{1}{2}\sin 2\theta + \frac{1}{2^2}\sin 3\theta + \ldots,$$

prove that
$$\lim_{n\to\infty}\left\{\frac{s_1 + s_2 + \ldots + s_n}{n}\right\} = \frac{4\sin\theta}{5 - 4\cos\theta}.$$

11. Prove that
$$\underline{\lim}\, s_n \leqslant \underline{\lim}\,\frac{s_1 + s_2 + \ldots + s_n}{n} \leqslant \overline{\lim}\, s_n.$$

12. If $p_1, p_2, \ldots, p_n, \ldots$ are arbitrary positive numbers such that $\sum_1^n p_\nu \to \infty$ as $n \to \infty$, prove that, if $a_n \to a$ as $n \to \infty$, the sequence $\{a_n'\}$ tends to the same limit a, where

$$a_n' = \frac{p_1 a_1 + p_2 a_2 + \ldots + p_n a_n}{p_1 + p_2 + \ldots + p_n}.$$

13. If $a_n \to \alpha$ and $b_n \to \beta$ as $n \to \infty$, prove that both

$$\frac{a_1 b_1 + a_2 b_2 + \ldots + a_n b_n}{n}$$

and
$$\frac{a_1 b_n + a_2 b_{n-1} + \ldots + a_n b_1}{n}$$

tend to $\alpha\beta$.

14. (i) If $\{b_n\}$ is a positive monotonic decreasing sequence, and if $a_n \to 0$, $b_n \to 0$ as $n \to \infty$, prove that

$$\lim_{n\to\infty}\frac{a_n}{b_n} = \lim_{n\to\infty}\frac{a_n - a_{n+1}}{b_n - b_{n+1}},$$

provided that the second limit exists.

(ii) If $b_{n+1} > b_n$ and $b_n \to \infty$ as $n \to \infty$, prove that

$$\lim_{n\to\infty}\frac{a_n}{b_n} = \lim_{n\to\infty}\frac{a_{n+1} - a_n}{b_{n+1} - b_n},$$

provided that the second limit exists.

CHAPTER III

LIMITS AND CONTINUITY

3·1. The concept of a function.

In § 2·4 reference has already been made to the concept of a function with regard to the integral variable n. By recalling what was stated there, the reader will see that, if we take as our definition the statement that *a function is a relation between real numbers*, the equation

$$y = \phi(x)$$

implies that, given any set of arguments x, to some or all of them there correspond values of y. The same value of y may correspond to more than one value of the argument x; when to each value of x there corresponds one, and only one, value of y the function is said to be *single-valued* or *uniform*.

The reader must guard against the idea, which is so easily acquired after reading some elementary text-books, that the concept of a function applies only to those functions which are capable of graphical representation. The concept of a function is of very wide application, and is not restricted even to those functions for which the relation between x and y is expressible by means of an analytical formula.

For example, let y denote the number of windows in house number x on the odd side of a certain given street in London. Then y is defined for a certain number of odd integral values of x, namely 1, 3, 5, ..., m, where m is the number of the last house on the odd side of this street.

In order that later terminology shall not be misunderstood, we now mention briefly some of the commonest types of function to be met with in Analysis.

(1) *Polynomials.*

A polynomial in x is of the form

$$a_0 x^n + a_1 x^{n-1} + \dots + a_n,$$

where the coefficients a_r are constants. This polynomial is of *degree* n.

(2) *Rational functions.*

A rational function is the ratio of two polynomials, so that if $P(x)$ and $Q(x)$ denote two polynomials, the general rational function $R(x)$ is given by

$$R(x)=\frac{P(x)}{Q(x)},$$

$P(x)$ and $Q(x)$ having no common factor.

(3) *Algebraic functions.*

The function $y=f(x)$ is an algebraic function of x, if y is the root of an equation of the nth degree in y whose coefficients are rational functions of x,

$$y^n+R_1y^{n-1}+\ldots+R_n=0.$$

We shall assume that the above equation is *irreducible*, that is to say it is incapable of resolution into factors whose coefficients are also rational functions of x.

For example the equation $\qquad y^4-x^2=0$

is reducible, since it implies that either $y^2+x=0$ or else $y^2-x=0$, and each of these equations is irreducible.

(4) *Transcendental functions.*

All functions which are not rational nor algebraic are called *transcendental*; this is a wide class of functions, and includes such well-known functions as $\sin x$, $\cos x$ and $\log x$, as well as many which are less familiar.

It is not difficult to prove that $\sin x$ and $\cos x$ are transcendental; in fact we can prove the more general results that no periodic function can be either a rational function or an algebraic function.

A function $f(x)$ is said to be *periodic* with period ω if $f(x)=f(x+\omega)$ for all values of x for which the function is defined. The functions $\sin x$ and $\cos x$ are periodic in 2π, as are also the other trigonometrical functions $\tan x$, $\cot x$, $\sec x$, $\operatorname{cosec} x$. (See reference to these functions in § 4·33.)

THEOREM 1. *No periodic function (unless it be a mere constant) can be a rational function.*

Suppose that $\qquad f(x)=\dfrac{P(x)}{Q(x)}, \qquad (Q(0)\neq 0),$

where P and Q are polynomials, and that $f(x)=f(x+\omega)$ for all values of x. Let $f(0)=a$, then the equation of the nth degree

$$P(x)-aQ(x)=0$$

is satisfied by $n+1$ values of x, viz. $0, \omega, 2\omega, \ldots, n\omega$.

Hence, by the fundamental theorem of algebra, $f(x)$ must be identically equal to a for all values of x.

THEOREM 2. *No periodic function can be an algebraic function.*

For, let the algebraic function be defined by the equation

$$y^n + R_1 y^{n-1} + \dots + R_n = 0 \quad\dots\dots(1),$$

where the R's denote rational functions of x. This equation can be written

$$P_0 y^n + P_1 y^{n-1} + \dots + P_n = 0,$$

where the P's denote polynomials in x. Let $f(0) = a$, as above, then

$$P_0 a^n + P_1 a^{n-1} + \dots + P_n = 0$$

for all values of x: hence $y = a$ satisfies equation (1) for all values of x, and one set of values of the algebraic function reduces to a constant. Now divide (1) by $y - a$ and repeat the argument; divide again by $y - a_1$ and so on until it has been repeated n times. We conclude that the algebraic function has, for any value of x, the same n values $a, a_1, a_2, \dots, a_{n-1}$; in other words it is composed of n constants.

It is also of some interest to prove that $\log x$ is a transcendental function. By assuming that the reader is already familiar with elementary differential calculus, it is possible to give a simple proof that $\log x$ is not a rational function.

THEOREM 3. *The function $\log x$ is not a rational function.*

It will be seen in § 4·33 that $\log x$ satisfies the equation

$$\frac{d}{dx}(\log x) = \frac{1}{x} \qquad (x > 0).$$

Let X and Y denote two polynomials having no common factor, so that if $\log x$ be rational,

$$\log x = \frac{X}{Y} \quad\dots\dots(1),$$

whence, by differentiation*,

$$\frac{Y^2}{x} = YX' - XY' \quad\dots\dots(2).$$

Thus, Y must be divisible by x a certain number of times (say m), and consequently X cannot be divisible by x since X and Y are prime to each other. Write $Y = Zx^m$, where Z is a polynomial not divisible by x.

Now $$Y' = mZx^{m-1} + Z'x^m,$$

and, on substituting in (2),

$$Z^2 x^{2m-1} = ZX'x^m - mZXx^{m-1} - XZ'x^m,$$

that is, $$mXZ = x(ZX' - XZ') - Z^2 x^m,$$

which is impossible, for the right-hand side is divisible by x, but the left-hand side is not. Hence equation (1) cannot hold, so that $\log x$ is not a rational function of x.

It is proved in Examples X, 10 that $\log x$ is not an algebraic function of x. When this has been proved, $\log x$ will have been shewn to be a transcendental function.

* We use the symbol X' to denote $\frac{d}{dx}X$.

The reader should realise that besides the transcendental functions mentioned above there are others, perhaps less important, but nevertheless of frequent occurrence. An example will suffice at this point. Consider the function defined as follows:

$$y = 2 \text{ when } x \text{ is rational,}$$
$$y = 0 \text{ when } x \text{ is irrational.}$$

The graph consists of two series of points arranged upon the lines $y=2$ and $y=0$; and although these would not be visibly distinguishable from two continuous straight lines, in reality an indefinite number of points is missing from each line*.

3·11. Functions of more than one variable.

So far only functions representable by the relation $y=\phi(x)$ have been considered, but the extension to the concept of a function of several variables is easily made.

If $y=\phi(x)$, the argument x is called the *independent variable* and y is called the *dependent variable*, for if values of x are arbitrarily assigned the corresponding values of y are determined by the functional relation.

If we consider a set of n independent variables $x, y, z, \ldots, t$ and one dependent variable u, the equation

$$u = \phi(x, y, z, \ldots, t) \quad \text{.....................(1)}$$

denotes the functional relation. In this case if $x_1, y_1, z_1, \ldots, t_1$ are the n arbitrarily assigned values of the independent variables, the corresponding value of the dependent variable u is determined by the functional relation.

The function represented by equation (1) is an *explicit* function, but where several variables are concerned it is rarely possible to obtain an equation expressing one of the variables explicitly in terms of the others. Thus most of the functions of more than one variable are *implicit* functions, that is to say we are given a functional relation

$$\Phi(x, y, z, \ldots, t) = 0 \quad \text{.....................(2),}$$

connecting the n variables $x, y, z, \ldots, t$, and it is not in general possible to solve this equation to find an *explicit* function which expresses one of these variables, say x, in terms of the other $n-1$

* Functions of this type will be considered in more detail in § 3·6.

variables. Even when only two variables are involved it may be impossible to find an explicit algebraical expression for one variable y in terms of the other variable x. A simple illustration of this is afforded by the equation

$$y^5 - y - x = 0.$$

Although x can be found explicitly in terms of y by writing

$$x = y^5 - y,$$

no algebraical expression for y in terms of x can be found.

Further details about implicit functions will be considered later; and when we have to discuss differentiability of functions of several variables the reader will see the importance of distinguishing carefully between dependent and independent variables.

3·12. Function of a function.

In the equation $y = \phi(x)$, suppose that x, instead of being the independent variable subject to choice, is dependent upon some other independent variable t, for which the functional relation is $x = f(t)$. If values of t, say $t_1, t_2, \ldots, t_m$, are assigned arbitrarily we obtain values for x from the relation $x = f(t)$: when these values of x are substituted in $y = \phi(x)$ we obtain values for y which are dependent on the arbitrary choice of values for t.

If we write

$$y = \phi\{f(t)\},$$

the symbolism denotes that y is a *function of a function.*

This kind of relation is easily extended to the case of several variables.

3·2. Limits of functions.

For simplicity we confine our attention first of all to functions of a single real variable.

The reader should have no difficulty in realising that there is a theory of limits for functions of a real variable similar to that for limits of sequences which was discussed in the preceding chapter.

The sequence $\{s_n\}$ is a function of the positive integral variable n, where n assumes only the integral values $1, 2, 3, \ldots, p, \ldots$, and we were concerned with the behaviour of s_n as n is indefinitely increased.

If $f(x)$ denotes any given function of x, where x denotes any real number for which $f(x)$ is defined, we shall also discuss the behaviour of $f(x)$ as x increases indefinitely, ranging through *all real values.*

Clearly the approach to infinity in which the variable assumes the integral values 1, 2, 3, ... only is a particular case of the more general method of approach in which the variable assumes all real values. Hence, if a definite limit exists when the variable assumes all real values, the same limit must be obtained if a special set of these values be selected (say the positive integral values) and the variable restricted to range through this special set only. It is important to realise that the converse is not *necessarily* true*.

3·21. Definitions.

For the sake of precision it is convenient here to consider the exact meaning of some of the statements which we shall frequently have occasion to use.

A *point* is an ordered aggregate of real numbers. For example (x, y, z) represents, for different values of x, y and z, a set of points in three-dimensional space.

A *function*, in the general mathematical sense, is a relation between numbers and points. For instance $z = f(x, y)$ expresses a relation between the point (x, y) and the number z, which is called the *value* of $f(x, y)$.

An *interval* is a collection of points. Thus, the points (x, y, z) defined by the three inequalities

$$a \leqslant x \leqslant b, \quad a' \leqslant y \leqslant b', \quad a'' \leqslant z \leqslant b'' \quad \text{............(1)}$$

all belong to the interval which these inequalities define.

The interval defined by the inequalities (1) is called a *closed* interval, for the end-points are included among the points (x, y, z) which compose the interval.

The interval defined by the three inequalities

$$a < x < b, \quad a' < y < b', \quad a'' < z < b''$$

is called an *open* interval, the end-points not being included among the points which compose the interval.

A *neighbourhood* is a special kind of interval. If (a_1, a_2, a_3) be any point, the interval $\{a_1 - h, a_1 + h;\ a_2 - h, a_2 + h;\ a_3 - h, a_3 + h\}$,

* See Examples III, 20.

whether open or closed, is described as a neighbourhood of the point (a_1, a_2, a_3).

Roughly speaking, a neighbourhood is a "cubical" interval with its centre at the point.

By the phrase "(x, y, z) is *at* the point (a_1, a_2, a_3)" we mean that $x = a_1$, $y = a_2$, $z = a_3$.

By the phrase "(x, y, z) is *near* the point (a_1, a_2, a_3)" we mean that (x, y, z) lies within some neighbourhood of the point (a_1, a_2, a_3). When we wish to imply that the number h which defines the extent of the neighbourhood may be as small as we please, the point (x, y, z) is said to be *sufficiently near* the point (a_1, a_2, a_3).

Let $f(x)$ be a function defined and bounded in the linear interval (a, b), then, if M and m are the bounds of $f(x)$, the number $M - m$ is called the *oscillation* of the function $f(x)$ in the interval.

Although all the above definitions have been formulated or illustrated in terms of three-dimensional space, they are applicable equally to space of any number n of dimensions.

A point in n-dimensional space is the ordered aggregate of real numbers $(x_1, x_2, \ldots, x_n)$. Similarly the other definitions can be given in a form which is applicable to space of any finite number of dimensions.

3·22. Finite limit at a finite point.

In the preceding chapter the only kind of limit which was contemplated was the limit of a sequence s_n as n was indefinitely increased. A function may approach a definite limit when the variable tends to some *finite* value, as well as when the variable tends to infinity.

The definition of the existence of a limit for a function of one variable $f(x)$ as x approaches a finite value a will now be given.

$f(x)$ is said to have the limit b as x tends to a when, given ϵ, a positive number η, depending on ϵ, can be found such that whenever

$$|x - a| \leqslant \eta, \quad |f(x) - b| < \epsilon.$$

If the above criterion is satisfied we write

$$\lim_{x \to a} f(x) = b,$$

and $f(x)$ may be said to converge to (or approach) the limit b as x tends to a.

In some cases the variable x may approach the value a from one side only.

When $x \to a$ through values of x greater than a, x approaches a from the right (or from above). Similarly when $x \to a$ through values of x less than a, x approaches a from the left (or from below). If when either of these methods of approach is considered a definite limit exists, the following notation is adopted*:

If a right-hand limit exists as x approaches a from above, this limit is denoted by

$$\lim_{x \to a+0} f(x) \text{ or } f(a+0).$$

If a left-hand limit exists, the limit is denoted by

$$\lim_{x \to a-0} f(x) \text{ or } f(a-0).$$

The statement that $\lim_{x \to a} f(x) = b$ involves the equations

$$\lim_{x \to a-0} f(x) = b = \lim_{x \to a+0} f(x).$$

When $f(x)$ approaches a finite limit b as x tends to a we write

$$f(x) \to b \text{ as } x \to a.$$

Similarly, if a right-hand limit b_1 or a left-hand limit b_2 exists as x approaches a from above or from below, we write

$$f(x) \to b_1 \text{ as } x \to a+0,$$

or

$$f(x) \to b_2 \text{ as } x \to a-0.$$

The reader should observe that there is no reason why b_1 and b_2 should in all cases be equal, nor is it necessary that either b_1 or b_2 should be equal to $f(a)$. There are cases in which the three numbers b_1, b_2, $f(a)$ all exist and are equal, and any two of them may be equal while the third is different. Again, it may happen that some or all of these three numbers may not exist. The importance of these remarks will be seen later when continuity is considered; and several illustrative examples are given in § 3·31 below.

If $f(x)$ does not approach a definite limit as $x \to a$, it may diverge to ∞ or diverge to $-\infty$. This is expressed formally as follows.

* The late Dr Leathem has suggested that the notation $x \searrow a$, $x \nearrow a$ would prove useful in these cases. See *The Mathematical Theory of Limits* (1925), 35.

Another notation sometimes used is $x \rightharpoondown a$ and $x \rightharpoonup a$.

If when any positive number G, however large, is chosen, a positive number $\eta(G)$ *can be found such that* $f(x) > G$ *when* $|x-a| \leqslant \eta$, *then we say that*

$$\lim_{x\to a} f(x) = \infty \text{ or } f(x) \to \infty \text{ as } x \to a.$$

Similarly the formal definition of divergence to $-\infty$ can be given.

If $f(x)$ be a function which possesses a graph, the curve $y=f(x)$ possesses an asymptote parallel to the axis of y at the point $x=a$.

3·23. General limit theorems.

Analogy with the definition of the limit of a sequence $\{s_n\}$ as $n\to\infty$ suggests that there will be similar limit theorems to those proved in § 2·561. This is in fact the case, and the proofs of the first three of the following theorems will be left as exercises for the reader.

Suppose that as $x\to a$, $f(x)\to\alpha$ *and* $g(x)\to\beta$, *then*

(1) $f(x)+g(x)\to\alpha+\beta$,

(2) $f(x)\,g(x)\to\alpha\beta$,

(3) $\dfrac{1}{f(x)}\to\dfrac{1}{\alpha}$ *provided that* $\alpha\neq 0$.

(4) *If* $f(u)\to f(\beta)$ *as* $u\to\beta$, *then*

$$\lim_{x\to a} f\{g(x)\} = f\{\lim_{x\to a} g(x)\}.$$

Proof of (4). Since $f(u)\to f(\beta)$ as $u\to\beta$, given ϵ, there corresponds a number $\eta_1(\epsilon)$, such that, when $|g(x)-\beta|\leqslant\eta_1$,

$$|f\{g(x)\}-f(\beta)|<\epsilon \quad \text{.....................(1).}$$

But $g(x)\to\beta$ as $x\to a$, so that corresponding to the above number η_1, a number η (depending on η_1, and therefore on ϵ) can be found such that, when $|x-a|\leqslant\eta$,

$$|g(x)-\beta|<\eta_1 \quad \text{.........................(2).}$$

By combining (1) and (2) we have, when $|x-a|\leqslant\eta$,

$$|f\{g(x)\}-f(\beta)|<\epsilon.$$

Hence $$\lim_{x\to a} f\{g(x)\} = f(\beta) = f\{\lim_{x\to a} g(x)\}.$$

The reader will see later that the condition laid down above for the validity of (4) is equivalent to the statement that the function $f(u)$ must be *continuous** at the point $u=\beta$. Thus it is only true to state that

$$\lim f(g) = f(\lim g)$$

if f be a continuous function.

3·24. Finite limit as x increases indefinitely.

The function $f(x)$ is said to approach the limit b as x tends to infinity if given ϵ, there exists a positive number X such that $|f(x)-b|<\epsilon$ whenever $x \geqslant X$.

When the above condition is fulfilled, we write either

$$\lim_{x\to\infty} f(x) = b \text{ or } f(x)\to b \text{ as } x\to\infty.$$

The reader may easily modify the above definition to include the case in which x tends to infinity through negative values, that is when $x\to-\infty$.

If the function $f(x)$ possesses a graph, the above definition corresponds to the existence of an asymptote parallel to the axis of x; the asymptote is the line $y=b$.

All the different cases of tendency to a limit are included in a single general definition which can be stated as follows:

Any endless progress of x generally determines a corresponding endless progress of $f(x)$. If there is related to this progress a number b, such that if we select any positive number ϵ as small as we please, there is always a corresponding definite stage in the progress of $f(x)$ after which it is always the case that $|f(x)-b|<\epsilon$, then $f(x)$ in this progress tends to a limit.

As we have already seen, in the case of functions of the positive integral variable n, tendency to a finite limit or to infinity are not the only possibilities. In fact monotonic sequences are the only ones which possess this property. The reader should observe that finite or infinite oscillation is possible for functions of the real variable x, and it will be found to be a helpful exercise to formulate precise definitions corresponding to these cases.

The general limit theorems of the preceding paragraph all apply to the case in which x tends to infinity. By contrasting the definitions of the existence of a limit as $x\to a$ and as $x\to\infty$, the reader will see that, with slight modifications, the proofs are almost

* For the definition of a continuous function see § 3·32 below.

identical. In fact, the proofs of these theorems in the case where $x \to \infty$ differ only slightly from the proofs of the fundamental limit theorems for sequences given in § 2·561.

3·3. Continuity.

The notion of continuity is a direct consequence of the concept of a limit. The special class of functions known as *continuous functions* possesses many important properties which will be investigated in this chapter. It is very important to separate the properties of continuous functions from those of functions in general. In many elementary text-books, in which none but continuous functions are contemplated, the properties of continuous functions are assumed and applied freely to every function considered. This leads to no difficulty so long as continuous functions alone are considered, but it is easy to construct cases of failure by considering a discontinuous function for which the property in question fails to hold. Examples which are given later will illustrate this point clearly.

For simplicity we confine ourselves mainly to functions of a single real variable, but the extensions to functions of more than one variable will be suggested in § 3·5. It may, however, be stated that the proofs of most of the theorems which follow can be adapted, usually with only slight changes in terminology, to the case of functions of more than one variable.

3·31. Continuous functions.

Roughly speaking, continuity means the identity of limits with values. Continuity is the property of a point, and the question for our consideration is as follows. Given any function $f(x)$ defined in the range $a \leqslant x \leqslant b$, say, then at the point x_0, where $a < x_0 < b$, is the function $f(x)$ continuous or not?

We can extend the concept of continuity and say that a function $f(x)$ is continuous *in an interval* if it is continuous at every point in the interval. Essentially, however, continuity is the property of a point, and the concept is only extended to an interval in the way just indicated.

Several examples will now be considered.

Example 1. *Consider the function defined as follows:*

$$f(x)=\frac{x^2-a^2}{x-a} \qquad \text{when } 0\leqslant x<a,$$

$$f(x)=a \qquad \text{when } x\geqslant a.$$

Discuss the behaviour of this function at the point $x=a$.

Now, *except when $x=a$,*

$$\frac{x^2-a^2}{x-a}=x+a,$$

but as x approaches the value a from below, it is clear that the value of the fraction $(x^2-a^2)/(x-a)$ becomes more and more nearly equal to $2a$, in fact

$$\lim_{x\to a-0}\frac{x^2-a^2}{x-a}=2a \quad \text{..............................(1).}$$

It must be clearly realised that the *value* of the fraction $(x^2-a^2)/(x-a)$ when $x=a$ is indeterminate or meaningless, for the ratio 0/0 is undefined.

The function $f(x)$, however, has the value a when $x=a$, for this value has been assigned to it *by definition*, hence, *when $x=a$,*

$$f(x)=a \quad \text{..................................(2).}$$

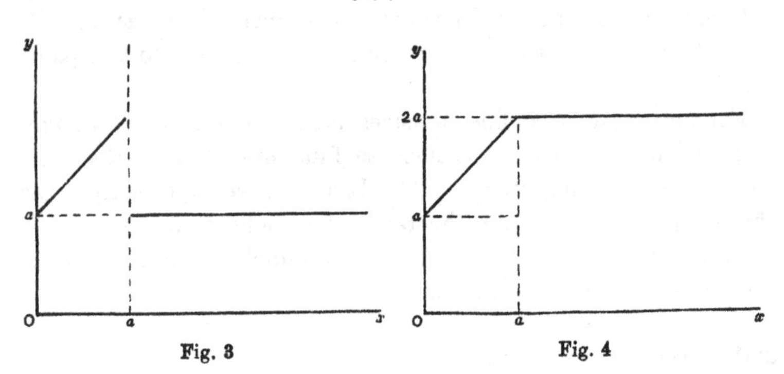

Fig. 3 Fig. 4

Examination of the graph of the function $f(x)$ in Fig. 3 at once suggests a discontinuity* when $x=a$. As we shall see, this is ensured analytically by the inequality of the limit (1) and the value (2).

Example 2. *Consider the function*

$$\phi(x)=\frac{x^2-a^2}{x-a} \qquad \text{when } 0\leqslant x<a,$$

$$\phi(x)=2a \qquad \text{when } x\geqslant a.$$

The graph of $\phi(x)$ is illustrated in Fig. 4, and it suggests no discontinuity when $x=a$.

* The primitive geometrical concept of continuity for a function which possesses a graph is that the function is continuous if its graph be an unbroken curve. A point at which there is a sudden break in the curve is thus a point of discontinuity.

The limit (1) of Example 1 is again the same, but in this case *when* $x=a$

$$\phi(x)=2a \quad \text{.................................(3)},$$

so that the limit of $\phi(x)$ as $x \to a$ is the same as the value of $\phi(x)$ at $x=a$, both being equal to $2a$. In this case there is no sudden break in the graph.

Example 3.
$$\psi(x)=\frac{x^2-a^2}{x-a} \quad \textit{when } 0 \leqslant x < a,$$
$$\psi(x)=a \quad \textit{when } x=a,$$
$$\psi(x)=2a \quad \textit{when } x > a.$$

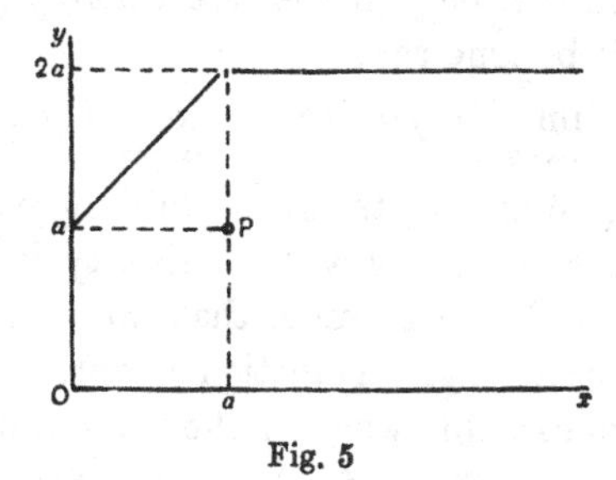

Fig. 5

In this case
$$\lim_{x\to a-0}\psi(x)=2a, \quad \psi(a)=a, \quad \lim_{x\to a+0}\psi(x)=2a,$$
and there is again a discontinuity when $x=a$ due to the isolated point P in Fig. 5.

3·32. Formal definitions of continuity.

The reader who has carefully examined the preceding examples will have realised how easily discontinuities may be recognised in a graph. In fact for those functions which admit of simple graphical representation, this is certainly the easiest way of finding points of discontinuity. We now give the formal definitions of continuity, from which definite conclusions can be drawn for any function, whether it be capable of graphical representation or not. In any case the graphical method illustrated above only *indicates* points of discontinuity of functions which can be represented graphically; it does not *prove* anything.

DEFINITION 1. *The function* $f(x)$ *is said to be continuous when* $x=x_0$, *if* $f(x)$ *possesses a definite limit as* x *tends to the value* x_0 FROM EITHER SIDE, *and each of these limits is equal to* $f(x_0)$:

$$\lim_{x\to x_0-0} f(x)=f(x_0)=\lim_{x\to x_0+0} f(x).$$

DEFINITION 2. *The function* $f(x)$ *is continuous when* $x = x_0$, *if given* ϵ, *a number* $\eta(\epsilon)$ *can be found, such that, whenever* $|x - x_0| \leqslant \eta$,

$$|f(x) - f(x_0)| < \epsilon.$$

The equivalence of the two definitions is at once obvious if the second be compared with the definition of a limit in § 3·22. The remark in that section about the right-hand and left-hand limits as x tends to a not necessarily being equal will now be understood. We have now seen that only in the case where $f(x)$ is continuous when $x = a$ will it be true that

$$\lim_{x \to a-0} f(x) = f(a) = \lim_{x \to a+0} f(x).$$

For most simple functions the graph will suggest at what points in a given range of values of x discontinuity is to be expected. It must however be borne in mind that the analytical method of deciding whether or not a given function is continuous at a specified point x_0 (say) is to examine whether the three numbers

$$\lim_{x \to x_0-0} f(x), \quad f(x_0), \quad \lim_{x \to a+0} f(x)$$

are or are not equal. Cases frequently arise in which one or more of these numbers do not exist. If $f(x)$ assumes, when $x = x_0$, the indeterminate form $0/0$ we must say that $f(x_0)$ *does not exist.* In such a case there cannot be continuity at $x = x_0$, even though $f(x_0 - 0)$ and $f(x_0 + 0)$ both exist and are equal.

The function $(x^2 - a^2)/(x - a)$ is defined for all real values of x *except* $x = a$. Since its value when $x = a$ is indeterminate, nothing more can be said about its continuity when $x = a$. However, a value for the function when $x = a$ may be assigned *by definition.* It will be seen that the function defined as follows:

$$f(x) = \frac{x^2 - a^2}{x - a} \qquad \text{when } x \gtrless a,$$

$$f(x) = A \qquad \text{when } x = a,$$

is a discontinuous function at the point $x = a$ for all values of A except $A = 2a$.

The function defined above when $A = 2a$ is said to be continuous at the point $x = a$ *by completing its definition.*

Continuity in an interval. A function $f(x)$ is said to be continuous in the closed interval (a, b) if it is continuous for every value of x in $a < x < b$, and if $f(a + 0)$ exists and is equal to $f(a)$, and $f(b - 0)$ exists and is equal to $f(b)$.

It is easily deduced from the theorems on limits that the sum, product, difference or quotient of two functions which are continuous at a certain point are themselves continuous at that point (except that, in the case of the quotient, the denominator must not vanish at the point in question).

Further it is true that a continuous function of a continuous function is a continuous function (§ 3·23, 4).

Examples. The polynomial

$$P(x) \equiv a_0 x^n + a_1 x^{n-1} + \ldots + a_n$$

is continuous for all values of x.

The rational function $$R(x) = \frac{P(x)}{Q(x)}$$

is continuous in any interval which does not include values of x which make $Q(x) = 0$.

The functions $\sin x$ and $\cos x$ are continuous for all values of x; $\log x$ is continuous for $x > 0$; e^x is continuous for all values of x; $\tan x$ is continuous in any range not including any of the points $x = (n + \frac{1}{2})\pi$ $(n = 0, 1, 2, \ldots)$.

3·4. Properties of continuous functions.

THEOREM 1. THE FUNDAMENTAL THEOREM ON INTERVALS.

If $I_1, I_2, \ldots, I_n, \ldots$ are linear intervals, each of which is contained in the preceding, and the lengths of which tend to zero as n tends to infinity, then there exists a unique point ξ such that every neighbourhood of ξ contains all but a finite number of the given intervals.

Let I_n be (a_n, b_n), then since I_n contains I_{n+1},

$$a_n \leqslant a_{n+1}, \quad b_n \geqslant b_{n+1}.$$

Hence,

$$a_1 \leqslant a_2 \leqslant \ldots \leqslant a_n \leqslant \ldots \qquad \text{(1)},$$

$$b_1 \geqslant b_2 \geqslant \ldots \geqslant b_n \geqslant \ldots \qquad \text{(2)}.$$

The sequence (1) of left-hand end-points is monotonic increasing and every term is less than b_1, hence, by § 2·58,

$$a_n \to \alpha \leqslant b_1.$$

Similarly the sequence (2) is monotonic decreasing, and

$$b_n \to \beta \geqslant a_1.$$

If we shew that $\alpha = \beta$, we have proved the existence of the unique point ξ.

We have

$$|a_n - \alpha| < \epsilon \text{ for } n \geqslant \nu_1,$$

$$|b_n - \beta| < \epsilon \text{ for } n \geqslant \nu_2,$$

and since the length of I_n tends to zero as $n \to \infty$,

$$|a_n - b_n| < \epsilon \text{ for } n \geqslant \nu_3.$$

Now

$$|\alpha - \beta| = |\alpha - a_n + a_n - b_n + b_n - \beta|$$
$$\leqslant |\alpha - a_n| + |a_n - b_n| + |b_n - \beta| < 3\epsilon$$

for $n \geqslant \nu$, where $\nu = \max(\nu_1, \nu_2, \nu_3)$.

Since α and β are independent of n, it follows that α and β coincide and $\alpha = \beta = \xi$. Clearly h can be so chosen that the neighbourhood $(\xi - h, \xi + h)$ strictly contains the interval I_m, and hence $I_{m+1}, I_{m+2}, \ldots$; and however small h is chosen there will be a value of n such that the length of I_n is small enough for I_n to be strictly contained in the neighbourhood $(\xi - h, \xi + h)$. Hence every neighbourhood of ξ contains all but a finite number of the intervals I.

3·41. THEOREM 2. BOREL'S THEOREM*.

Let (a, b) be a closed linear interval; suppose that we are given a set J of closed intervals such that every point P of (a, b) is an interior point† *of at least one of these intervals $J(P)$; then a* FINITE *number of these closed intervals*

$$J_1, J_2, \ldots, J_m$$

can be chosen which possess the same property:—every point P of (a, b) is an interior point of at least one of them: in other words the interval (a, b) can be completely covered by a FINITE *number of intervals of the set J.*

A set of intervals j_r can be associated with the intervals J as follows. Each j_r is an interval which contains at least one point (not an end-point) P_r of (a, b), and no point of j_r lies outside the interval $J(P_r)$. The intervals j_r will be called *suitable* intervals, or briefly, intervals (S).

If the whole interval (a, b) is suitable, the theorem is proved. If not, bisect (a, b); if either or both of these intervals is not

* Also called the theorem of Borel-Lebesgue or the Heine-Borel theorem. The proof given by Heine that every continuous function is uniformly continuous contained the germ of this theorem (*Journal für Math.* LXXIV (1871), 188). It appears to have been first explicitly stated and proved for a linear interval (a, b) by Borel, *Ann. de l'École Norm.* (3), XII (1895), 50.

† Except when P coincides with a or b, when it is an end-point.

an (S), bisect it or them*. The process of bisecting intervals which are unsuitable either will terminate or it will not. If it does terminate the theorem is proved, for (a, b) will have been divided into a finite number of suitable intervals $j_1, j_2, \ldots, j_m$, say, and the set of associated intervals $J_1, J_2, \ldots, J_m$ will be the *finite* set required.

Suppose that the process of bisection does not terminate, and let an interval which *can* be divided into suitable intervals by the process of bisection be called an interval (B).

By hypothesis (a, b) itself is not a (B), therefore one at least of the bisected portions of (a, b) is not a (B). Choose the one which is not (if neither is a (B) choose the left-hand one); bisect this and select again that bisected part which is not a (B), and so on. This process of bisection gives an unending sequence of intervals

$$I_1, I_2, I_3, \ldots, I_n, \ldots,$$

such that

(i) each is contained in the preceding,

(ii) the length of I_n tends to zero as $n \to \infty$,

(iii) the interval I_n is not an (S).

The same argument as was used in Theorem 1 shews that these intervals converge to a unique limit point ξ, which is certainly a point of (a, b).

By hypothesis there is a number h such that every point within a distance h from ξ is a point of one of the intervals J. If n be chosen so large that the length of I_n is less than h, then ξ is an interior or an end-point of I_n and the distance of every point of I_n from ξ is less than h.

Hence I_n is an (S), which contradicts (iii).

The hypothesis that the process of bisecting intervals does not terminate involves a contradiction: hence the process does terminate and the theorem is proved.

3·411. Note on the preceding theorems.

Both the preceding theorems are of an abstract nature, and the reader may wonder what is their importance. The importance of

* A suitable interval is not to be bisected, for one of the parts into which it is divided might not be suitable.

these theorems cannot be over-emphasised, for they are fundamental. Without appealing to one or other of these it is impossible to prove the theorem which follows concerning the oscillation of a continuous function, a theorem which will be seen to have fundamental importance in the theory of integration.

It is essential that the reader should realise that a point has now been reached beyond which we cannot proceed unless the first or second of the above theorems has been proved. This will be clear when it is seen how all the theorems which follow depend upon those which have preceded them. Since the important theorems of the differential calculus, including Taylor's theorem, depend for their proof upon the properties of continuous functions, the preceding theorems are therefore also fundamental to the differential calculus.

The reader will see that Theorem 1 is sufficient for our immediate requirements, and the properties of continuous functions which will now be proved can be demonstrated without appeal to Borel's theorem. Theorem 1 is a theorem on intervals which is less general than Theorem 2, but Borel's theorem has important applications in other branches of Pure Mathematics*, and it is one of the important theorems in Analysis. There are several other theorems of the same type which are outside the scope of this book; theorems of this kind are sometimes spoken of as "covering theorems."

3·42. THEOREM 3. *If $f(x)$ is continuous in the closed interval (a, b), then, given ϵ, the interval can always be divided up into a finite number of sub-intervals such that*

$$|f(x')-f(x'')|<\epsilon,$$

when x' and x'' are any two points in the same sub-interval.

Suppose that the theorem is false. Then, however we sub-divide (a, b) there must be at least one of the sub-intervals for which the theorem is false.

* It was pointed out by Baker, *Proc. London Math. Soc.* (1), XXXV, 459, and (2), I, 24, that Goursat's proof of Cauchy's fundamental theorem in the theory of functions of a complex variable (*Trans. Amer. Math. Soc.* I (1900), 15) practically contains Borel's theorem.

Borel's theorem is also of fundamental importance in the theory of sets of points; in particular in the theory of measure of a linear set. See de la Vallée Poussin, *Intégrales de Lebesgue* (1916), § 17.

Apply the process of repeated bisection; and at each stage choose a bisected part for which the theorem is false. In this way we obtain a set of intervals (a_n, b_n) satisfying the conditions of Theorem 1, and whose end-points therefore converge to a unique limit point ξ. Also each interval (a_n, b_n) is such that the theorem is false for it. Suppose, for definiteness, that ξ does not coincide with a or b.

Since $f(x)$ is continuous when $x=\xi$, there is a value of η such that, when $|x-\xi| \leqslant \eta$,

$$|f(x)-f(\xi)| < \epsilon.$$

If n be chosen so large that $b_n - a_n$ is less than η, then the interval (a_n, b_n) is contained entirely within the interval

$$(\xi-\eta, \xi+\eta),$$

for ξ is such that

$$a_n \leqslant \xi \leqslant b_n.$$

Thus, if x' and x'' are any two points in (a_n, b_n),

$$|f(x')-f(\xi)| < \epsilon, \quad |f(x'')-f(\xi)| < \epsilon.$$

Hence,
$$|f(x')-f(x'')| \leqslant |f(x')-f(\xi)| + |f(\xi)-f(x'')|$$
$$< 2\epsilon.$$

The supposition of falsity leads to a contradiction, hence the theorem must be true.

The argument is easily modified if ξ coincides either with a or with b.

3·421. THEOREM 4. *A function which is continuous in an interval is bounded therein.*

Suppose that the interval (a, b) is divided into sub-intervals which satisfy the conditions of Theorem 3. Let the dividing points be

$$x_0 = a, \ x_1, \ x_2, \ \ldots, \ x_{n-1}, \ x_n = b.$$

Now
$$|f(x)| \leqslant |f(a)| + |f(x)-f(a)|$$
$$< |f(a)| + \epsilon,$$

when $0 < x-a \leqslant x_1 - a$, so that

$$|f(x_1)| < |f(a)| + \epsilon \quad \ldots\ldots\ldots\ldots\ldots\ldots\ldots(1).$$

Similarly, when $0 < x - x_1 \leqslant x_2 - x_1$,

$$|f(x)| \leqslant |f(x_1)| + |f(x)-f(x_1)|$$
$$< |f(x_1)| + \epsilon$$
$$< |f(a)| + 2\epsilon,$$

by (1). Hence, $|f(x_2)| < |f(a)| + 2\epsilon.$

By proceeding in this way we get, when $0 < x - x_{n-1} \leqslant b - x_{n-1}$,

$$|f(x)| < |f(a)| + n\epsilon \quad \text{.....................(2).}$$

Thus, since inequality (2) is clearly satisfied in the whole interval (a, b), the theorem is proved.

3·43. Uniform continuity.

We have seen that, if $f(x)$ is continuous in (a, b)*, then, given ϵ, we can find a number η such that, whenever $|x' - x| \leqslant \eta$,

$$|f(x') - f(x)| < \epsilon.$$

We know that η depends upon the chosen value of ϵ, but η is also *a function of* x, for η will in general be different at different points of the interval (a, b).

This is at once clear from Fig. 6, for if AB is divided into equal parts each of length ϵ, the corresponding sub-division of (a, b) is such that the value of η is not the same for all points x in (a, b).

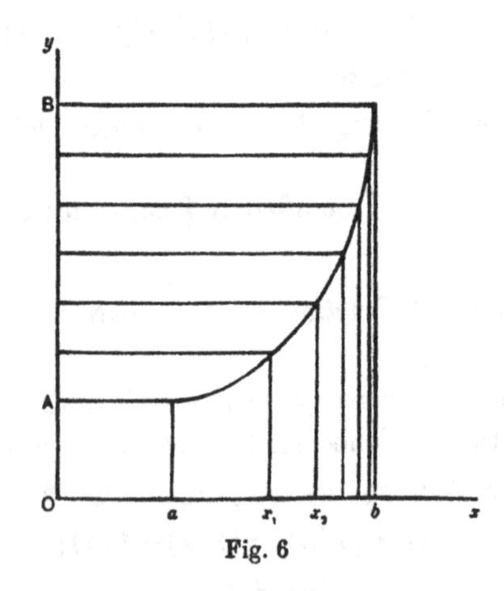

Fig. 6

If it is possible to find a positive number h, such that, when ϵ has been chosen,

$$|f(x) - f(x')| < \epsilon,$$

whenever $$|x' - x| \leqslant h(\epsilon),$$

* The interval (a, b) is taken to mean the *closed* interval unless the contrary is stated.

the number h being INDEPENDENT OF x, then $f(x)$ is said to be *uniformly continuous* in (a, b).

In general, the value of η depends upon the particular pair of values x and x'. The important point in the definition of uniform continuity is that the number h must be *independent of* x. Whenever such a number h can be found,

$$|f(x)-f(x')|<\epsilon$$

for every pair of values x and x' in (a, b) whose distance apart does not exceed h.

THEOREM 5. *A function which is continuous in an interval is uniformly continuous in the interval*.*

Let the given interval (a, b) be divided up into sub-intervals (a, x_1), (x_1, x_2), ..., (x_{n-1}, b), such that for any two points x, x' in the same sub-interval,

$$|f(x)-f(x')|<\tfrac{1}{2}\epsilon.$$

Let h be a positive number not exceeding the least of the numbers

$$x_1-a,\quad x_2-x_1,\quad \ldots,\quad b-x_{n-1}.$$

Now choose any two points x', x'' in (a, b) such that $|x'-x''|$ does not exceed h.

If these two points belong to the same sub-interval, then

$$|f(x')-f(x'')|<\tfrac{1}{2}\epsilon.$$

If they do not belong to the same sub-interval, they certainly lie one in each of two consecutive intervals; suppose x_r is the dividing point such that

$$x_{r-1}<x'<x_r<x''<x_{r+1},$$

then
$$|f(x')-f(x'')|\leqslant|f(x')-f(x_r)|+|f(x_r)-f(x'')|$$
$$<\tfrac{1}{2}\epsilon+\tfrac{1}{2}\epsilon=\epsilon.$$

Hence, given ϵ, there exists a positive number h, such that

$$|f(x')-f(x'')|<\epsilon,$$

when x' and x'' are *any* two values of x in (a, b) such that

$$|x'-x''|\leqslant h.$$

Note. It is essential to the above theorem that (a, b) should be a *closed* interval, so that $f(x)$ is continuous at every point in $a\leqslant x\leqslant b$. For example consider the function $\sin\frac{1}{x}$ in $(0, 1)$.

* This theorem on uniform continuity is due to Heine, *loc. cit.*

Now $\sin\frac{1}{x}$ is continuous in $0<x\leqslant 1$, but it is discontinuous when $x=0$. The function $\sin\frac{1}{x}$ is not uniformly continuous in $(0, 1)$; it is of course uniformly continuous in $(\delta, 1)$ where $\delta>0$.

3·431. Note on the concept of uniformity.

The above theorem shews that it is unnecessary, for a function of a real variable x, to draw any distinction between functions which are uniformly and those which are non-uniformly continuous in the continuous domain of x, for all continuous functions are uniformly continuous.

The concept of uniformity, however, is of fundamental importance in Analysis, and should be clearly understood.

Given ϵ, suppose that $f(x, \xi)$ is a function of two variables x and ξ, which, for every point x of a given closed interval (a, b), satisfies the inequality

$$|f(x, \xi)|<\epsilon \quad\text{..................................(1)},$$

when ξ is given any one of a certain set of values denoted by ξ_x, the particular set of values depending on the particular value of x under consideration. If a set ξ_0 can be found, such that every member of the set ξ_0 is a member of all the sets ξ_x, then the function $f(x, \xi)$ is said to satisfy the inequality (1) *uniformly* for all points x in (a, b). If a given function $\phi(x)$ possesses some property for every value of ϵ in virtue of the inequality (1), the function $\phi(x)$ is then said to *possess the property uniformly*.

One of the most important ideas of Analysis is that of *uniform convergence*. Consider the sequence $s_n(x)$, which is a function of n and of x. In this case the ξ of the preceding statement is the integral variable n.

Suppose that $s_n(x)\to s(x)$ as $n\to\infty$ for all values of x in a given interval (a, b). *If, given* ϵ, *a number* m *can be found such that*

$$|s_n(x)-s(x)|<\epsilon$$

for all values of $n\geqslant m(\epsilon)$ *where* m *is* INDEPENDENT OF x, *then* $s_n(x)$ *is said to converge* UNIFORMLY *to* $s(x)$ *in the interval* (a, b).

In general, m depends on ϵ and on x; if $m(\epsilon, x)$ denotes the *least* m for which the above holds, and if, for fixed ϵ, and for all values of x in (a, b),

$$m(\epsilon, x)\leqslant\mu(\epsilon),$$

then μ can be chosen instead of m, and the convergence of $s_n(x)$ to $s(x)$ is uniform in (a, b).

Uniformity of convergence is a highly important concept, and it will be considered in detail in the last chapter. We have seen that continuity in an interval implies uniformity of continuity, hence the latter concept is not an important one. Convergence however does *not* imply uniformity of convergence, and so uniformity of convergence is a highly important concept.

3·44. THEOREM 6. *A continuous function attains its bounds.*

If $f(x)$ is continuous in (a, b) and M and m are its upper and lower bounds, it will be proved that there are at least two points x_1 and x_2 in (a, b) such that

$$f(x_1) = M, \quad f(x_2) = m.$$

Suppose that M is not attained; then $M - f(x)$ does not vanish at any point of (a, b); hence $(M - f(x))^{-1}$ is a continuous function, and therefore bounded. If $G > 0$ be its upper bound, we have

$$\frac{1}{M - f(x)} \leqslant G,$$

so that $$M - f(x) \geqslant \frac{1}{G},$$

that is, $$f(x) \leqslant M - \frac{1}{G};$$

but this contradicts the fact that M is the upper bound of $f(x)$ in (a, b).

Hence M must be attained.

Similarly it may be proved that m is attained.

COROLLARY. In virtue of the above theorem we can now state Theorem 3 in the following useful form:

If $f(x)$ is continuous in the interval (a, b), then (a, b) can be subdivided into a finite number of partial intervals in each of which the oscillation of $f(x)$ is less than any given ϵ.

3·45. THEOREM 7. *If $f(x)$ is continuous in (a, b) and $f(a)$ and $f(b)$ differ in sign, then $f(x)$ vanishes at least once between a and b.*

To fix the ideas, suppose that $f(a) < 0$ and $f(b) > 0$. Since $f(x)$ is continuous, it will be negative in the neighbourhood of a and positive in the neighbourhood of b. The set of values of x between a and b which make $f(x)$ positive is bounded below by a, and hence possesses an exact lower bound k: clearly $a < k < b$.

From the definition of the lower bound, the values of $f(x)$ must be negative or zero in $a \leqslant x < k$. Since $f(x)$ is continuous when $x = k$,

$$\lim_{x \to k-0} f(x) = f(k).$$

Hence $f(k)$ is also negative or zero. We shall shew that $f(k)$ cannot be negative; for if $f(k)=-c$, where c is positive, then there exists a positive number $\eta(c)$ such that

$$|f(x)-f(k)|<c \quad \text{when} \quad |x-k|\leqslant\eta,$$

since $f(x)$ is continuous when $x=k$. The function $f(x)$ would then be negative for those values of x in (a, b) which lie between k and $k+\eta$, which contradicts the fact that k is the lower bound of the set of values of x between a and b which make $f(x)$ positive.

It follows that $f(k)=0$, and the theorem is therefore proved.

COROLLARY. *If $f(a)$ and $f(b)$ are unequal, and $f(x)$ is continuous in (a, b), $f(x)$ assumes at least once every value between $f(a)$ and $f(b)$.*

Let G be *any* number between $f(a)$ and $f(b)$. The continuous function $\phi(x)=f(x)-G$ has opposite signs when $x=a$ and when $x=b$. Hence, by the theorem, $\phi(x)$ vanishes at least once between a and b, so that for this value of x, $f(x)=G$.

3·5. Continuity of functions of more than one variable.

Consider, for definiteness, the case of three variables x, y, z. The extensions of the preceding theorems to functions of more than one variable are rendered more simple by the following method of reducing multiple limits to simple limits.

If $x\to a$, $y\to b$, $z\to c$, expressed symbolically as

$$(x, y, z)\to(a, b, c) \quad \text{.....................(1)},$$

and we define ω by the equation

$$\omega=|x-a|+|y-b|+|z-c|,$$

then the statement that ω tends to zero is equivalent to the three statements which are symbolically expressed by (1).

If η be defined by the equation

$$\eta=\{(x-a)^2+(y-b)^2+(z-c)^2\}^{\frac{1}{2}},$$

the statement that η tends to zero is also equivalent to (1).

The geometrical interpretation of ω and η is that ω is the length of the shortest path from (a, b, c) to (x, y, z) which is composed solely of segments parallel to the axes, while η is of course the distance between the points (a, b, c) and (x, y, z).

Although the geometrical analogy does not extend to more than three dimensions, the above notions can be extended to a function of n variables if $(x_1, x_2, \ldots, x_n)$ denote the coordinates of a point in n-dimensional hyper-space.

3·51. Continuity for a function of two variables.

For definiteness, the definitions of a limit and of continuity will now be given for a function of two variables x and y. The reader will see that they are the obvious extensions of the definitions already given for functions of one variable.

DEFINITION OF A LIMIT. *The function $f(x, y)$ is said to tend to the limit l as x tends to the value a and y tends to the value b if, given ϵ, a positive number $\mu(\epsilon)$ can be found such that*

$$|f(x, y) - l| < \epsilon,$$

for all values of x and y such that

$$|x-a| \leqslant \mu \text{ and } |y-b| \leqslant \mu.$$

Thus for all points (x, y) within the square of centre (a, b) and whose sides are of length 2μ, the values of $f(x, y)$ will differ from l by less than ϵ.

Clearly we could substitute for the square a circle of centre (a, b) and in this case the statement that

$$\sqrt{\{(x-a)^2 + (y-b)^2\}} \leqslant \eta,$$

replaces the statements $|x-a| \leqslant \mu$, $|y-b| \leqslant \mu$.

DEFINITION OF CONTINUITY. *The function $f(x, y)$ is said to be continuous at the point (a, b) if, given ϵ, we can find a positive number $\mu(\epsilon)$ such that*

$$|f(x, y) - f(a, b)| < \epsilon$$

whenever $|x-a| \leqslant \mu$ and $|y-b| \leqslant \mu$.

The alternative definition of continuity is that $f(x, y)$ is continuous at the point (a, b) if $f(x, y) \to f(a, b)$ when $x \to a$ and $y \to b$ *in any manner*.

The reader should observe that to assert the continuity of a function of two variables is to assert more than its continuity with respect to each variable separately. Clearly, if $f(x, y)$ is continuous in both variables in accordance with the above definition, it will also be continuous with respect to x (or y) when any fixed value is given to y (or x). The following example will shew that the converse is not necessarily true.

Example. Consider the function

$$f(x, y) = \frac{2xy}{x^2+y^2} \qquad \text{when } x \text{ and } y \text{ are not both zero,}$$

$$f(0, 0) = 0.$$

If $y=\beta$ the function $2x\beta/(x^2+\beta^2)$ is continuous for all values of x, including the value $x=0$, and similarly the function $2\alpha y/(\alpha^2+y^2)$ is continuous for all values of y, including the value $y=0$.

Write $x=\eta\cos\theta$, $y=\eta\sin\theta$, then the statements $x\to 0$, $y\to 0$ are equivalent to the statement that $\eta\to 0$. But $f(x, y)=\sin 2\theta$, which is independent of η and may have any value between -1 and 1.

Since the value of the function at the point (0, 0) is zero, the limit and the value are not the same at the origin, so the function is not continuous in both variables in any domain which includes the origin.

The explanation is as follows. When we assign a fixed value β to y the consideration of the continuity of the function $2x\beta/(x^2+\beta^2)$ involves only the behaviour of this function along the axis of x. A similar remark applies to the function of y obtained by giving x a fixed value α. Continuity in both variables involves that the function $f(x, y)$ shall tend to the same limit when the point (a, b) is approached along *any* radius of the circle whose centre is (a, b) and which passes through the point (x, y); that is to say, ***the limit must be independent of*** θ.

3·6. Some special functions.

An illustration has already been given in § 3·1 of a type of function which has a good deal of importance in Analysis. The example given there was the function defined as follows:

$$y=2 \quad \textit{when } x \textit{ is rational},$$
$$y=0 \quad \textit{when } x \textit{ is irrational}.$$

The reader who has only been accustomed to functions defined by analytical expressions may be inclined to regard such functions as the one given above as artificial and unimportant. This is not however the case, and functions of this type are frequently met with in Analysis.

Also, functions which at first sight seem to have no possible analytical expression sometimes have quite simple ones. We shall consider a few examples.

(1) Let

$$\begin{aligned} y&=1 \quad \textit{when } x>0,\\ y&=0 \quad \textit{when } x=0,\\ y&=-1 \quad \textit{when } x<0. \end{aligned}$$

The graph is illustrated in Fig. 7.

Analytical expressions for this function are

$$y=\frac{2}{\pi}\lim_{n\to\infty}\text{arc tan}\, nx=\frac{2}{\pi}\int_0^\infty \frac{\sin xt}{t}\,dt.$$

This function is important in the Theory of Numbers; it is sometimes called "signum x" and it is written

$$y=\text{sgn}\, x.$$

(2) Let

$$y=1 \quad \textit{when } x \neq 0,$$
$$y=0 \quad \textit{when } x=0.$$

An analytical expression for this function is

$$y=\lim_{n\to\infty}\frac{nx}{1+nx}.$$

(3) Even the function referred to at the beginning of this section may be represented by the somewhat complicated analytical formula

$$y=2\{1-\lim_{n\to\infty}\operatorname{sgn}(\sin^2 n!\,\pi x)\}.$$

It is a special case of Dirichlet's function, and, as will be seen later in § 7·32, it is a function which cannot be integrated by Riemann's method. The graph of this function cannot be represented in a diagram.

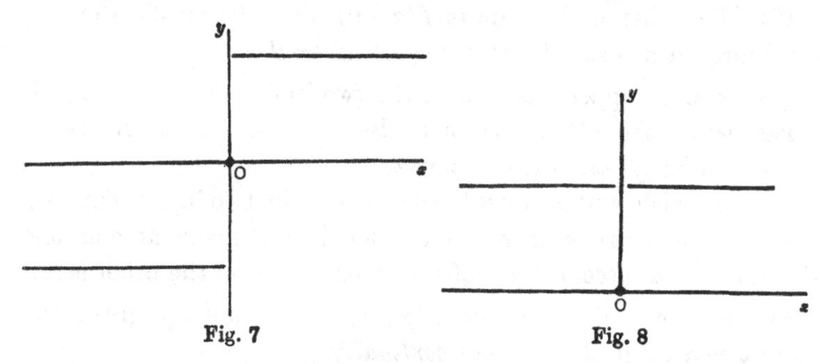

Fig. 7 Fig. 8

3·61. Classification of functions.

The functions considered above may be described as "functions defined as limits." A classification of functions has been made by Baire*, in which continuous functions constitute class 0, discontinuous functions which are limits of continuous functions constitute class 1, the limits of functions of class 1 which are neither of class 1 nor of class 0 constitute class 2, and so on.

The function given in example (2) above is a function of Baire's class 1. The function $nx/(1+nx)$ is, for all values of n, a continuous function of x and is therefore of class 0. The function $y=\lim_{n\to\infty} nx/(1+nx)$ is discontinuous when $x=0$, and is therefore of class 1. The graph is illustrated in Fig. 8.

3·7. Classification of discontinuities.

The points of discontinuity of a function may be classified as follows:

(1) If $x=\alpha$ be the point under consideration, then if both the limits $f(\alpha-0)$, $f(\alpha+0)$ exist and have different values, the point α is said to be a point of discontinuity *of the first kind*, or a point of *ordinary* discontinuity.

* See the Borel tract by de la Vallée Poussin, *Intégrales de Lebesgue*, § 33 *et seq.*

The difference between the greatest and least of the three numbers $f(\alpha+0)$, $f(\alpha-0)$, $f(\alpha)$ is the *saltus*, or measure of discontinuity of the function at the point α.

If $f(\alpha)=f(\alpha-0)$ while $f(\alpha)\neq f(\alpha+0)$, the function is said to be *ordinarily discontinuous at α on the right.* A similar definition can be given of an ordinary discontinuity on the left.

If it happens that $f(\alpha+0)$ and $f(\alpha-0)$ have equal values which differ from $f(\alpha)$, the discontinuity at α is said to be *removable,* since by altering the functional value at the one point α the function can be made continuous at the point.

(2) If neither of the limits $f(\alpha+0)$, $f(\alpha-0)$ exists, the discontinuity at α is said to be *of the second kind.*

(3) It may happen that one of the two limits $f(\alpha+0)$, $f(\alpha-0)$ exists, while the other does not. Such a point α is sometimes called a *point of mixed discontinuity.*

If $f(\alpha)$ exists and is equal to that one of the two limits $f(\alpha+0)$, $f(\alpha-0)$ which exists, then the function is continuous at α on one side and has a discontinuity of the second kind on the other side.

(4) If either of the limits $f(\alpha\pm 0)$ is indefinitely great, the point α is a point of *infinite discontinuity.*

(5) When $f(x)$ oscillates at α on one side or the other, α is said to be a *point of oscillatory discontinuity.* The oscillation is *finite* when $f(x)$ is bounded in some neighbourhood of α; it is *infinite* when there is no neighbourhood of α in which $f(x)$ is bounded.

3·8. Semi-continuous functions*.

The condition of continuity of a function $f(x)$ at a point x, namely that, given ϵ, an open interval $(x-h, x+h)$ exists such that for any point x' in it, $|f(x')-f(x)|<\epsilon$, can be divided into two separate conditions,

$$\text{(i)}\quad f(x')<f(x)+\epsilon,$$

and

$$\text{(ii)}\quad f(x')>f(x)-\epsilon.$$

It is possible that at a point x one of these conditions may be satisfied and not the other. This fact gives rise to the concept of semi-continuity.

If $f(x)$ be a function defined in a given interval (a, b) and if, corresponding to every chosen value of ϵ, an open neighbourhood $(x-h, x+h)$ of a particular point x can be determined such that, for every point x' in this open interval

* The concept of semi-continuity is due to Baire, *Annali di Mat.* (3 A), III (1899). For further information the reader should consult Hobson, *Functions of a Real Variable*, I (1921), 290.

the condition $f(x') < f(x) + \epsilon$ is satisfied; then the point x is said to be a point of *upper semi-continuity* of the function $f(x)$.

If an open neighbourhood of the point x can be determined for each ϵ, such that $f(x') > f(x) - \epsilon$, then the point x is said to be a point of *lower semi-continuity*.

It is clearly necessary that both the above conditions shall be satisfied in order that x may be a point of continuity of $f(x)$.

If every point of the interval (a, b) in which $f(x)$ is defined is a point of upper semi-continuity, then the function $f(x)$ is said to be an *upper semi-continuous function* in (a, b).

A similar definition applies to a lower semi-continuous function.

3·9. Continuity in an infinite interval.

Suppose that $f(x)$ is continuous in the interval $x \geqslant a$, where a is some definite positive number, and that $\lim_{x\to\infty} f(x)$ exists.

If we write $\xi = a/x$, the interval $x \geqslant a$ becomes $0 < \xi \leqslant 1$; with the values of ξ in $0 < \xi \leqslant 1$ associate the values of $f(x)$ at the corresponding points in $x \geqslant a$, and assign as the value of the function when $\xi = 0$ the number $\lim_{x\to\infty} f(x)$. We thus obtain a function of ξ which is continuous in the closed interval $(0, 1)$. On this understanding we can speak of the continuity of a function $f(x)$ in the closed interval (a, ∞) and we denote by $f(\infty)$ the value of $\lim_{x\to\infty} f(x)$.

To a function $f(x)$ continuous in (a, ∞) we can extend some of the results already proved for functions which are continuous in a finite interval (a, b). In particular, such a function $f(x)$ is bounded, attains its bounds, and takes at least once every value between these bounds.

The function $\dfrac{x^2}{1+x^2}$ is continuous in $(0, \infty)$; it does not attain its upper bound, unity, when $x \geqslant 0$, but $f(\infty) = \lim_{x\to\infty} \{x^2/(1+x^2)\} = 1$, so that its upper bound is attained at the point $x = \infty$.

EXAMPLES III.

1. If $f(x) < \phi(x)$ for $a - h < x < a + h$ while $f(x) \to l_1$ and $\phi(x) \to l_2$ as $x \to a$, shew that $l_1 \leqslant l_2$.

2. If $\lim_{x\to a} f(x) = l$, shew that $\lim_{x\to a} |f(x)| = |l|$. Prove that, unless $l = 0$, the converse does not hold.

3. Find the limits of the following functions*:

(i) $\dfrac{\sin 2x}{x} + \dfrac{x}{\sin 2x} + \dfrac{1-x}{1+x}$ as $x \to 0$,

(ii) $x^2 - x + 2 + \dfrac{x-1}{\sin(x-1)} + \cos(x-1)$ as $x \to 1$,

* Assume that $\lim_{x\to 0} \dfrac{\sin x}{x} = 1$, see p. 98.

(iii) $\dfrac{1+\cos \pi x}{\tan^2 \pi x}$ as $x \to 1$,

(iv) $\sec x - \tan x$ as $x \to \frac{1}{2}\pi$,

(v) $\dfrac{\sqrt{(1+x)}-\sqrt{(1+x^2)}}{\sqrt{(1-x^2)}-\sqrt{(1-x)}}$ as $x \to 0$.

4. Prove that the function $f(x) = x \sin \dfrac{1}{x}$ $(x \neq 0)$,

$$f(0)=0$$

is continuous for every value of x; and sketch its graph.

Is $f(x)$ uniformly continuous in the range $(0, 1/\pi)$?

5. Shew that the function $f(x)$, which is equal to 0 when $x=0$, to $\frac{1}{2}-x$ in $0<x<\frac{1}{2}$, to $\frac{1}{2}$ when $x=\frac{1}{2}$, to $\frac{3}{2}-x$ when $\frac{1}{2}<x<1$ and to 1 when $x=1$, is discontinuous when $x=0$, $x=\frac{1}{2}$, $x=1$.

Illustrate by a sketch of the graph in the interval (0, 1).

6. Shew that the function defined as follows:

$$\phi(x) = (x^2/a) - a \qquad 0<x<a,$$
$$\phi(x) = 0 \qquad x=a,$$
$$\phi(x) = a - (a^3/x^2) \qquad x>a$$

is continuous at $x=a$. Sketch its graph. Is $\phi'(x)$ continuous when $x=a$?

7. Find the sum to n terms of the series

$$\frac{x}{x+1} + \frac{x}{(x+1)(2x+1)} + \frac{x}{(2x+1)(3x+1)} + \ldots;$$

by making $n \to \infty$, find the sum-function $s(x)$ of the infinite series. Shew that $s(x)$ is a discontinuous function in the neighbourhood of the origin.

8. Construct an example to illustrate that if $f(x)$ be continuous at every point of the infinite interval $x \geqslant a$, $f(x)$ is not necessarily uniformly continuous. What happens if, in addition, $f(x)$ tends to a limit as $x \to \infty$?

9. For the function $\phi(x) = x \log \sin^2 x$ when $x>0$,

$$\phi(0)=0,$$

discuss the continuity on the right at the origin.

10. (i) Discuss the continuity in the neighbourhood of the origin of the function

$$f(x, y) = \frac{2xy^2}{x^3+y^3} \quad \text{when } x \text{ and } y \text{ are not both zero}$$
$$f(0, 0) = 0.$$

(ii) Consider the same problem for the function

$$\phi(x, y) = \frac{2xy}{\sqrt{(x^2+y^2)}}, \quad \phi(0, 0)=0.$$

11. Deduce *from Borel's theorem* that, if $f(x)$ is continuous in (a, b), the interval can be divided into a finite number of sub-intervals in each of which

the oscillation of $f(x)$ is less than any assigned ϵ. Shew also that, given ϵ, we can find a number η such that if (a, b) be divided into sub-intervals of length less than η, then the oscillation of $f(x)$ in each of them will be less than ϵ.

12. Examine whether the conditions of Borel's theorem are satisfied for the interval $(0, 1)$ in the following cases: (i) if the interval is divided into n equal parts; (ii) if with each point ξ of $(0, 1)$ an interval $(\xi-\delta, \xi+\delta)$ is associated, where $0<\delta<1$, and with the points 0 and 1 we associate the intervals $(0, \delta)$, $(1-\delta, 1)$ respectively; (iii) with each of the rational points p/q we associate the interval $\left(\frac{p}{q}-\frac{\delta}{q^3}, \frac{p}{q}+\frac{\delta}{q^3}\right)$ (in each case rejecting the parts of the intervals associated with 0 and 1 which lie outside $(0, 1)$).

13. Shew that the function

$$y=\lim_{n\to\infty}\frac{f(x)+n\phi(x)\sin^2\pi x}{1+n\sin^2\pi x}$$

is equal to $f(x)$ when x is an integer, but is equal to $\phi(x)$ in every other case. Define the same function by another analytical expression.

14. Sketch the graph of the function $y=f(x)$, where

$$f(x)=\lim_{n\to\infty}\frac{\log(2+x)-x^{2n}\sin x}{1+x^{2n}}$$

in the interval $0\leqslant x\leqslant\frac{1}{2}\pi$; and explain why the function does not vanish anywhere in this interval, although $f(0)$ and $f(\frac{1}{2}\pi)$ differ in sign.

15. Find the points of discontinuity of the function defined as follows in $(0, 1)$: $f(x)=(\log m)^{-1}$ when $x=(2n+1)/2^m$, where n takes all integral values such that $2n+1<2^m$, and m ranges from 1 to ∞; otherwise $f(x)=0$.

16. Classify the points of discontinuity of the following functions:

(i) $f(x)=\sin x/x$ when $x\neq 0$, $f(x)=A$ when $x=0$.

(ii) $f(x)=\dfrac{1}{x-a}\operatorname{cosec}\dfrac{1}{x-a}$.

(iii) $f(x)=1/(1-e^{1/x})$.

(iv) $f(x)=pnx$, where pnx denotes the positive or negative excess of x over the nearest integer; when x exceeds an integer by $\frac{1}{2}$ let $pnx=0$.

17. Prove that, if $f(x)$ is an upper semi-continuous function in (a, b), then $f(x)$ has an upper bound, and it attains that upper bound somewhere in (a, b).

18. *The upper limit of the function $f(x)$ (bounded in $0\leqslant x<1$) as $x\to 1$ ($\overline{\lim}_{x\to 1} f(x)$) may be defined as $\lim_{y\to 1} g(y)$, where $g(y)$ is the upper bound of $f(x)$ in $y\leqslant x<1$.*

Define similarly $\underline{\lim} f(x)$ as $x\to 1$.

Prove that if $y=(1-x^2)\sin 1/x$, $\underline{\lim}\, y=-1$, $\overline{\lim}\, y=1$ as $x\to 0$.

19. If
$$y=\lim_{n\to\infty}\frac{x^n\left(a+\sin\dfrac{1}{x-1}\right)+b+\sin\dfrac{1}{x-1}}{x^n+1},$$
prove that, as $x\to 1$,
$$R\,\overline{\lim}\,y=a+1,\qquad L\,\overline{\lim}\,y=b+1,$$
$$R\,\underline{\lim}\,y=a-1,\qquad L\,\underline{\lim}\,y=b-1.$$

20. Shew that the function
$$f(x)=\frac{x\tan x+2x-1}{x+1}$$
does not tend to a unique limit as $x\to\infty$ *through all real values*, but if x ranges through the sequence of values $x=n\pi+\frac{\pi}{4}$ $(n=0,\ 1,\ 2,\ \ldots)$, then $f(x)\to 3$.

21. Prove that $\quad \lim_{x\to\infty}\left(1+\dfrac{1}{x}\right)^x=\lim_{u\to 0}(1+u)^{1/u}=e.$

[It must be proved that the function $\left(1+\dfrac{1}{x}\right)^x$ approaches the same limit when x tends to infinity ranging through *all real values*, as it does when x ranges through the particular sequence of values $1,\ 2,\ 3,\ \ldots,\ n,\ \ldots.$

If
$$n\leqslant x<n+1,$$
$$1+\frac{1}{n}\geqslant 1+\frac{1}{x}>1+\frac{1}{n+1},$$
and so
$$\left(1+\frac{1}{n}\right)^{n+1}>\left(1+\frac{1}{x}\right)^x>\left(1+\frac{1}{n+1}\right)^n,$$
and so on. See § 2·8.

The second limit is equivalent to the first if $u=1/x$.]

22. Outline a proof of Borel's theorem for a two-dimensional region.

[In the notation of § 3·41, the interval $(a,\ b)$ is replaced by a closed two-dimensional region, the interval $J\,(P)$ by a circle of centre P, and the interval j_r by a square with sides parallel to the axes.]

23. If $a_1,\ a_2,\ \ldots,\ a_p$ are all positive and if
$$\mu_n=(a_1{}^n+a_2{}^n+\ \ldots\ +a_p{}^n)/p,$$
prove that the sequence $\{\mu_{n+1}/\mu_n\}$ steadily increases; and deduce that the same is true of $\mu_n{}^{1/n}$.

24. For the function $f(x)$ of Example 4, find an upper bound of $|f'(x)|$ in $\frac{1}{2}\epsilon\leqslant x\leqslant 1$ and hence shew that, if the interval $(-1,\ 1)$ be divided into n sub-intervals in each of which the oscillation of $f(x)$ does not exceed ϵ, then n need not exceed $4\epsilon^{-2}+2$.

CHAPTER IV

DIFFERENTIAL CALCULUS

4·1. Introduction.

Throughout this treatment prominence will be given to the concept of a "differential." This important concept is either ignored or else badly and inadequately treated in almost every English text-book. The concept of a "differential," which is a development of the work of Leibniz, has, with few exceptions, been fully treated only in Continental text-books*. Although for functions of one variable the concept of a differential is not so important as it becomes when dealing with functions of several variables, it is essential at the outset to distinguish carefully between derivatives and differentials, and between the processes of derivation and differentiation. In subsequent chapters the reader will see how the formal treatment of differentiability of functions of more than one variable can be rendered both more simple and more concise by the use of the differential notation.

4·11. Derivatives.

Let $y=f(x)$ denote a single-valued function of x in a given interval (a, b), and let x be any point in this interval. If x be given a positive or negative increment $\Delta x = h$, then the corresponding increment of y, which we denote by Δy, will be

$$f(x+h)-f(x).$$

The ratio of these increments is

$$\frac{\Delta y}{\Delta x}=\frac{f(x+h)-f(x)}{h}.$$

If this ratio tends to a definite limit, as h tends to zero, this limit is called the *derivative* of $f(x)$ at the point x, and it is usually

* See, for example, de la Vallée Poussin's *Cours d'Analyse Infinitésimale*, I (1921), Chs. I and III, in which the theory of differentials is given its due importance in the development of the subject.

For a historical account of the dispute between Newton and Leibniz and its consequences, see Rouse Ball's *History of Mathematics* (1912), 356–362.

denoted by $f'(x)$. Other accepted notations for the derivative* are y', $D_x y$, or Dy.

If h be restricted to have only *positive* values, and if a definite limit exists when h approaches zero *through positive values* only, this limit may be called the *right-hand derivative* of $f(x)$ at the point x and it will be denoted by $Rf'(x)$.

If h approaches zero through negative values only, the *left-hand derivative* $Lf'(x)$ is similarly defined.

The function $f(x)$ only possesses a unique derivative $f'(x)$ if

$$Rf'(x) = Lf'(x).$$

If the function $f(x)$ possesses a unique derivative at every point of the open interval (a, b), and further, a right-hand derivative at a and a left-hand derivative at b, then $f(x)$ is said to be DERIVABLE† *in the interval* (a, b).

We now prove three elementary but very important theorems.

(1) *Every function $f(x)$ which possesses a* FINITE *derivative for a given value of x must be a continuous function of x at this point.*

Since $f(x)$ possesses a finite derivative at the point x, say α,

$$\lim_{h \to 0} \frac{f(x+h) - f(x)}{h} = \alpha,$$

hence

$$|f(x+h) - f(x)| = |h| (|\alpha| + \epsilon),$$

where $\epsilon \to 0$ as $h \to 0$. Since α is finite we have

$$|f(x+h) - f(x)| \to 0 \text{ as } |h| \to 0.$$

Hence $f(x)$ is continuous at the point x.

(2) *If $f(x)$ be a constant, its derivative is zero.*

For

$$\lim_{h \to 0} \frac{f(x+h) - f(x)}{h} = \lim \frac{0}{h} = 0.$$

(3) *If $f(x) = x$, its derivative is unity.*

For

$$\lim_{h \to 0} \frac{(x+h) - x}{h} = \lim \frac{h}{h} = 1.$$

* The reader who is accustomed also to use the notation $\frac{dy}{dx}$ should note that in this treatment we cannot *yet* allow it to be used to denote the *derivative* of $f(x)$.

† The reader is warned not to use the term *differentiable* as a synonym for the term *derivable*. The definition of differentiability is given in § 4·12.

4·12. Differentials.

A function $f(x)$ is said to be *differentiable* at the point x, if it is finite and determinate in the neighbourhood of this point, and if when x is given the increment Δx, which may be assigned arbitrarily, the increment Δy can be expressed in the form

$$\Delta y = A\Delta x + \epsilon\Delta x \quad\text{.......................(1)},$$

where A is independent of Δx, and $\epsilon \to 0$ as $\Delta x \to 0$.

In this case the first term on the right-hand side of (1) is called the *differential of y*, and it is denoted by dy (or by df).

Thus $$dy = A\Delta x,$$

so that $$\Delta y = dy + \epsilon\Delta x.$$

By making $\Delta x \to 0$ we deduce from equation (1) that

$$\lim \frac{\Delta y}{\Delta x} = A;$$

and hence that, if $f(x)$ is differentiable, $f'(x)$ exists and has a finite and definite value A.

Also, by the definition of the derivative, if $f'(x)$ has a definite value A, equation (1) holds.

Thus, the necessary and sufficient condition that the function $y = f(x)$ should be differentiable at the point x is that it possesses a finite definite derivative at this point.

When this is so, we have

$$dy = f'(x)\,\Delta x \quad\text{.......................(2)},$$

and so the *differential of a function* is the product of its derivative and an (arbitrary) increment Δx of the independent variable x.

Now suppose that $f(x) = x$; we have seen that $f'(x) = 1$ and equation (2) becomes

$$dx = \Delta x;$$

and so we define the differential of the INDEPENDENT *variable* to be the same as the (arbitrary) increment of that variable.

Equation (2) now becomes

$$dy = f'(x)\,dx \quad\text{.......................(3)},$$

or, dividing by dx, $$\frac{dy}{dx} = f'(x) \quad\text{.........................(4)}.$$

We have thus proved that the derivative of a function $f(x)$, of one variable x, is the ratio of the differential of the function and the differential of the variable. This now enables us to employ the expression for the derivative which is most frequently used, and it is due to Leibniz.

The reader will now see that, from the point of view of the differential notation, the statement which is frequently made in elementary books that $\frac{dy}{dx}$ is *not a ratio*, but a symbol denoting the operation $\frac{d}{dx}(y)$, is quite misleading. Of course it is allowable to *define* $\frac{dy}{dx}$ to have whatever meaning is most suitable for our purpose, but it is not easy to give any justification for writing an equation such as $dy=\phi(x)\,dx$ unless dy and dx have been defined. When $\frac{dy}{dx}$ is defined as synonymous with $f'(x)$, then no meaning can be given to dx and dy standing alone, unless dx and dy are subsequently *defined.*

4·121. Geometrical interpretation of a differential.

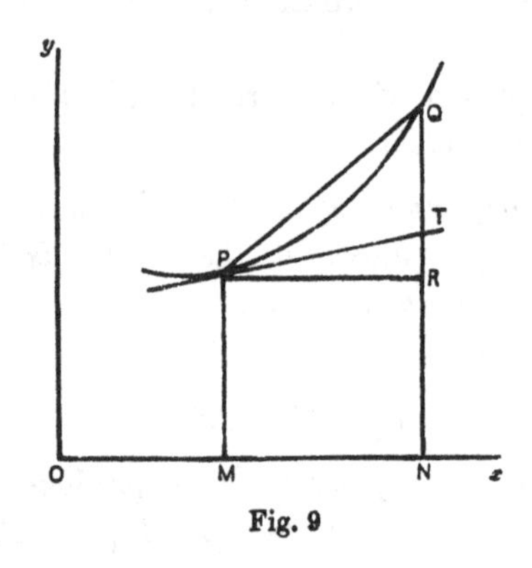

Fig. 9

Let P and Q be the points (x, y), $(x+\Delta x, y+\Delta y)$ on the curve $y=f(x)$. The derivative $f'(x)$ is the slope of the tangent to the curve at P, so that

$$f'(x)=\tan T\hat{P}R=\frac{RT}{PR}=\frac{RT}{MN}.$$

Now by equation (4) above, $\frac{dy}{dx}=f'(x)$,

and $MN=\Delta x$ (which for the *independent* variable is the same as dx), so that $dy=RT$, while $\Delta y=RQ$. This illustrates the important fact that Δy and dy are not the same, for

$$\Delta y-dy=RQ-RT=TQ.$$

This geometrical consideration suggests that the nearer Q is to P, that is the smaller Δx (or dx) is taken, the smaller does the difference between Δy and dy become. It assists us to understand that, as $\Delta x \to 0$, we may expect that

$$\frac{\Delta y}{\Delta x} \to \frac{dy}{dx}.$$

4·2. Infinitesimals.

In the older forms of the Infinitesimal Calculus the fallacious theory that numbers, called infinitesimals, existed, which, whenever convenient, could be neglected, has led to a good deal of confusion of ideas in the presentation of the foundations of the differential and integral calculus.

By assuming that every section of the aggregate of real numbers defines a single real number, it is implicitly assumed that *if a and b are any two positive real numbers such that $a < b$, then a positive integer n can be found such that $na > b$.* This is the *theorem of Eudoxus* (usually called the *principle of Archimedes*). It is not difficult to prove that the aggregate of real numbers satisfies the theorem of Eudoxus*, and an important consequence of this fact is that so-called infinitesimal numbers do not exist within the aggregate. Every positive number ϵ, being such that an integer n can be found so that $n\epsilon > 1$ is a "finite" number, in the sense in which *finite numbers* were distinguished from *infinitesimals* in the older forms of the Infinitesimal Calculus.

A correct theory of infinitesimals can be based upon the following definition:

An infinitesimal is a VARIABLE *whose limit is zero.*

A variable α is said to be infinitesimal with respect to another variable β when the ratio α/β tends to zero. Two infinitesimals α and β are said to be *of the same order* if a positive constant κ exists such that $|\alpha| \leqslant \kappa |\beta|$. The above definitions are frequently expressed concisely by writing

$$\alpha = o(\beta), \quad \alpha = O(\beta)$$

respectively†.

A particular infinitesimal α is chosen, called the *principal infinitesimal*, and by means of it all other infinitesimals may be

* For a proof, see Hobson, *Functions of a Real Variable*, I, 41.

† The "O" notation is only completely unambiguous if we know the limit to which the variable tends. Thus the function $2x + 3x^2$ is $O(x)$ when $x \to 0$, but $O(x^2)$ when $x \to \infty$.

classified. Thus, an infinitesimal β which is such that $\beta = O(\alpha)$ is said to be *of the first order*. If $\beta = O(\alpha^r)$, then β is an infinitesimal *of order r*.

If a function be expressible as a sum of a number of infinitesimals of different orders, the one of lowest order is called the *principal part*.

Example. Consider the function $(x+y-a)^2 - b^2$, where $x \to a$ and $y \to b$. Then $x-a$ and $y-b$ are the principal infinitesimals. It is clear that the given function is an infinitesimal since its limit, as $x \to a$ and $y \to b$, is zero. By writing

$$(x+y-a)^2 - b^2 = 2b\{(x-a)+(y-b)\} + \{(x-a)+(y-b)\}^2$$

the first term on the right-hand side is the principal part, and the second term on the right is an infinitesimal of order 2.

4·21. Theorems on infinitesimals.

THEOREM 1. *The limit of the ratio of two infinitesimals α and β is unaltered when two other infinitesimals α_1 and β_1 are respectively substituted for them so long as α/α_1 and β/β_1 both tend to unity.*

For
$$\frac{\alpha_1}{\beta_1} = \frac{\alpha}{\beta} \cdot \frac{\alpha_1}{\alpha} \cdot \frac{\beta}{\beta_1},$$

and by the fundamental limit theorems,

$$\lim \frac{\alpha_1}{\beta_1} = \lim \frac{\alpha}{\beta} \lim \frac{\alpha_1}{\alpha} \lim \frac{\beta}{\beta_1}$$

$$= \lim \frac{\alpha}{\beta}.$$

THEOREM 2. *If the ratio of two infinitesimals α and α_1 tends to unity, then their difference δ is infinitesimal with respect to each of them; and conversely.*

For
$$\delta = \alpha - \alpha_1$$

may be written
$$\frac{\delta}{\alpha_1} = \frac{\alpha}{\alpha_1} - 1,$$

but the right-hand side tends to zero, and hence $\delta = o(\alpha_1)$.

Conversely since
$$\frac{\alpha}{\alpha_1} = 1 + \frac{\delta}{\alpha_1},$$

and $\delta = o(\alpha_1)$, then $\alpha/\alpha_1 \to 1$.

Similarly it may be shewn that $\delta = o(\alpha)$, and that if $\delta = o(\alpha)$, $\frac{\alpha_1}{\alpha} \to 1$.

4·22. Application to differentials.

For the function $y = f(x)$, *if* Δx *be infinitesimal, and if the derivative* $f'(x)$ *is finite and different from zero at a given point* x, *then* Δy *and* dy *are equivalent infinitesimals.*

By the definition of the derivative

$$\frac{\Delta y}{\Delta x} = f'(x) + \epsilon,$$

where $\epsilon \to 0$ as $\Delta x \to 0$.

Hence, by using the formula $dy = f'(x) . \Delta x$, we get

$$\frac{\Delta y}{dy} = 1 + \frac{\epsilon}{f'(x)},$$

since, by hypothesis, $f'(x)$ does not vanish.

Now ϵ tends to zero as Δx tends to zero, and so

$$\frac{\Delta y}{dy} \to 1.$$

Thus, *when* Δx *is infinitesimal*, Δy and dy are two infinitesimals which may be substituted for each other.

4·3. Rules for derivation.

Let u and v be two given functions of x, and suppose that $\frac{du}{dx}$ and $\frac{dv}{dx}$ exist and are finite, then it is easily shewn that

(i) if $y = u \pm v$ then $\frac{dy}{dx} = \frac{du}{dx} \pm \frac{dv}{dx}$,

(ii) if $y = uv$ then $\frac{dy}{dx} = u\frac{dv}{dx} + v\frac{du}{dx}$,

(iii) if $y = \frac{u}{v}$ then $\frac{dy}{dx} = \left(v\frac{du}{dx} - u\frac{dv}{dx}\right)\Big/ v^2$, $(v \neq 0)$.

To illustrate the method it will suffice to prove (iii). Let Δy, Δu and Δv be the increments of the functions y, u, v corresponding to the increment Δx of x. Then

$$\frac{\Delta y}{\Delta x} = \frac{\dfrac{u + \Delta u}{v + \Delta v} - \dfrac{u}{v}}{\Delta x} = \frac{v\dfrac{\Delta u}{\Delta x} - u\dfrac{\Delta v}{\Delta x}}{v(v + \Delta v)},$$

and by taking the limit when $\Delta x \to 0$ it follows that*

$$\frac{dy}{dx} = \left(v\frac{du}{dx} - u\frac{dv}{dx}\right)\Big/ v^2.$$

4·31. Notation for the derivative.

There is much to be said for adopting a notation such as $D_x y$ for the derivative of a function $y = f(x)$, for when this is done it emphasises further the distinction between the derivative of a function and the concept of differentials.

In this book, however, we shall use the notation $\frac{dy}{dx}$, thereby conforming with the usual custom of English writers.

The reader should observe, however, that $\frac{dy}{dx}$ has the meaning which has been assigned to it in § 4·12, namely that it is the ratio of the differential dy of the dependent variable to the differential dx of the independent variable. We have proved in that section that if $f'(x)$ exists finitely it is equal to the ratio $dy : dx$, and that whenever $f(x)$ is differentiable the ratio $dy : dx$ is equal to the derivative $f'(x)$ which then certainly exists finitely.

It should be observed that the rules for derivation proved in the previous section become at once rules for differentiation by multiplying each of the equations in (i), (ii) and (iii) by the differential dx.

4·32. Rules for derivation†.

We now prove two theorems.

1. *Let $y = f(x)$ and $x = \phi(t)$ be functions such that $f'(x)$ is finite at a certain point and $\phi'(t)$ is finite at the corresponding point, then y is a function of t having a finite derivative at that point given by the formula*

$$\frac{dy}{dt} = \frac{dy}{dx} \cdot \frac{dx}{dt}.$$

* Note that $v + \Delta v \neq 0$ when Δx is sufficiently small, for the function v is continuous and not equal to zero at the point x; thus division by $v + \Delta v$ is always permissible, and also $\Delta v \to 0$ as $\Delta x \to 0$.

† The reader is expected to be familiar with the simple standard forms and the technique of derivation. See, for example, Gibson's *Elementary Treatise on the Calculus* (1919), Ch. VI.

By hypothesis, to each value of t in the given interval (α, β) for t there is a value of x in the corresponding interval (a, b), and to this x there corresponds a value of y. Hence y may be regarded as a function of t, say

$$y = f(x) = F(t) \quad \text{.......................(1)}$$

and

$$x = \phi(t), \quad x + \Delta x = \phi(t + \Delta t) \quad \text{..............(2).}$$

Now, however Δt approaches zero, $\phi(t + \Delta t) \rightarrow \phi(t)$, since $\phi(t)$ is continuous by § 4·11 (1). Hence, by (2), $\Delta x \rightarrow 0$ as $\Delta t \rightarrow 0$.

With the conditions stated above, from the identity

$$\frac{\Delta y}{\Delta t} = \frac{\Delta y}{\Delta x} \cdot \frac{\Delta x}{\Delta t}$$

the result follows by a direct appeal to the fundamental limit theorems.

Note on the above theorem.

If $x = \phi(t)$ be a function such that $\Delta x = 0$ for some point in every neighbourhood of the point t, the case needs more careful examination. It will be best to consider two illustrative examples*.

Example 1. Let

$$x = t \sin 2m\pi t.$$

The function $\sin 2m\pi t$ is periodic in $1/m$; if m be very large and fixed, the value of x will oscillate a great many times in the neighbourhood of the origin. Where the graph cuts the axis of t, that is when

$$\Delta t = \pm\frac{1}{2m}, \ \pm\frac{1}{m}, \ \pm\frac{3}{2m}, \ \ldots,$$

we have $\Delta x = 0$. However large m be taken, a number δ can be chosen such that there is an interval for t, $0 < t < \delta$, in which $\Delta x \neq 0$, for all that is necessary is to choose $\delta < 1/(2m)$.

Example 2. Consider the function

$$x \begin{cases} = t^2 \sin \pi/t & \text{when } t \neq 0, \\ = 0 & \text{when } t = 0. \end{cases}$$

In this case we cannot determine an interval for t near the origin throughout which $\Delta x \neq 0$; for, however small δ be taken, the value of x oscillates infinitely often in $0 < t < \delta$, so that for an infinity of points in this interval $\Delta x = 0$. The graph of the function is shewn in Fig. 10.

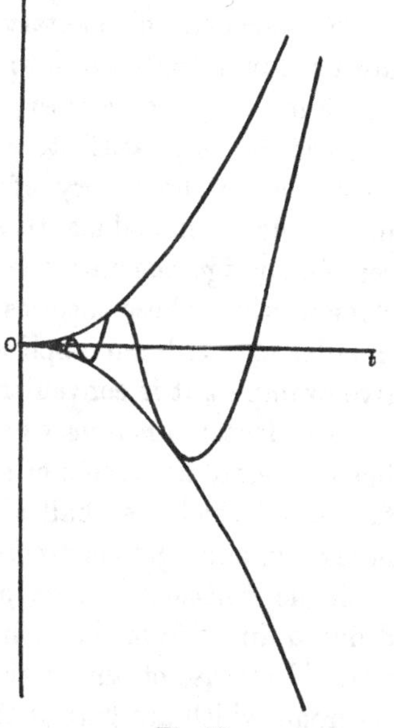

Fig. 10

* See Pierpont, *The Theory of Functions of Real Variables* (Ginn, 1905), I, 233.

In most cases in which the rule is required in practice, the functions f and ϕ will be found to satisfy the conditions of the above theorem.

2. *Let $y = f(x)$ be single-valued, monotonic and continuous, and let $x = g(y)$ be the inverse function, then, if $f'(x)$ be finite and not zero,*

$$g'(y) = 1/f'(x).$$

Since $f(x)$ is monotonic, the identity

$$\frac{\Delta x}{\Delta y} = \frac{1}{\dfrac{\Delta y}{\Delta x}}$$

nowhere involves division by 0, and the theorem is proved by proceeding to the limit.

The fundamental existence theorem on implicit functions is proved in § 10·1. For the present we shall assume that the inverse function exists whenever the above theorem is used.

4·33. The derivatives of the elementary transcendental functions.

The so-called "elementary functions" $\log x$, e^x, $\sin x$, and $\cos x$ are usually introduced into elementary text-books by methods which involve an appeal to geometrical intuition.

Since the only satisfactory definitions of these functions involve an appeal to the theory of integration and infinite series, it is impossible to introduce these functions by methods which are completely rigorous until the subject has been developed up to the stage at which these theories have been fully discussed. However, in order to enrich our applications, and to provide useful illustrative examples, it is convenient to be able to use these elementary transcendental functions as early in the book as possible. Accordingly we state here some of the most important properties of these functions, which we shall always assume to be known wherever necessary, but rigorous proofs of most of them must be postponed until the functions can be properly defined, and their properties deduced direct from the definitions. Care must of course be taken to avoid the use of any of these functions in the proofs of general theorems which are part of the fundamental structure of Analysis.

(1) *The function $\log x$.*

It is assumed that the reader is familiar with the usual discussion of the function $\log x$ which is given in elementary treatises, and that he knows the form of the graph of this function*. The function $\log x$ may be *defined* in more than one way, but the most satisfactory definition is by means of an integral, as follows. If $x > 0$,

$$\log x = \int_1^x \frac{dt}{t}.$$

The function is so defined† in § 10·11, and its elementary properties are there deduced from the definition.

In Algebra the logarithm is defined in the following way: *if a and b are two real numbers such that $a > 0$ and $b > 1$, then the number x, such that $b^x = a$, is called the logarithm of a to the base b and is written*

$$x = \log_b a.$$

For all theoretical considerations the particular real number e, of which one definition has already been given‡, is chosen as the base, but for practical calculations the common or Briggian logarithms, of which the base is 10, are used. We shall assume that the reader is familiar with the rules for working with logarithms, and so he will already know such properties of logarithms as the following:

$$\log_b (a_1 a_2) = \log_b a_1 + \log_b a_2,$$

$$\log_b 1 = 0, \quad \log_b (1/a) = -\log_b a, \quad \log_b b = 1,$$

$$\log_b a^m = m \log_b a.$$

The problem of indefinite integration (which is the inverse problem of derivation), when applied to the power x^n, requires the solution of the differential equation

$$\frac{dy}{dx} = x^n \quad \text{..........................(1).}$$

[It is clear that, except when $n = -1$, the solution is

$$y = x^{n+1}/(n+1) + C,$$

* See, for example, Gibson, *loc. cit.* § 29.

† Although in this book definite integrals are not defined until Ch. VII, the reader should observe that since a definite integral only depends upon the concept of a bound, which was introduced in Ch. II, the definite integral, and consequently the definition of $\log x$, might have appeared earlier in the book.

‡ See § 2·8. Whenever $\log x$ is written the base e is implied. It is only when we contemplate the use of any other base that the base will be definitely indicated (as in the equation $x = \log_b a$ above).

where C is an arbitrary constant; but the solution of the equation

$$\frac{dy}{dx}=\frac{1}{x} \quad \text{..............................(2)}$$

is an exception to the general rule. This indicates that the primitive of equation (2) is a function of a different type from that of equation (1). This function is called $\log x$. We have already seen in § 3·1 that the function $\log x$ so defined cannot be a rational function; and it can also be shewn, as in Examples X, 10, that it cannot be an algebraic function either. It must therefore be a transcendental function.

Many of the important properties of the function $\log x$ are suggested intuitively to us from geometrical considerations, and so, until these properties are proved direct from the definition in § 10·11, we shall assume the following results:

(i) $\log x$ is defined only for $x>0$, and it is a continuous function of x at every point in the interval $x>0$;

(ii) $\log x$ steadily increases from $-\infty$ to ∞ as x ranges from 0 to ∞, and $\log 1=0$.

(iii) The function satisfies the fundamental laws of operation

$$\log(xy)=\log x+\log y,$$
$$\log(1/x)=-\log x.$$

(2) *The function e^x.*

It will be seen in § 10·12 that when $\log x$ has been defined as indicated above, the obvious way to define the exponential function is as the inverse of the logarithmic function, so that if

$$y=\log x, \quad x=e^y.$$

The fundamental existence theorem on implicit functions is proved in § 10·1, and by an appeal to this theorem many properties of the exponential function can be deduced from those of the logarithmic function.

In § 2·8 the number e has been defined as

$$\lim_{n\to\infty}\left(1+\frac{1}{n}\right)^n;$$

and if y is real and positive we have proved that

$$\lim_{n\to\infty}\left(1+\frac{y}{n}\right)^n$$

is the sum-function of the convergent power series

$$1+y+\frac{y^2}{2!}+\frac{y^3}{3!}+\ldots.$$

The identification of the value of the above limit with the function e^y, which is the inverse of the logarithmic function, must be postponed until later*.

If we assume the existence of this inverse function, that is to say that if

$$y=\log x, \quad x=e^y,$$

then since

$$\frac{dy}{dx}=\frac{1}{x},$$

we have, by § 4·32, 2,

$$\frac{dx}{dy}=x=e^y.$$

Hence the derivative of the exponential function is the function itself. More generally, if $x=e^{ky}$, where k is any constant,

$$\frac{dx}{dy}=ke^{ky}.$$

(3) *The general power* a^x.

The function a^x, where a is positive and x is rational, is a function which is defined in elementary Algebra.

Let x be a positive rational number, say p/q: then the positive value y of the power $a^{p/q}$ is given by $y^q=a^p$ and it follows that

$$q\log y=p\log a, \quad \log y=(p/q)\log a=x\log a,$$

and so

$$y=e^{x\log a}.$$

If x is irrational the last equation may be taken as the definition† of a^x. Thus

$$\frac{d}{dx}(a^x)=\frac{d}{dx}(e^{x\log a})=\log a\,.\,e^{x\log a}=a^x\log a.$$

Since a^x is always expressible in the form $e^{x\log a}$, the properties of the function a^x are similar to those of e^x.

(4) *The circular functions.*

We have already made use of the functions $\sin x$ and $\cos x$, the two fundamental *circular functions*, and in doing so it has been

* See § 10·13.

† Notice that both a and a^x are positive. Another method of defining a^x when x is irrational is given in § 6·3.

tacitly assumed that the reader is familiar with the usual method of defining these functions which is adopted in books on elementary trigonometry. It is not yet possible to define these functions without any reference to geometrical intuition as an element of proof. The definition which we shall give of the circular functions $\sin x$ and $\cos x$, based solely on the concept of a real number, implies a knowledge of the theory of power series; and this theory is not discussed until the last chapter*.

There are difficulties in every elementary mode of introducing the circular functions, but these difficulties can be satisfactorily overcome by defining the functions $\sin x$ and $\cos x$ to be the sum-functions of the absolutely convergent power series

$$x - \frac{x^3}{3!} + \frac{x^5}{5!} - \dots,$$

$$1 - \frac{x^2}{2!} + \frac{x^4}{4!} - \dots,$$

for all values of x, and then developing their properties from this definition.

In order to be able to take full advantage of the use of the functions $\sin x$ and $\cos x$ for the purpose of illustration, it is convenient to be able to use the results that *the derivatives of* $\sin x$ *and* $\cos x$ *are respectively* $\cos x$ *and* $-\sin x$.

These results cannot be rigorously proved unless we assume the validity of term-by-term derivation of a power series (see § 13·6). The usual elementary demonstration† based upon the inequalities $\sin \theta < \theta < \tan \theta$ and upon the result that $\lim_{\theta \to 0} \frac{\sin \theta}{\theta} = 1$, cannot be entirely freed from geometrical intuition; and in any case it involves an appeal to the concept of the length of a curve‡ (length of an arc of a circle).

The definitions of $\sin x$ and $\cos x$ by means of the above power series, and the deduction of their properties from the definition, are considered in Chapter XIII.

* Another method of defining the circular functions begins by defining the inverse tangent of x as the integral $\int_0^x \frac{dt}{1+t^2}$. See Hardy's *Pure Mathematics* (1928), § 217.

† See, for example, Gibson's *Calculus*, pp. 77 and 129.

‡ See § 8·6.

4·34. Note on some special functions.

(1) An example similar to Example 2 in § 4·32 is worth closer examination, for it reveals points of considerable importance.

Let
$$f(x) = x^2 \sin 1/x, \quad x \neq 0,$$
$$f(0) = 0.$$

Since $\lim_{x \to 0} x^2 \sin 1/x = 0$, the function is certainly continuous at the origin. Let us consider the derivative of this function.

If $x \neq 0$, by the ordinary rules we get

$$f'(x) = 2x \sin 1/x - \cos 1/x.$$

When $x = 0$, since $\sin 1/x$ and $\cos 1/x$ have no meaning, the ordinary rules of derivation do not apply. We cannot, however, conclude from this that $f'(0)$ does not exist. In fact,

$$f'(0) = \lim_{h \to 0} \frac{f(h) - f(0)}{h} = \lim_{h \to 0} \frac{h^2 \sin 1/h}{h} = 0.$$

We therefore see that, although the function $f'(x)$ does not tend to any limit as $x \to 0$ (for $\cos 1/x$ oscillates infinitely often near the origin), yet $f'(0)$ has a perfectly definite value 0.

Thus $f'(a)$ may have a definite value without being equal to $\lim_{x \to a} f'(x)$. In fact, $f'(a) = \lim_{x \to a} f'(x)$ only if $f'(x)$ is *continuous* at the point $x = a$.

(2) A very curious function was discovered by Weierstrass which is continuous for every value of x, while it has *no derivative anywhere**.

This function is defined by the sum of the infinite series

$$\sum_{n=0}^{\infty} a^n \cos b^n \pi x,$$

where b is an odd integer, $0 < a < 1$, and $ab > 1 + \frac{3}{2}\pi$.

We merely mention this as one among several other extraordinary functions which have been discovered more recently. It should help

* This function appears to have been first mentioned by Du Bois-Reymond in 1874. The reader should consult a paper on "Infinite Derivatives" by G. C. Young, *Quart. Journ. of Math.* XLVII (1916), 127. See also Hardy, *Trans. Amer. Math. Soc.* XVII (1916).

to convince the reader that even among *continuous* functions there are many which cannot be represented by a graph.

4·4. The theorems of the differential calculus.

The theorems which will now be proved are of great importance and are of frequent application in the subsequent development of the subject. The proofs of the theorems themselves are not difficult, but rigorous proofs can only be given by appealing to the properties of continuous functions, and the proofs of these properties depend upon the abstract theorem known as Borel's theorem (or something equivalent to it). The difficulty of Borel's theorem is one of the main obstacles which prevents the giving of a rigorous *elementary* account of the subject of Analysis.

4·41. Rolle's theorem.

Let $f(x)$ be a function subject to the conditions,

(i) *$f(x)$ is a continuous function in the interval $a \leqslant x \leqslant b$,*

(ii) *$f'(x)$ exists in the open interval $a < x < b$,*

(iii) *$f(a) = f(b)$;*

then there is a point c, such that $a < c < b$, at which $f'(c) = 0$.

By the theorems on continuous functions, $f(x)$ is bounded in (a, b) and it attains its bounds. Hence, either $f(x)$ is constant throughout (a, b), in which case the theorem is obvious, or else one or other of the bounds of $f(x)$ is different from $f(a)$ and is attained at $x = c$. In this case we shall shew that $f'(c) = 0$.

To fix the ideas, let the bound attained at c be the upper bound; then

$$f(c \pm h) \leqslant f(c),$$

where h is positive.

Hence
$$\frac{f(c \pm h) - f(c)}{h} \leqslant 0.$$

That is
$$\frac{f(c + h) - f(c)}{h} \leqslant 0 \quad \text{.....................(1)},$$

$$\frac{f(c - h) - f(c)}{-h} \geqslant 0 \quad \text{.....................(2)}.$$

Now make $h \to 0$ and (1) gives $f'(c) \leqslant 0$, (2) gives $f'(c) \geqslant 0$; hence

$$f'(c) = 0.$$

4·42. The mean-value theorem*.

Let $f(x)$ be a function subject to the following conditions:

(i) *$f(x)$ is single-valued and continuous in $a \leqslant x \leqslant b$,*

(ii) *$f'(x)$ exists in the open interval $a < x < b$;*

then there is a point c, such that $a < c < b$, for which

$$f(b) - f(a) = (b-a) f'(c).$$

Except for the fact that $f(b)$ is not equal to $f(a)$, the conditions are the same as those in Rolle's theorem. Define the function $\psi(x)$ such that

$$\psi(x) = f(x) - A \,.\, x \quad \text{.....................(1)},$$

where A is a constant to be so chosen that $\psi(b) = \psi(a)$.

Clearly $\psi(x)$, the difference between the functions $f(x)$ and $A \,.\, x$ which both satisfy conditions (i) and (ii), now satisfies all the conditions of Rolle's theorem; hence

$$\psi'(c) = 0,$$

for some value of c such that $a < c < b$.

From (1) we get $\quad 0 = \psi'(c) = f'(c) - A$.....................(2).

Now A is given by the equation

$$f(b) - A \,.\, b = f(a) - A \,.\, a \quad \text{..................(3)}.$$

By equating the values of A from (2) and (3) the theorem is proved.

Note. The reader may observe that although $f'(x)$ is not restricted to be necessarily *finite* at every point of (a, b), the point c must be one at which $f'(c)$ is finite.

4·43. Geometrical interpretations.

(1) Rolle's theorem is almost obvious geometrically, for if $f(x)$ be a continuous function which possesses a graph, and $f(a) = f(b)$, the diagram in Fig. 11 indicates the existence of a point or points between a and b at which $f'(x)$ vanishes, that is at which the tangent to the curve $y = f(x)$ is parallel to the axis of x.

The conditions of the theorem involve that the graph of $f(x)$ possesses a tangent at every point of the interval (a, b) (save

* This theorem is variously called "the law of the mean," "the formula of finite increments" (*formule des accroissements finis*) and "the *first* mean-value theorem," to distinguish it from the so-called "*second* mean-value theorem." See Examples IV, 11.

possibly at the end-points A and B). At some point C the tangent is parallel to the x-axis. Since $f'(x)$ may be infinite, the graph may have points of inflexion with vertical tangents, as at P.

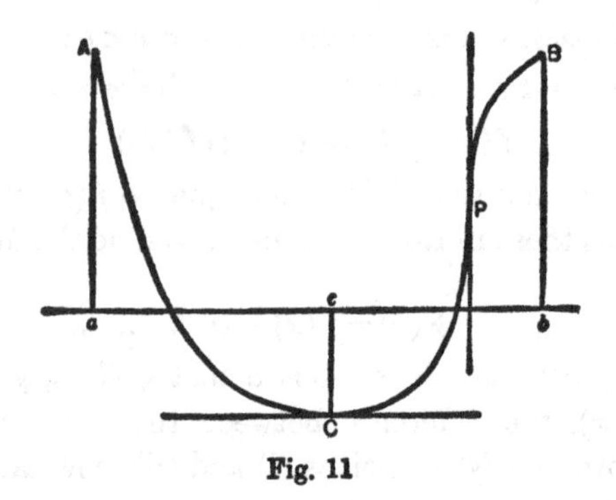

Fig. 11

The curve must not have a vertical cusp or angular point, for at such a point $f'(x)$ *does not exist*, and condition (ii) is therefore not fulfilled.

The reader should observe that although it is impossible *to draw a curve* with an *infinite* number of oscillations or which *does not* have a tangent at $x = a$ or at $x = b$, neither of these cases need be excluded in Rolle's theorem.

When $f(x)$ is a *polynomial*, and $f(a) = f(b) = 0$, we deduce the useful result that between any two roots of the equation $f(x) = 0$, there is at least one root of $f'(x) = 0$.

(2) The geometrical interpretation of the mean-value theorem is also simple.

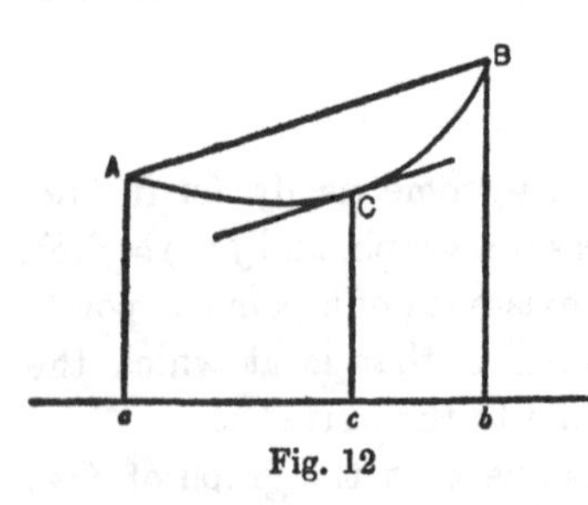

Fig. 12

In the figure let ACB be the graph of $f(x)$ in (a, b) and let the chord AB make an angle α with the x-axis, then

$$\tan\alpha = \frac{f(b) - f(a)}{b - a}.$$

By the theorem we also have

$$\tan\alpha = f'(c).$$

Thus, at some point c within (a, b) the tangent to the curve $y = f(x)$ is parallel to the chord AB.

4·44. Cauchy's formula.

Let the two functions $f(x)$ and $\phi(x)$ satisfy the conditions of the mean-value theorem, and also suppose that $\phi'(x) \neq 0$ anywhere in (a, b), then, there is a point c, such that $a < c < b$, for which

$$\frac{f(b)-f(a)}{\phi(b)-\phi(a)} = \frac{f'(c)}{\phi'(c)}.$$

Define the function $\psi(x)$ by the equation

$$\psi(x) = f(x) - A\phi(x) \quad \text{..................(1)},$$

where A is a constant, which is to be chosen so that $\psi(b) = \psi(a)$.

Thus
$$A = \frac{f(b)-f(a)}{\phi(b)-\phi(a)} \quad \text{......................(2)},$$

and since $\phi'(x) \neq 0$ anywhere in (a, b), $\phi(b) \neq \phi(a)$*, so that A is always finite and determinate.

The function $\psi(x)$ satisfies all the conditions of Rolle's theorem and so

$$\psi'(c) = 0, \quad a < c < b.$$

Hence
$$0 = \psi'(c) = f'(c) - A\phi'(c) \quad \text{..................(3)},$$

and Cauchy's formula follows by equating the values of A from (2) and (3).

Note on the theorem.

The reader should observe that Cauchy's formula cannot be deduced by applying the mean-value theorem to the functions $f(x)$ and $\phi(x)$ separately and dividing the results. For, with the same conditions on $f(x)$ and $\phi(x)$ as stated above, we obtain by the mean-value theorem

$$\frac{f(b)-f(a)}{\phi(b)-\phi(a)} = \frac{f'(c_1)}{\phi'(c_2)},$$

where $a < {c_1 \atop c_2} < b$, but c_1 is not *necessarily* equal to c_2.

Cauchy's formula is more general than the mean-value theorem, and reduces to the latter when $\phi(x) = x$.

4·45. Taylor's theorem.

If $f(x)$ be a single-valued continuous function of x, the object of Taylor's theorem is to obtain an expansion for $f(a+h)$ in

* If $\phi(b) = \phi(a)$ the function $\phi(x)$ would satisfy in (a, b) *all* the conditions of Rolle's theorem, and $\phi'(x)$ would therefore vanish at some point in (a, b).

ascending powers of h, up to the term of any given order n, in the form

$$f(a+h)=f(a)+hf'(a)+\frac{h^2}{2!}f''(a)+\dots$$
$$+\frac{h^{n-1}}{(n-1)!}f^{(n-1)}(a)+R_n(h).$$

Various forms of the "remainder" $R_n(h)$ may be found, of which the simplest is the expression due to Lagrange. In a subsequent section one method of proof is given by which it is possible to determine several forms of the "remainder of order n in Taylor's theorem," $R_n(h)$. Lagrange's form of the remainder can be obtained in a simple way by successive applications of Cauchy's formula.

Lagrange's expression for the remainder.

Let $f(x)$ be a single-valued function of x, and suppose that (i) *$f(x)$ and all its derivatives up to the $(n-1)$th are continuous in $a \leqslant x \leqslant a+h$, and* (ii) *$f^{(n)}(x)$ exists in $a<x<a+h$, then, if the expression*

$$f(a+h)-f(a)-hf'(a)-\dots-\frac{h^{n-1}}{(n-1)!}f^{(n-1)}(a)$$

be denoted by $R_n(h)$ or $\psi(h)$,

$$\psi(h)=\frac{h^n}{n!}f^{(n)}(a+\theta h),\ \textit{where } 0<\theta<1.$$

Let $\chi(h)=h^n/n!$, then, as is easily verified,

$$\psi(0)=\psi'(0)=\dots=\psi^{(n-1)}(0)=0,$$
$$\chi(0)=\chi'(0)=\dots=\chi^{(n-1)}(0)=0.$$

Now
$$\frac{\psi(h)}{\chi(h)}=\frac{\psi(h)-\psi(0)}{\chi(h)-\chi(0)}=\frac{\psi'(h_1)}{\chi'(h_1)},$$

where $0<h_1<h$, by Cauchy's formula.

Similarly, if $0<h_2<h_1$,

$$\frac{\psi'(h_1)}{\chi'(h_1)}=\frac{\psi'(h_1)-\psi'(0)}{\chi'(h_1)-\chi'(0)}=\frac{\psi''(h_2)}{\chi''(h_2)},$$

and continuing the process we get finally, if $0<h_n<h_{n-1}$,

$$\frac{\psi(h)}{\chi(h)}=\frac{\psi^{(n)}(h_n)}{\chi^{(n)}(h_n)}.$$

Clearly $0<h_n<h$, so that we can write $h_n=\theta h$, where $0<\theta<1$.

Hence $$\frac{\psi(h)}{\chi(h)}=\frac{\psi^{(n)}(\theta h)}{\chi^{(n)}(\theta h)}=\frac{f^{(n)}(a+\theta h)}{1},$$

and so $$\psi(h)=\frac{h^n}{n!}f^{(n)}(a+\theta h).$$

4·451. Other forms of the remainder in Taylor's theorem.

With the same conditions as above, write

$$R_n(h)=h^p.P(h),$$

where $P(h)$ is a function of h to be determined.

Let $a+h=b$, and consider the function

$$F(x)=f(x)+(b-x)f'(x)+\dots$$
$$+\frac{(b-x)^{n-1}}{(n-1)!}f^{(n-1)}(x)+(b-x)^p P(h).$$

The function $F(x)$ satisfies all the conditions of Rolle's theorem, for

(i) *$F(x)$ is continuous in $a\leqslant x\leqslant b$*, since it is the sum of $n+1$ continuous functions of x.

(ii) *$F'(x)$ exists in $a<x<b$*, for the highest order derivative of $f(x)$ involved in it is $f^{(n)}(x)$.

(iii) $F(a)=F(b)$, for

$$F(a)=f(a)+hf'(a)+\dots+\frac{h^{n-1}}{(n-1)!}f^{(n-1)}(a)+h^pP(h)$$
$$=f(a+h),$$
$$F(b)=f(b)=f(a+h).$$

Hence there is a point c, such that $a<c<b$, for which $F'(c)=0$.

Now $$F'(x)=\frac{(b-x)^{n-1}}{(n-1)!}f^{(n)}(x)-p(b-x)^{p-1}P(h),$$

and so $$p(b-c)^{p-1}P(h)=\frac{(b-c)^{n-1}}{(n-1)!}f^{(n)}(c).$$

Hence, if $0<\theta<1$,

$$P(h)=\frac{(b-c)^{n-p}}{p.(n-1)!}f^{(n)}(c)=\frac{h^{n-p}(1-\theta)^{n-p}}{p.(n-1)!}f^{(n)}(a+\theta h).$$

In the special cases

(*a*) $p=n$, we get Lagrange's form of the remainder,

$$R_n(h)=\frac{h^n}{n!}f^{(n)}(a+\theta h)\quad\dots\dots\dots\dots\dots\dots\text{(L)};$$

(*b*) $p=1$, we get Cauchy's form,

$$R_n(h)=\frac{h^n(1-\theta)^{n-1}}{(n-1)!}f^{(n)}(a+\theta h) \quad \text{...........(C).}$$

4·46. Maclaurin's theorem.

This theorem is really a special case of Taylor's theorem and is proved in the same way. If we put $a=0$ we obtain Maclaurin's theorem for the expansion of $f(h)$ in ascending powers of h up to the term of order n as follows:

$$f(h)=f(0)+hf'(0)+\dots+\frac{h^{n-1}}{(n-1)!}f^{(n-1)}(0)+R_n(h),$$

where the forms of the remainder analogous to (*a*) and (*b*) above are

$$\frac{h^n}{n!}f^{(n)}(\theta h) \quad \text{..........................(L),}$$

$$\frac{h^n(1-\theta)^{n-1}}{(n-1)!}f^{(n)}(\theta h) \quad \text{.................(C).}$$

The infinite series known as Taylor's and Maclaurin's series are discussed in the next chapter, where the question of their convergence will be investigated.

4·5. Indeterminate forms.

The function $f(x)/\phi(x)$ for which $f(a)=\phi(a)=0$ is a function whose value when $x=a$ "is *indeterminate* of the form 0/0." Frequently, however, it happens that $\lim\limits_{x\to a}\{f(x)/\phi(x)\}$ is definite, and limits of this kind are most easily evaluated by an application of Cauchy's formula.

Many of the most important limits can be evaluated by the methods which will now be discussed. Before employing these methods the theory must be clearly understood, and so we shall consider several different cases in detail. In each case suppose that $f(a)=\phi(a)=0$.

(1) *Let* $f(a)=\phi(a)=0$, *but suppose that* $f'(a)$ *and* $\phi'(a)$ *both exist and have a definite ratio**.

Now $$\frac{f(x)}{\phi(x)}=\frac{f(x)-f(a)}{\phi(x)-\phi(a)}=\frac{\{f(x)-f(a)\}/(x-a)}{\{\phi(x)-\phi(a)\}/(x-a)},$$

so that $$\lim_{x\to a}\frac{f(x)}{\phi(x)}=\frac{f'(a)}{\phi'(a)}.$$

* For this case Cauchy's formula is not required.

(2) *Let $f'(x)$ and $\phi'(x)$ exist near, but not necessarily at, a; if $f'(x)/\phi'(x)$ tends to a limit l as $x \to a$, then the limit of $f(x)/\phi(x)$ exists, and is also equal to l.*

By Cauchy's formula, if $a < \xi < x$,

$$\frac{f(x)}{\phi(x)} = \frac{f(x) - f(a)}{\phi(x) - \phi(a)} = \frac{f'(\xi)}{\phi'(\xi)}.$$

Hence, assuming that the limits in question exist,

$$\lim_{x \to a+0} \frac{f(x)}{\phi(x)} = \lim_{\xi \to a+0} \frac{f'(\xi)}{\phi'(\xi)} = l.$$

Since we have here assumed that $x > a$ we have taken right-hand limits only, the point a being approached from above.

If $x < a$, we have, if $x < \xi_1 < a$,

$$\frac{f(x)}{\phi(x)} = \frac{f(a) - f(x)}{\phi(a) - \phi(x)} = \frac{f'(\xi_1)}{\phi'(\xi_1)},$$

and taking left-hand limits

$$\lim_{x \to a-0} \frac{f(x)}{\phi(x)} = \lim_{\xi_1 \to a-0} \frac{f'(\xi_1)}{\phi'(\xi_1)} = l.$$

We have therefore proved that if $f'(x)/\phi'(x)$ possesses a unique finite limit l when x approaches a, then the limit of $f(x)/\phi(x)$ is also equal to l.

If $f'(x)/\phi'(x)$ possesses a limit on the right (or left) only, then the right- (or left-) hand limit of $f(x)/\phi(x)$ exists, and these limits must be equal.

(3) *Suppose that both $f(x)/\phi(x)$ and $f'(x)/\phi'(x)$ assume the indeterminate form $0/0$ when $x = a$.*

In this case we apply Cauchy's formula again to the function $f'(x)/\phi'(x)$. By the same argument as in (2),

$$\lim_{x \to a} \frac{f(x)}{\phi(x)} = \lim_{\xi \to a} \frac{f'(\xi)}{\phi'(\xi)},$$

provided that the right-hand side exists.

If $f'(a) = \phi'(a) = 0$ but $f''(a)/\phi''(a)$ is determinate, then by (1) the right-hand side does exist and is equal to $f''(a)/\phi''(a)$: and by (2) we have also

$$\lim_{x \to a} \frac{f'(x)}{\phi'(x)} = \lim_{x \to a} \frac{f''(x)}{\phi''(x)}$$

whenever the right-hand side exists.

In general, if $f^{(\nu)}(a)$ and $\phi^{(\nu)}(a)$ both vanish for $\nu < m$, we have, by repeated applications of (2),

$$\lim_{x\to a}\frac{f(x)}{\phi(x)}=\lim_{x\to a}\frac{f^{(m-1)}(x)}{\phi^{(m-1)}(x)}$$

$$=\lim_{x\to a}\frac{f^{(m)}(x)}{\phi^{(m)}(x)} \text{ by (2)}$$

$$=\frac{f^{(m)}(a)}{\phi^{(m)}(a)} \text{ by (1)}$$

whenever the right-hand side exists.

The reader should note that unless at each stage $f^{(\nu)}(a)/\phi^{(\nu)}(a)$ assumes the form 0/0, the above process cannot be applied.

4·51. The indeterminate form ∞/∞.

Let $f(x)$ and $\phi(x)$ be two functions each of which tends to infinity as x tends to a definite value a: suppose that both $f'(x)$ and $\phi'(x)$ exist (except at the point $x=a$) and that in the neighbourhood of $x=a$, $f'(x)$ and $\phi'(x)$ are finite and not simultaneously zero.

Under these conditions the same rules can be applied to evaluate the limit of $f(x)/\phi(x)$ when x tends to a as were used when the indeterminate form was 0/0.

Let x_1 and x be two values sufficiently near a to ensure that $f'(x)$ and $\phi'(x)$ exist in the interval (x_1, x) and have no values in that interval at which they simultaneously vanish. Then by Cauchy's formula, if $x_1 < \xi < x$,

$$\frac{f(x)-f(x_1)}{\phi(x)-\phi(x_1)}=\frac{f'(\xi)}{\phi'(\xi)},$$

and so

$$\frac{f(x)}{\phi(x)}\cdot\frac{1-f(x_1)/f(x)}{1-\phi(x_1)/\phi(x)}=\frac{f'(\xi)}{\phi'(\xi)};$$

that is

$$\frac{f(x)}{\phi(x)}=\frac{f'(\xi)}{\phi'(\xi)}\cdot\frac{1-\phi(x_1)/\phi(x)}{1-f(x_1)/f(x)} \quad\text{................(1).}$$

Suppose that $f'(x)/\phi'(x)$ has a finite limit l as $x\to a$; then if x_1 and x be taken sufficiently near to a, ξ can be made as near as we please to a and then $f'(\xi)/\phi'(\xi)$ will be as near as we please to the value l. Further, without violating the above condition, if x_1 be kept fixed and x be made to approach the value a the second fraction on the right-hand side of (1) tends to the limit unity. It therefore follows, since $\xi \to a$ as $x \to a$, that

$$\lim_{x \to a} \frac{f(x)}{\phi(x)} = l.$$

Example. Let $f(x) = \log 1/x$ and $\phi(x) = 1/x$, then when $x = 0$, $f(x)/\phi(x)$ is indeterminate of the form ∞/∞.

$$\frac{f'(x)}{\phi'(x)} = \frac{-1/x}{-1/x^2} = x.$$

Hence
$$\lim_{x \to 0} \frac{f(x)}{\phi(x)} = \lim_{x \to 0} \frac{f'(x)}{\phi'(x)} = 0.$$

4·52. Other indeterminate forms.

Indeterminate forms of other types occasionally arise, but in general they can always be reduced to the fundamental form 0/0. Consequently only a brief note on the commonest cases is given. In each case the limits are as $x \to a$.

(1) If $f(x) \to 0$, $\phi(x) \to \infty$, and we have to consider the function $f(x).\phi(x)$; the product $f\phi$ is here of the form $0.\infty$, but by writing

$$f\phi = \frac{f}{1/\phi}$$

the form becomes of the type 0/0.

(2) If $f(x) \to \infty$, $\phi(x) \to \infty$, and we have to consider the function

$$f(x) - \phi(x);$$

since
$$f - \phi = \left(\frac{1}{\phi} - \frac{1}{f}\right) \Big/ \frac{1}{f\phi},$$

this form again reduces to 0/0.

(3) If $f(x) \to 0$ and $\phi(x) \to 0$ and we have to consider the function

$$[f(x)]^{\phi(x)},$$

we proceed as follows.

If $f > 0$, write $y = f^{\phi}$, then

$$\log y = \phi \log f = \frac{\phi}{1/\log f},$$

which is of the form 0/0.

If $\log y$ tends to a limit, say c, then since e^x is a continuous function of x when $x > 0$,

$$\lim y = \lim f^{\phi} = e^c.$$

4·53. Criticism.

The reader should clearly understand that the use of 0 and ∞ which has been made in the preceding sections is purely symbolical.

It is convenient, for brevity, to speak of "the indeterminate form $0 . \infty$" for instance. Stated fully this indicates that two functions $f(x)$ and $\phi(x)$ are such that $f(x) \to 0$ and $\phi(x) \to \infty$ as $x \to a$, although possibly $\lim\limits_{x \to a} f(x) . \phi(x)$ may be finite and definite.

Some writers describe the "true value" of the function

$$g(x) = \frac{f(x)}{\phi(x)}$$

when $f(a) = \phi(a) = 0$, to be what is really $\lim\limits_{x \to a} g(x)$, and this is not actually a *value* of the function $g(x)$ at all.

Even when it is chosen to adopt the convention that whenever $g(x)$ is indeterminate it shall be *defined* to have the value $\lim g(x)$ as x approaches the value a, say, the term "true value" is still misleading, and it is better to dispense with it altogether. At best it is a survival from the time when the theory of indeterminate forms was not properly understood.

4·6. Young's form of Taylor's theorem.

If $f^{(n)}(a)$ exists finitely, then

$$f(a+h) = f(a) + hf'(a) + \frac{h^2}{2!} f''(a) + \ldots + \frac{h^n}{n!} \{f^{(n)}(a) + \epsilon_h\},$$

where $\epsilon_h \to 0$ as $h \to 0$.

Let $\chi(h) = h^n/n!$ and let $\lambda(h)$ be the function

$$f(a+h) - f(a) - hf'(a) - \ldots - \frac{h^n}{n!} f^{(n)}(a).$$

As in § 4·45 we can shew that

$$\lambda(0) = \lambda'(0) = \ldots = \lambda^{(n)}(0) = 0,$$

and
$$\chi(0) = \chi'(0) = \ldots = \chi^{(n-1)}(0) = 0 ; \ \chi^{(n)}(0) = 1.$$

By the theory of indeterminate forms we have

$$\lim_{h \to 0} \frac{\lambda(h)}{\chi(h)} = \frac{\lambda^{(n)}(0)}{\chi^{(n)}(0)} = \frac{0}{1} = 0;$$

and, by the definition of $\lambda(h)$, $\lambda(h)/\chi(h)$ is the ϵ_h of the enunciation. This proves the theorem.

Note. The above theorem is useful in practical applications such as the theory of contact of plane curves and differential geometry. It assumes less than the Lagrange form of the remainder.

4·7. The uniqueness of the Taylor expansion.

If $f^{(n)}(a)$ be finite, Taylor's theorem states that $f(a+h)$ can be expanded in the form

$$A_0 + A_1 h + A_2 h^2 + \ldots + A_{n-1} h^{n-1} + M h^n,$$

where the coefficients A_r are constant with respect to h and M is bounded when $h \to 0$.

Such an expansion is possible in only one way, for if we suppose that another similar expansion is possible, then

$$A_0 + A_1 h + \ldots + A_{n-1} h^{n-1} + M h^n \equiv a_0 + a_1 h + \ldots + a_{n-1} h^{n-1} + m h^n.$$

Since M and m are bounded it follows by making $h \to 0$ that $A_0 = a_0$.

Similarly, by dividing by h and again making $h \to 0$, we get $A_1 = a_1$, and so on. The expressions are therefore identical, and so the Taylor expansion of $f(a+h)$ in ascending powers of h is unique.

By writing $a + h = x$ the Taylor expansion assumes the form

$$f(x) = A_0 + A_1 (x-a) + A_2 (x-a)^2 + \ldots + M (x-a)^n.$$

This is the Taylor expansion of $f(x)$ *about the point* $x = a$. By choosing another base point α, say, a *different expansion* for $f(x)$ in powers of $x - \alpha$ can be found. What has been proved is that there is only one Taylor expansion about a given point a as base.

4·8. Leibniz's theorem on the nth derivative of a product.

Let $y = uv$, where u and v are two given functions of x, then by using suffixes $1, 2, \ldots, n, \ldots$ to denote the first, second, ..., nth, ... derivatives respectively it is easily seen that

$$\begin{aligned} y_1 &= uv_1 + u_1 v, \\ y_2 &= uv_2 + 2u_1 v_1 + u_2 v, \\ y_3 &= uv_3 + 3u_1 v_2 + 3u_2 v_1 + u_3 v, \end{aligned}$$

which suggests that the law of the coefficients is the same as in the binomial theorem. *Leibniz's theorem states that*

$$y_n = u_n v + n u_{n-1} v_1 + \frac{n(n-1)}{2!} u_{n-2} v_2 + \ldots + n u_1 v_{n-1} + u v_n.$$

The theorem may be proved by induction either by assuming it for $n = m$ and proving it for $n = m + 1$, or as follows.

Let

$$y_n = A_0 u_n v + A_1 u_{n-1} v_1 + A_2 u_{n-2} v_2 + \ldots + A_n u v_n \quad \ldots(1),$$

where the coefficients A_r are to be determined.

Let $u = e^{px}$, $v = e^x$, so that $y = e^{(1+p)x}$, then

$$y_n = (1+p)^n e^{(1+p)x},$$

$$u_r v_{n-r} = p^r e^{px} . e^x = p^r e^{(1+p)x}.$$

On substituting in (1) we get

$$(1+p)^n \equiv A_0 + A_1 p + A_2 p^2 + \ldots + A_n p^n,$$

and this proves that the coefficients A_r are the binomial coefficients.

4·9. Extreme values of functions of one variable.

In every elementary treatise on the Calculus the theory of maxima and minima (extreme values) of a function $f(x)$ is considered, if only from geometrical considerations. For a function $f(x)$ such that $f'(x)$ and $f''(x)$ both exist in the interval $a \leqslant x \leqslant b$, it is shewn that the values of x at which $f(x)$ is stationary are the solutions of $f'(x) = 0$, and provided that $f''(x) \neq 0$ at any of these points the stationary value is a maximum or a minimum according as $f''(x)$ is less than or greater than zero at the point in question.

This well-known elementary criterion is a particular case of the more general theorem which, together with several other theorems on maxima and minima, is considered here.

All the points at which $f'(x) = 0$ give *stationary* values of $f(x)$. Values of $f(x)$ which are either maxima or minima will be termed *extreme values* (or *extremes*).

4·91. Definition.

Let $f(x)$ be defined in the interval $a \leqslant x \leqslant b$, and let c be any interior point of (a, b). If a sufficiently small positive number η can be found such that

$$\Delta f = f(x) - f(c)$$

preserves the same sign for all values of x such that $|x - c| < \eta$, then $f(x)$ is said to have an extreme value at the point $x = c$.

The extreme value is a *maximum* if Δf is negative and a *minimum* if Δf is positive.

The use of the terms "maximum" and "minimum" in this connection must not be confused with the maximum (absolutely greatest) or the minimum (absolutely least) value assumed by the function $f(x)$ in the interval (a, b).

If $f(x)=x(1+\sin^2 1/x)$ when $x>0$, $f(0)=0$,
the function oscillates between the two straight lines $y=x$ and $y=2x$. Clearly 0 is the absolutely least value of $f(x)$, but in any interval $(0, \delta)$, where δ is positive and as small as we please, the function $f(x)$ has an infinite number of maxima and minima.

4·92. Criteria for extreme values.

Let $f(x)$ be defined in $a \leqslant x \leqslant b$ and let c be an interior point of (a, b).

Suppose that,

(i) *$f^{(n)}(c)$ exists and is not zero,*

(ii) $f'(c)=f''(c)=\ldots=f^{(n-1)}(c)=0$,

then $f(x)$ has no extreme value at $x=c$ if n be odd. If n be even $f(x)$ has a maximum at $x=c$ if $f^{(n)}(c)<0$, or a minimum if $f^{(n)}(c)>0$.

By § 4·6, under the conditions stated above, if $f^{(n)}(c)$ is finite,

$$\Delta f = f(c+h) - f(c) = \frac{h^n}{n!}\{f^{(n)}(c) + \epsilon_h\},$$

and since h may be positive or negative, when n is odd Δf does not preserve the same sign in the neighbourhood of the point c, and so $f(x)$ has no extreme at that point.

If n is even, the sign of Δf is the same as the sign of $f^{(n)}(c)$; hence

if $f^{(n)}(c)<0$, there is a maximum at $x=c$,
if $f^{(n)}(c)>0$, there is a minimum at $x=c$.

If $f^{(n)}(c)=\pm\infty$ we write

$$f(c+h)-f(c)=\frac{h^{n-1}}{(n-1)!}f^{(n-1)}(c+\theta h)$$

and observe that $f^{(n-1)}(c+\theta h)$ has the same sign as $hf^{(n)}(c)$.

The ordinary elementary criterion for an extreme corresponds to the above when $n=2$. If $f''(c)=0$, by the theorem we see that the discrimination may be made by considering derivatives of higher orders. Thus if $f'(c)=f''(c)=f'''(c)=0$, $f^{\text{iv}}(c)\neq 0$, the sign of $f^{\text{iv}}(c)$ enables discrimination to be made. If $f'(c)=0$, $f''(c)=0$, $f'''(c)\neq 0$ the point $x=c$ is not a point at which $f(x)$ has an extreme. In this case there is a point of inflexion at $x=c$*.

* See example (1), § 4·94.

4·93. THEOREM. *If $f'(x)$ exists in (a, b) the points at which $f(x)$ has an extreme value are among the solutions of the equation $f'(x)=0$.*

If $f(x)$ has a minimum at c, then, for $h>0$,

$$f(c+h)-f(c)>0, \quad f(c-h)-f(c)>0$$

and so

$$\frac{f(c+h)-f(c)}{h}>0 \text{ and } \frac{f(c-h)-f(c)}{-h}<0 \quad \ldots\ldots(1).$$

By making $h \to 0$ and taking the limit of each of the expressions in (1), we get

$$f'(c) \geqslant 0 \text{ and } f'(c) \leqslant 0,$$

hence $f'(c)=0$.

COROLLARY. The reasoning of the theorem also shews that *if $f(x)$ has an extreme at $x=c$, then $f'(c)=0$ provided that $f'(c)$ exists.*

The reader should compare this result with Rolle's theorem.

4·94. Examples for illustration of two important points.

(1) *The function $f(x)$ need not have an extreme at every point at which $f'(x)=0$.*

Let $y=(x-1)^3$, then clearly $f'(x)=0$ when $x=1$. The point $x=1$ is not however one at which the function has an extreme, for

$$f''(1)=0, \quad f'''(1)=6 \neq 0.$$

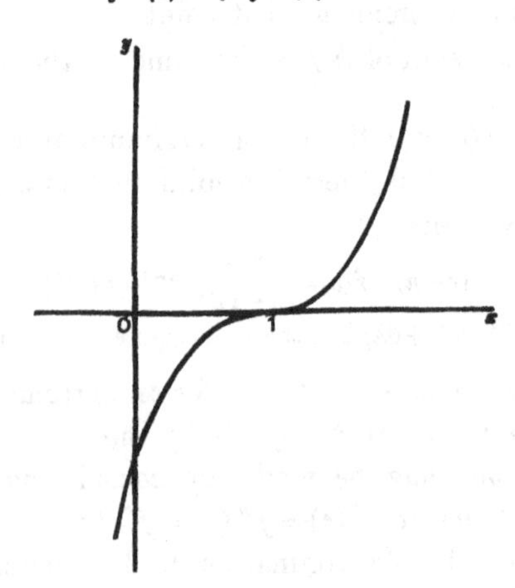

Fig. 13

It is seen from the graph in Fig. 13 that there is a point of inflexion at $x=1$. This point of inflexion is a *stationary* value of $f(x)$, for $f'(x)=0$ at this point, but it is not an *extreme* value of $f(x)$.

(2) *All the extreme values may not be given by the solutions of the equation $f'(x)=0$; there may be extremes at points where $f'(x)$ does not exist.*

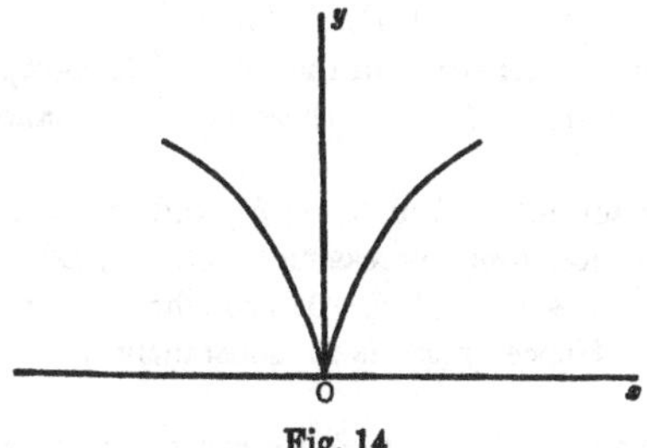

Fig. 14

Let $y=x^{\frac{2}{3}}$, then from the graph we see that the origin is a point at which $f(x)$ has a minimum, but the derivative of the function does not exist at the origin. In this case

$$Rf'(0)=\infty, \quad Lf'(0)=-\infty.$$

This case is an illustration of the theorem which follows.

4·95. THEOREM. *If $f(x)$ is continuous at and near the point c, and $f'(x)$ is finite near c, then $f(x)$ has a minimum at c if*

$$Rf'(c)=\infty, \quad Lf'(c)=-\infty \quad \text{..............(1)};$$

$f(x)$ has a maximum at c if

$$Rf'(c)=-\infty, \quad Lf'(c)=\infty \quad \text{..............(2)}.$$

To fix the ideas let (1) hold, then, if $h>0$,

$$\lim_{h\to 0}\frac{f(c+h)-f(c)}{h}=\infty, \quad \lim_{h\to 0}\frac{f(c-h)-f(c)}{-h}=-\infty;$$

thus there exists a positive number δ such that, for $h<\delta$,

$$f(c\pm h)-f(c)>0.$$

Hence $f(x)$ has a minimum at $x=c$.

The case of a maximum is proved similarly.

4·96. THEOREM. *If $f(x)$ is continuous at and near c, and $f'(x)$ is finite or infinite near c but never zero throughout any sub-interval of $(c-\delta, c+\delta)$; then $f(x)$ has a minimum at c provided that*

$$f'(x)\geqslant 0 \text{ on the right of } c \text{ and } f'(x)\leqslant 0 \text{ on the left of } c, \text{ ......(1)};$$

$f(x)$ has a maximum at c if the inequality signs are interchanged.

Let h be positive, then in the interval $c<x\leqslant c+h$ we have, from conditions (1), $f'(x)\geqslant 0$. We shew that in this interval $f(x)$ is an *increasing* function (in the strict sense).

Suppose that $c \leqslant x_1 < x_2 \leqslant c+h$, then by the mean-value theorem

$$f(x_2) = f(x_1) + (x_2 - x_1) f'(\xi),$$

where $x_1 < \xi < x_2$; but $x_2 - x_1 > 0$ and $f'(\xi) \geqslant 0$, hence

$$f(x_2) \geqslant f(x_1),$$

so that $f(x)$ is *monotonic increasing* in the interval $(c, c+h)$.

It remains to shew that $f(x)$ is a *constantly increasing* function in this interval.

Let (α, β) be any sub-interval of $(c, c+h)$, and suppose that $f(\alpha) = f(\beta)$; then, since $f(x)$ is a monotonic increasing function, $f(x) = f(\alpha)$ for all points in (α, β), and this implies that $f'(x) = 0$ throughout (α, β) which is contrary to our hypothesis. Hence $f(x)$ is a constantly increasing function in $c < x \leqslant c+h$.

Similarly it may be proved that $f(x)$ is a constantly decreasing function in $c-h \leqslant x < c$.

Hence $f(x)$ satisfies the conditions for a minimum at $x = c$.

The case where $f(x)$ has a maximum at c is proved similarly.

EXAMPLES IV.

1. Shew that, when $x \to 0$, x^n and $\sin^n ax$ are infinitesimals of the same order when a has any constant value, zero excepted.

2. Prove that, if $\alpha = x(5 + 2\cos 1/x)$ and $\beta = \sin x$, then α and β are infinitesimals of the same order when $x \to 0$.

Shew that the same is true for the infinitesimals α and β if

$$\alpha = \operatorname{cosec} 2mx - \tfrac{1}{2}\operatorname{cosec} mx, \quad \beta = x.$$

3. Prove that, when $x \to 0$,

(i) $\sin 2mx - 2\sin mx$ is infinitesimal of the same order as x^3;

(ii) $(x - x^2)^2 + \lambda(x + x^2)^2$ is infinitesimal of the same order as x^2 unless $\lambda = -1$. What is the result when $\lambda = -1$?

4. Calculate the right-hand and left-hand derivatives at the origin for the functions

$$\text{(i)}\quad f(x) = \begin{cases} x \arctan 1/x & (x \neq 0) \\ 0 & (x = 0) \end{cases},$$

$$\text{(ii)}\quad \phi(x) = \begin{cases} x/(1 + e^{1/x}) & (x \neq 0) \\ 0 & (x = 0) \end{cases}.$$

5. Prove that the function $f(x) = x \sin 1/x$ $(x \neq 0)$ and which has the value 0 when $x = 0$ is continuous for all values of x, but has no derivative at the origin.

6. Discuss the existence of $f'(x)$ and $f''(x)$ at the origin for the function $f(x) = x^2 \sin 1/x$ $(x \neq 0)$ and which has the value 0 at the origin.

7. For the function $f(x) = x^2$ in $0 \leqslant x < 1$, $f(1) = 0$, the derivative $f'(x)$ does not vanish anywhere in the interval $(0, 1)$. Why does Rolle's theorem fail for $f(x)$ in this interval?

Give cases of failure of Rolle's theorem for other reasons; where possible illustrate by diagrams.

8. Discuss the applicability of Rolle's theorem in (0, 2) to the function $f(x)=2+(x-1)^{\frac{2}{3}}$. Illustrate your answer by a rough sketch.

9. Shew by the mean-value theorem that

(i) $h<\log 1/(1-h)<h/(1-h)$, where $0<h<1$,

(ii) $1+\dfrac{x}{2\sqrt{(1+x)}}<\sqrt{(1+x)}<1+\dfrac{x}{2}$, where $-1<x<0$.

10. The function $f(x)$ is differentiable in $a\leqslant x\leqslant b$, and $f(a)=f(b)=0$; shew, by dividing the range (a, b) into two equal parts and applying the mean-value theorem to each part, that there is at least one point ξ in (a, b) for which

$$|f'(\xi)|>\frac{4}{(b-a)^2}\int_a^b f(x)\,dx.$$

Verify the theorem for the function $\sin^2 x$ in $(0, \pi)$.

[For this example it is assumed that the reader is already acquainted with a definite integral. The integral calculus is discussed in Chapter VII.]

11. If, in the interval $a\leqslant x\leqslant b$, $f(x)$, $f'(x)$ are continuous, and $f''(x)$ exists, prove that

$$f(b)=f(a)+(b-a)f'(a)+\tfrac{1}{2}(b-a)^2 f''(x_2), \text{ where } a<x_2<b.$$

Deduce that, if $x>0$,

$$\log(1+x)>x-\tfrac{1}{2}x^2.$$

Shew similarly that if h is small and positive

$$\tan\left(\frac{\pi}{4}+h\right)>1+2h+2h^2+\frac{8}{3}h^3.$$

12. Evaluate the limits, as $x\to 0$, of the functions:

(i) $\dfrac{e^x-e^{\sin x}}{x-\sin x}$, (ii) $\dfrac{\log(1+x+x^2)+\log(1-x+x^2)}{\sec x-\cos x}$,

(iii) $\dfrac{\log(1+x^2+x^4)}{x(e^x-1)}$, (iv) $\dfrac{e^{ax}-e^{-ax}}{\log(1+bx)}$.

13. Find the limits of the following functions:

(i) $\dfrac{e^{mx}-e^{ma}}{(x-a)^n}$ $(x\to a)$, (ii) $(\cos x)^{1/x}$ $(x\to 0)$,

(iii) $\left(\dfrac{1}{x}\right)^{\tan x}$ $(x\to 0)$, (iv) $(\cos x)^{\cot^2 x}$ $(x\to 0)$,

(v) $\log\left(2-\dfrac{x}{a}\right)\cot(x-a)$ $(x\to a)$,

(vi) $(\cos\theta/n)^{n^2}$ $(n\to\infty)$, (vii) $1/x^2-\operatorname{cosec}^2 x$ $(x\to 0)$,

(viii) $\{(x^3+ax^2)^{\frac{1}{3}}-(x^3-ax^2)^{\frac{1}{3}}\}$ $(x\to\infty)$,

(ix) $\{(a^n+b^n+c^n)/3\}^{1/n}$ $(n\to 0)$.

14. The functions $f(x)$ and $\phi(x)$ are such that $f(0)=\phi(0)=0$; by considering the case where $f(x)=x^2\sin 1/x$, $\phi(x)=\tan x$, shew that $\lim\limits_{x\to 0}\dfrac{f(x)}{\phi(x)}$ may exist when $\lim\limits_{x\to 0}\dfrac{f'(x)}{\phi'(x)}$ does not exist.

[The former limit is 0, the second does not exist.]

15. Evaluate $\lim\limits_{x\to n}\dfrac{1}{(n^2-x^2)^2}\left\{\dfrac{n^2+x^2}{nx}-2\sin\dfrac{n\pi}{2}\sin\dfrac{x\pi}{2}\right\}$

when n is an odd integer.

16. The functions $\phi(x)$ and $\phi'(x)$ are both continuous in $a\leqslant x\leqslant b$; and $\phi(a)=\phi(b)=0$. Prove that whatever number λ may be, then for at least one value of x between a and b

$$\phi(x)=\lambda\phi'(x).$$

17. If α and β lie between the least and greatest of a, b, c, prove that

$$\begin{vmatrix} f(a), & f(b), & f(c) \\ \phi(a), & \phi(b), & \phi(c) \\ \psi(a), & \psi(b), & \psi(c) \end{vmatrix} = k \begin{vmatrix} f(a), & f'(\alpha), & f''(\beta) \\ \phi(a), & \phi'(\alpha), & \phi''(\beta) \\ \psi(a), & \psi'(\alpha), & \psi''(\beta) \end{vmatrix}$$

where $k=\frac{1}{2}(b-c)(c-a)(a-b)$.

18. Prove that a steadily increasing function of a real variable x must tend to a finite limit or to ∞ as $x\to\infty$.

Shew that, if $b>0$, $\dfrac{\log b-\log x}{b-x}$

decreases steadily as x increases from 0 to b, and find its limit as $x\to b$.

19. If $\phi(x)=(x-[x])^2(1-x+[x])^2$, where $[x]$ denotes the greatest integer contained in x, prove that $\phi'(x)$ exists for all values of x in $-1<x<1$ and that, in particular, $\phi'(\frac{1}{4})\neq 0$, and $\phi'(u+\frac{1}{4})=\phi'(\frac{1}{4})$ for all integral u.

If
$$\begin{aligned} f(x)&=0 \text{ for } x=0,\\ &=x^2\phi\left(\frac{1}{x}\right) \text{ for } x\neq 0, \end{aligned}$$

prove that $f(x)$ is everywhere differentiable but that $f'(x)$ is not continuous on the right of $x=0$.

20. A function $f(x)$ possesses a derivative $f'(a)$ at $x=a$; prove that

$$\phi(h,k)\equiv\frac{f(a+h)-f(a-k)}{h+k}-f'(a)$$

tends to zero as h and k tend to zero simultaneously in any way by *positive* values. Prove also that we can remove the restriction that h and k are to be positive $(h+k\neq 0)$ if $f(x)$ possesses a continuous derivative in the neighbourhood of $x=a$.

21. Find the maxima and minima of the function $x^2(1-x)^3$, and sketch its graph from $x=-1$ to $x=2$.

22. (i) Use the theorem of § 4·95 to shew that the function

$$f(x)=1+x^{2/3}$$

has a minimum at $x=0$.

(ii) Consider whether the functions $\phi(x)$ and $\psi(x)$ have extreme values at the origin,

$$\phi(x)=\begin{cases}\dfrac{e^{1/x}-1}{e^{1/x}+1} & (x\neq 0),\\ 0 & (x=0);\end{cases}\qquad \psi(x)=\begin{cases}e^{-1/x^2} & (x\neq 0),\\ 0 & (x=0).\end{cases}$$

Sketch roughly the graphs of these functions in the neighbourhood of the origin.

[$\psi(x)$ is called Cauchy's function. Use the theorem in § 4·96.]

CHAPTER V

INFINITE SERIES

5·1. Introduction.

The present chapter contains an elementary discussion of the theory of infinite series, and some of the ordinary simple criteria for convergence. Its main object is to meet the needs of such readers as may be entirely unacquainted with the elementary theory of series. It is of course impossible in a single chapter to attempt to give any kind of *complete* account of a vast subject such as this upon which valuable books in English have already been written*; and since these exist there is no urgent necessity for dealing here with more than is necessary to make this book self-contained. This chapter includes all the elementary theorems and convergence tests which are needed to enable the reader to make an intelligent use of infinite series so far as they are required in the subsequent development of Analysis which is included within the scope of this book.

The concept of uniformity of convergence is discussed in Chapter XIII.

5·11. The sum of an infinite series.

The discussion of the infinite series

$$u_1 + u_2 + \ldots + u_n + \ldots\ldots\ldots\ldots\ldots\ldots\ldots\ldots(1)$$

can be reduced to the study of the behaviour of the sequence $\{s_n\}$, where

$$s_n = u_1 + u_2 + \ldots + u_n.$$

If $\{s_n\}$ converges to a limit s, then s is defined to be the *infinite sum* (or more briefly *the sum*) of the series (1).

If $s_n \to \pm\infty$ the infinite series (1) is *divergent*, and its sum (in this sense) does not exist.

When $\{s_n\}$ does not possess a unique limit, the series (1) is said to *oscillate*. It *oscillates finitely* if the sequence has finite lower

* For example, Bromwich's *Theory of Infinite Series* (1926) and Knopp's *Theory and Application of Infinite Series* (1928). The latter is a translation from the German.

and upper limits λ and Λ; it oscillates infinitely if $\lambda = -\infty$, $\Lambda = \infty$*.

THEOREM. *If a series converges, then its nth term tends to zero as n tends to infinity.*

This can be deduced at once from the general principle of convergence of a sequence $\{s_n\}$, for if the series Σu_n converges, then for a value of n large enough, and for all positive integral values of p,

$$|s_{n+p} - s_n| < \epsilon.$$

In particular this relation holds when $p = 1$, and $|s_{n+1} - s_n| = |u_{n+1}|$, so that $|u_{n+1}| < \epsilon$; in other words $u_{n+1} \to 0$ as $n \to \infty$.

COROLLARY. *If u_n does not tend to zero the series Σu_n cannot converge.*

The above corollary gives a useful one-sided test which is frequently used *to prove that a given series does not converge.*

The reader should observe that the condition that u_n must tend to zero is a *necessary* condition for the convergence of Σu_n, but it is *not sufficient.*

If $u_n = 1/n$, then clearly $u_n \to 0$, but $\Sigma 1/n$ diverges†.

5·12. General theorem on series of positive terms.

If all the terms of the series Σu_n are positive, it is evident that the sequence $\{s_n\}$ is monotonic increasing; and so $\{s_n\}$ must tend to a finite limit or tend to infinity. Hence a series of *positive* terms cannot oscillate. From this result we obtain the following

THEOREM. *If $\{s_n\}$ is bounded‡ the series of positive terms Σu_n must converge.*

If s_n is less than a constant K then s_n tends to a limit s which cannot exceed K. For since $\{s_n\}$ is monotonic increasing, it tends to its upper bound. If K be a rough upper bound the (exact) upper bound s will be less than K; and if K is the upper bound, s coincides with K. Hence we have

$$s_n \to s \leqslant K.$$

* See § 2·55. † See § 5·41 (*a*).

‡ The theorem is still true if $s_n < K$ for $n \geqslant \nu$ only, and not for all values of n. See note on Test I in § 5·2.

5·13. Absolute and conditional convergence.

When we consider series whose terms are not all of the same sign, there is an important distinction to be made between series which do and which do not remain convergent when all the terms are replaced by their absolute values. First of all we have the

THEOREM. *A series Σu_n is certainly convergent if the series (of positive terms) $\Sigma | u_n |$ is convergent. If $\Sigma u_n = s$ and $\Sigma | u_n | = S$, then $|s| \leqslant S$.*

Clearly, for a value of n large enough, and for all positive integral values of p,

$$|s_{n+p} - s_n| \leqslant |S_{n+p} - S_n| < \epsilon.$$

Hence Σu_n converges.

Also, $$|s_n| \leqslant |u_1| + |u_2| + \ldots + |u_n| < S,$$
and so, by the preceding theorem,

$$|s| \leqslant S.$$

By this theorem all convergent series are divided into two classes, and Σu_n belongs to one or the other according as $\Sigma | u_n |$ is or is not also convergent.

DEFINITION. If a convergent series Σu_n is such that $\Sigma | u_n |$ also converges, then the first series will be called *absolutely convergent*, and otherwise *non-absolutely convergent.*

If $\Sigma | u_n |$ diverges it may happen that Σu_n still converges, although it does not converge absolutely. For example, it is easy to prove that the series

$$1 - \tfrac{1}{2} + \tfrac{1}{3} - \tfrac{1}{4} + \ldots$$

is convergent*. It is certainly not absolutely convergent, because the series obtained when every term is replaced by its absolute value is the divergent† harmonic series

$$1 + \tfrac{1}{2} + \tfrac{1}{3} + \tfrac{1}{4} + \ldots.$$

The reader must refer to treatises on the theory of series for a detailed discussion of the validity of various operations on infinite series.

Since the sum of an infinite series is defined as a *limit*, it is essentially different from the sum of a finite number of terms,

* It satisfies the alternating series test of § 5·61.

† See § 5·41 (*a*).

which is obtained by adding these terms together. It is thus essential to guard against the assumption that any operation which is justifiable when applied to finite series can also be applied to an infinite series without further investigation.

The important property of an *absolutely convergent* series is that, roughly speaking, it may be treated as though it were a finite series: its terms may be deranged in any order, and grouped in various ways without affecting either its convergence or its sum. Also, if Σa_n and Σb_n are two absolutely convergent series, it can be proved that

$$(a_1+a_2+a_3+\ldots)(b_1+b_2+b_3+\ldots)$$
$$=a_1b_1+(a_2b_1+a_1b_2)+(a_3b_1+a_2b_2+a_1b_3)+\ldots.$$

With series which are non-absolutely convergent, however, great care must be exercised. A convergent series whose behaviour as to convergence can be altered by rearrangement, and for which therefore the order of the terms must be taken into account, is called *conditionally convergent*.

The series $$s=1-\tfrac{1}{2}+\tfrac{1}{3}-\tfrac{1}{4}+\ldots \qquad \ldots\ldots(1)$$
when rearranged so that each positive term is followed by two negative terms thus,

$$\sigma=1-\tfrac{1}{2}-\tfrac{1}{4}+\tfrac{1}{3}-\tfrac{1}{6}-\tfrac{1}{8}+\tfrac{1}{5}-\tfrac{1}{10}-\tfrac{1}{12}+\ldots \qquad \ldots\ldots(2)$$

no longer has the same sum. For

$$\sigma_{3n}=(1-\tfrac{1}{2})-\tfrac{1}{4}+(\tfrac{1}{3}-\tfrac{1}{6})-\tfrac{1}{8}+\ldots+\left(\frac{1}{2n-1}-\frac{1}{4n-2}\right)-\frac{1}{4n}$$
$$=\tfrac{1}{2}-\tfrac{1}{4}+\tfrac{1}{6}-\tfrac{1}{8}+\ldots+\frac{1}{4n-2}-\frac{1}{4n}$$
$$=\tfrac{1}{2}\left(1-\tfrac{1}{2}+\tfrac{1}{3}-\ldots+\frac{1}{2n-1}-\frac{1}{2n}\right)$$
$$=\tfrac{1}{2}s_{2n}.$$

Thus $\lim\limits_{n\to\infty}\sigma_{3n}=\tfrac{1}{2}s$; and since $\lim\sigma_{3n+1}=\lim\sigma_{3n+2}=\lim\sigma_{3n}$ the sum of the series (2) is $\tfrac{1}{2}s$.

It can be shewn that a non-absolutely convergent series can, by suitable rearrangement of the order of its terms, be made to converge to any arbitrarily prescribed sum, or even to diverge.

5·2. Criteria for convergence of series of positive terms.

We shall write $$U_n=u_1+u_2+\ldots+u_n,$$
and if the series Σu_n converges, we denote its sum by U so that
$$U=\lim_{n\to\infty}U_n.$$

Test I. The direct comparison test.

(*a*) *Let* $u_1 + u_2 + u_3 + \dots$ *be the series to be tested, and suppose that* $c_1 + c_2 + c_3 + \dots$ *is a series known to be convergent; then if* $u_n \leqslant kc_n$ *for all* n, *where* k *is a positive constant,* Σu_n *is convergent.*

Since Σc_n converges, $\lim C_n = C$, where C is finite, and

$$C_n < C.$$

Also
$$U_n \leqslant kC_n < kC,$$

hence U_n tends to a limit U which cannot exceed kC; in other words Σu_n converges.

(*b*) *If* $d_1 + d_2 + d_3 + \dots$ *is a known divergent series, and if* $u_n \geqslant kd_n$, *then* Σu_n *diverges.*

The proof is left to the reader.

Note. It would be sufficient in the above tests if $u_n \leqslant kc_n$ or $u_n \geqslant kd_n$ *for all values of* n *greater than* ν, where ν is fixed; for in studying the convergence of series it is permissible to discard any finite number of terms at the beginning of the series and to consider the new series which results, for

$$\begin{aligned} U_n &= (u_1 + u_2 + \dots + u_\nu) + u_{\nu+1} + \dots + u_n \\ &= a + U_{n,\nu}, \end{aligned}$$

where a is a finite constant. Hence U_n must tend to a finite limit or to infinity according as $U_{n,\nu}$ does, and conversely.

Test II. Comparison by limits.

If the ratio u_n/v_n *tends to a finite (non-zero) limit as* $n \to \infty$, *then* Σu_n *will converge or diverge according as* Σv_n *converges or diverges.*

Suppose that $\dfrac{u_n}{v_n} \to l$, then, for $n \geqslant \nu$,

$$l - \epsilon < \frac{u_n}{v_n} < l + \epsilon,$$

and so, for $n \geqslant \nu$,

$$\text{(i)} \quad u_n > v_n(l - \epsilon),$$

and
$$\text{(ii)} \quad u_n < v_n(l + \epsilon);$$

hence by Test I, both results follow.

Test III. D'Alembert's ratio test.

If $\lim \dfrac{u_{n+1}}{u_n} = l$, *then the series* Σu_n *converges if* $l < 1$, *and diverges if* $l > 1$.

(i) *Let* l *be less than unity.* Choose β between l and 1, then, since the values of u_{n+1}/u_n when n is large enough differ from l by as little as we please, we have, when $n \geqslant m$,

$$\frac{u_{n+1}}{u_n} < \beta;$$

and since u_m is positive, it follows that

$$u_{m+1} < \beta u_m, \quad u_{m+2} < \beta^2 u_m, \ldots.$$

Thus, from and after the term u_m the terms of the series Σu_n do not exceed those of the convergent geometric series

$$u_m (1 + \beta + \beta^2 + \ldots);$$

hence Σu_n converges.

(ii) *Let l be greater than unity.* In this case we have, for $n \geqslant m$,

$$u_{n+1} > u_n,$$

so that u_n does not tend to zero and Σu_n cannot converge; and so, since it is a series of positive terms, it must diverge.

COROLLARY 1. From part (i) of the preceding proof we deduce that Σu_n *converges if* $\dfrac{u_{n+1}}{u_n} < \beta$, *where β is a* FIXED *constant which is less than unity.*

It is *not* sufficient for the convergence of Σu_n that

$$\frac{u_{n+1}}{u_n} < 1$$

merely; for if $u_n = 1/n$ we have, for all values of n,

$$\frac{u_{n+1}}{u_n} = \frac{n}{n+1} < 1,$$

but $\Sigma \dfrac{1}{n}$ diverges.

COROLLARY 2. *If $l = 1$ there is no criterion, save when u_{n+1}/u_n approaches the limit* 1 *from above. In this case Σu_n is divergent.*

For when $n \geqslant \nu$, and $\delta > 0$, $\qquad \dfrac{u_{n+1}}{u_n} = 1 + \delta,$

and so $u_{n+1} > u_n$ when $n \geqslant \nu$. Thus u_n does not tend to zero, and so Σu_n diverges.

Test IV. Cauchy's tests.

The series Σu_n converges or diverges according as $\overline{\lim}\,(u_n)^{1/n}$ is less than or greater than unity.

(i) Suppose that $\overline{\lim}\,(u_n)^{1/n} = \alpha < 1$. Choose β between α and 1, then, from the definition of an upper limit, for $n \geqslant m\,(\beta)$,

$$(u_n)^{1/n} < \beta,$$

that is

$$u_n < \beta^n;$$

and so Σu_n converges by comparison with the convergent geometric series $\Sigma\beta^n$, $(\beta < 1)$.

(ii) Suppose that $\overline{\lim}\,(u_n)^{1/n} = \alpha > 1$, then, by definition,

$$(u_n)^{1/n} > 1$$

for an infinity of values of n; the same is therefore true of u_n, so that u_n does not tend to zero, and Σu_n accordingly diverges.

Note. If $u_n^{1/n}$ has a *unique* limit when $n \to \infty$ the preceding results hold if we replace "$\overline{\lim}$" by "lim."

5·3. THEOREM. *If u_n is positive and if u_{n+1}/u_n tends to a limit l, then $u_n^{1/n}$ tends to the same limit.*

From some particular value of n onwards all the ratios

$$\frac{u_{n+1}}{u_n},\ \frac{u_{n+2}}{u_{n+1}},\ \ldots,\ \frac{u_{n+p}}{u_{n+p-1}},\ \ldots,$$

lie between $l - \epsilon$ and $l + \epsilon$, and so, for $n \geqslant n_0$ (say),

$$(l-\epsilon)^p < \frac{u_{n+p}}{u_n} < (l+\epsilon)^p;$$

and since u_n is positive,

$$u_n^{\frac{1}{n+p}}(l-\epsilon)^{\frac{p}{n+p}} < (u_{n+p})^{\frac{1}{n+p}} < u_n^{\frac{1}{n+p}}(l+\epsilon)^{\frac{p}{n+p}}.$$

Keep n fixed and make p tend to infinity; the extreme members tend respectively to $l - \epsilon$ and $l + \epsilon$, and so, when $m \geqslant m_0$,

$$l - 2\epsilon < u_m^{1/m} < l + 2\epsilon,$$

hence $u_n^{1/n} \to l$.

The converse however is not true. Consider the series

$$1 + a + ab + a^2b + a^2b^2 + \ldots + a^nb^{n-1} + a^nb^n + \ldots,$$

where a and b are unequal positive numbers. It is easily seen that the ratio of any term to the preceding is alternately a or b, whereas $u_n^{1/n} \to \sqrt{(ab)}$.

It follows from the theorem that d'Alembert's ratio tests are less general than Cauchy's tests, for the latter may succeed when the former fail.

5·4. Method of formation of convergent and divergent series.

One of the simplest methods of testing series is by comparison with other series which are known to be either convergent or divergent.

The following method of formation of convergent and divergent series provides, among its special cases, several useful auxiliary series which may be advantageously employed for comparison tests.

THEOREM. *Let D_n be any number which tends steadily to infinity as n tends to infinity, then the two series of positive terms*

$$(D_2 - D_1) + (D_3 - D_2) + \dots + (D_{n+1} - D_n) + \dots \quad \dots(1),$$

$$\left(\frac{1}{D_1} - \frac{1}{D_2}\right) + \left(\frac{1}{D_2} - \frac{1}{D_3}\right) + \dots + \left(\frac{1}{D_n} - \frac{1}{D_{n+1}}\right) + \dots \dots(2),$$

are respectively divergent and convergent.

Since $D_n \to \infty$ the same is true of the sum to n terms of the series (1),

$$s_n = D_{n+1} - D_1.$$

Hence the series (1) diverges.

The sum to n terms of the series (2) is clearly

$$\frac{1}{D_1} - \frac{1}{D_{n-1}},$$

and this tends to the finite limit $1/D_1$ as $n \to \infty$.

Special cases of the above theorem.

(i) Let $D_n = n^\lambda$, where λ is positive, then the series

$$\Sigma \left\{\frac{1}{n^\lambda} - \frac{1}{(n+1)^\lambda}\right\}$$

converges by (2).

(ii) Let $D_n = \log n$, then the series

$$\Sigma \{\log(n+1) - \log n\} = \Sigma \log\left(1 + \frac{1}{n}\right)$$

diverges by (1).

(iii) Let $D_n = \log \log n$, then the series

$$\Sigma \{\log\log(n+1) - \log\log n\} = \Sigma\, 1/(n+\theta)\log(n+\theta),$$

where $0 < \theta < 1$ by the mean-value theorem, is a *divergent* series by (1).

5·41. Useful auxiliary series.

From the special cases of the preceding theorem we deduce the following useful results:

(a) $\Sigma\, 1/n$ *is a divergent series;*

(b) $\Sigma\, 1/n^{1+\lambda}$, *where λ is positive, is a convergent series;*

(c) $\Sigma\, 1/n \log n$ *is a divergent series.*

These may all be proved by employing the comparison Test II.

The proof of (*a*) is immediate by observing that

$$\lim \frac{\log(1+1/n)}{1/n} = 1,$$

and so $\Sigma\, 1/n$ diverges by comparison with (ii).

To prove (*b*) we consider the limit of the ratio

$$\frac{1/n^{1+\lambda}}{\dfrac{1}{n^{\lambda}} - \dfrac{1}{(n+1)^{\lambda}}} = \frac{1/n^{1+\lambda}}{\lambda/(n+\theta)^{1+\lambda}}$$

by the mean-value theorem, where $0 < \theta < 1$.

Now $$\lim \left\{\frac{1}{\lambda}\left(\frac{n+\theta}{n}\right)^{1+\lambda}\right\} = \frac{1}{\lambda},$$

which is finite*, and so series (*b*) converges by comparison with (i).

To prove (*c*) we consider the limit of the ratio

$$\frac{n \log n}{(n+\theta)\log(n+\theta)}.$$

Since $$\frac{\log(n+\theta)}{\log n} = \frac{\log(1+\theta/n) + \log n}{\log n} \to 1,$$

and also $\dfrac{n+\theta}{n} \to 1$, it follows that

$$\lim \frac{n \log n}{(n+\theta)\log(n+\theta)} = 1;$$

and so series (*c*) diverges by comparison with (iii).

5·5. Kummer's criteria.

We now prove a set of general criteria for convergence and divergence of series, from which quite a number of the most useful tests can be derived by considering simple special cases. Since the proofs in the general case are not difficult this method is a better one than proving each of the special cases independently.

THEOREM. *Let D_n denote any positive number, and let*

$$\phi(n) \equiv D_n \frac{a_n}{a_{n+1}} - D_{n+1},$$

then (1) *the series of positive terms Σa_n converges if*

$$\phi(n) > \alpha > 0$$

* For λ is positive so that $\lambda \neq 0$.

for values of n which exceed a fixed number m, where α is a constant, and (2), *if $\Sigma 1/D_n$ diverges, then Σa_n diverges if, for all $n > m$,*

$$\phi(n) \leqslant 0.$$

Proof of (1). Since a_{n+1} is positive, if the condition above is satisfied

$$D_n a_n - D_{n+1} a_{n+1} > \alpha\, a_{n+1},$$

for $n > m$; and so

$$D_{m+1} a_{m+1} - D_{m+2} a_{m+2} > \alpha\, a_{m+2},$$

$$D_{m+2} a_{m+2} - D_{m+3} a_{m+3} > \alpha\, a_{m+3},$$

$$\dots\dots\dots\dots\dots\dots\dots\dots\dots\dots$$

$$D_{m+p} a_{m+p} - D_{m+p+1} a_{m+p+1} > \alpha a_{m+p+1}$$

hence, by addition,

$$D_{m+1} a_{m+1} - D_{m+p+1} a_{m+p+1} > \alpha\,(a_{m+2} + \dots + a_{m+p+1}),$$

so that, since $D_{m+p+1} a_{m+p+1}$ is positive,

$$s_{m+p+1} - s_{m+1} < D_{m+1} a_{m+1}/\alpha.$$

Hence, it follows that, if K is a constant,

$$s_{m+p+1} < K,$$

and so Σa_n converges.

Proof of (2). If the condition be satisfied in this case we have, for $n > m$,

$$D_n a_n - D_{n+1} a_{n+1} \leqslant 0,$$

and so $$D_n a_n \geqslant D_{n-1} a_{n-1} \geqslant \dots \geqslant D_{m+1} a_{m+1} = c,$$

where c is constant, and so

$$a_n \geqslant c/D_n;$$

hence Σa_n diverges by comparison with $\Sigma 1/D_n$.

5·51. Alternative forms of Kummer's criteria.

In practice it is frequently easier to test a given series by criteria involving limits than by tests given in the above form. It is easy to give alternative forms of the preceding results which involve limits. In their most general form they involve upper and lower limits, but in almost all practical cases arising in elementary work the function $\phi(n)$ will be found to have a unique limit as $n \to \infty$.

THEOREM. *To prove that*

(1′) Σa_n *is convergent if* $\underline{\lim}\,\phi(n) > 0$,

(2′) Σa_n *is divergent if* $\overline{\lim}\,\phi(n) < 0$.

The theorem is proved if we shew that (1) and (1′) are exactly equivalent and that (2′) is included in (2).

If (1′) is true, $\underline{\lim}\, \phi(n) = \beta > 0$,

and from the definition of a lower limit, for $n > m$,

$$\phi(n) > \beta - \tfrac{1}{2}\beta,$$

that is, $\phi(n) > \tfrac{1}{2}\beta$,

and so $\frac{1}{2}\beta$ is the number which replaces α in (1).

If (1) is true, $\phi(n) > \alpha$ for $n > m$,

and so $\underline{\lim}\, \phi(n) \geqslant \alpha > 0$,

which is (1′).

From (2′) we have, if β is positive,

$$\overline{\lim}\, \phi(n) = -\beta < 0;$$

and so, for $n > m$, $\phi(n) < -\beta + \tfrac{1}{2}\beta$,

that is, for $n > m$, $\phi(n) < 0$,

which is (2).

5·52. Special cases of Kummer's tests.

Let us assume that $\phi(n)$ has a unique limit as n tends to infinity, then the criteria of the preceding section become equivalent to the statement that Σa_n *converges or diverges according as* $\lim \phi(n)$ *is greater than or less than zero.*

(1) Let $D_n = 1$ and we obtain d'Alembert's ratio tests.

(2) Let $D_n = n$ and we obtain the tests usually called Raabe's tests, that Σa_n *converges or diverges according as the limit of*

$$n\left(\frac{a_n}{a_{n+1}} - 1\right)$$

is greater than or less than unity.

(3) Gauss's test. *If a_n/a_{n+1} is expanded in powers of $1/n$ so that**

$$\frac{a_n}{a_{n+1}} = 1 + \frac{\mu}{n} + O\left(\frac{1}{n^2}\right),$$

then Σa_n converges if $\mu > 1$ and diverges if $\mu \leqslant 1$.

* The O notation has already been explained in § 4·2. Here, of course, $n \to \infty$. A more general form of the test may be given in which $O\left(\frac{1}{n^{1+\lambda}}\right)$, where $\lambda > 0$, replaces $O\left(\frac{1}{n^2}\right)$. In most elementary cases the form given above suffices.

(i) Suppose that $\mu \neq 1$, then

$$n\left(\frac{a_n}{a_{n+1}} - 1\right) = \mu + O\left(\frac{1}{n}\right),$$

so that

$$\lim n\left(\frac{a_n}{a_{n+1}} - 1\right) = \mu,$$

and by Raabe's test Σa_n converges if $\mu > 1$, and diverges if $\mu < 1$.

(ii) If $\mu = 1$, consider Kummer's tests, when $D_n = n \log n$.

$$\begin{aligned}\phi(n) &= n \log n \frac{a_n}{a_{n+1}} - (n+1)\log(n+1) \\ &= n \log n + \log n + O\left(\frac{\log n}{n}\right) - (n+1)\left\{\log n + \log\left(1 + \frac{1}{n}\right)\right\} \\ &= O\left(\frac{\log n}{n}\right) - (n+1)\log\left(1 + \frac{1}{n}\right),\end{aligned}$$

hence $\lim \phi(n) = -1$ since $\frac{\log n}{n} \to 0$ and $(n+1)\log\left(1 + \frac{1}{n}\right) \to 1$.

Thus by D_n test (2′), it follows that Σa_n diverges when $\mu = 1$.

Note 1. An alternative form of the Gauss test reads as follows:

If $\frac{a_{n+1}}{a_n} = 1 - \frac{\mu}{n} + O\left(\frac{1}{n^2}\right)$, *the series* Σa_n *converges if* $\mu > 1$, *and diverges if* $\mu \leqslant 1$.

Note 2. The Gauss test* is very useful in practice, and is usually found to be the best test to apply when d'Alembert's ratio test fails because

$$\lim (u_{n+1}/u_n) = 1.$$

It is more delicate than Raabe's test, and it will therefore certainly test every series for which the latter test is decisive; and furthermore Gauss's test gives a definite result when Raabe's test fails because $\lim n\left(\frac{a_n}{a_{n+1}} - 1\right) = 1$.

The Gauss test is usually easy to apply, and in every case in which a_n/a_{n+1} is the ratio of two polynomials in n, the expansion in powers of $1/n$ is readily obtained. In spite of this the test is rarely given in elementary books on Algebra in which elementary convergence tests are proved, although the direct proof of it is almost as easy as the direct proof of Raabe's test which is almost always given.

5·6. Series in general, not necessarily of positive terms.

Throughout the preceding sections all the series in question have been *series of positive terms*, and it is only to such series that the preceding tests are applicable. It is beyond the scope of this

* Gauss established this criterion in 1812, *Werke*, III, 140.

book to give a detailed discussion of series of arbitrary terms, and for further information the reader must refer to treatises on the theory of series. In § 5·13 reference has been made to the importance of the concept of absolute convergence. If Σu_n be a series of arbitrary terms and $\Sigma \mid u_n \mid$ converges, Σu_n is then said to be absolutely convergent. Thus, for absolutely convergent series all the simple criteria for convergence of series of positive terms are applicable. It is accordingly much easier to recognise the convergence of a series of arbitrary terms, if that series is absolutely convergent, than it is if the series is non-absolutely convergent. There are several tests which can be applied to non-absolutely convergent series, but the only case which we consider here is the important case of *alternating series*, that is, series whose terms are alternately positive and negative.

An important class of series consists of the power series* $\Sigma a_n x^n$, in which a_n is positive: when x takes a negative value these become series with alternately positive and negative terms. If the series $\Sigma \mid a_n \mid . \mid x \mid^n$ converges in a certain range of values for x, say in $-1 < x < 1$, then in that range $\Sigma a_n x^n$ is absolutely convergent. When we consider $\mid x \mid = 1$, the reader will see that if $x = -1$ the series becomes $\Sigma (-)^n a_n$, which is an alternating series. Accordingly some simple criterion for the convergence of alternating series is needed to enable a full discussion for the convergence of a real power series to be given.

Although the general theory of series of complex terms is not considered at all in this book, it may be well to mention that if Σu_n is an arbitrary *series of complex terms*, the series $\Sigma \mid u_n \mid$ is still a series of *positive terms*. Thus an *absolutely* convergent series Σu_n of complex terms is as simple to discuss as a series of positive terms.

5·61. Alternating series test.

THEOREM. *The series*

$$u_1 - u_2 + u_3 - u_4 + \dots$$

in which (i) $u_n \to 0$ *as* $n \to \infty$, *and* (ii) $u_m \geqslant u_{m+1}$ *for all values of* m, *is convergent.*

* Power series are discussed in detail in Chapter XIII. We are assuming here that a_n and x are *real* numbers.

Now

$$s_{2n} = (u_1 - u_2) + (u_3 - u_4) + \dots + (u_{2n-1} - u_{2n}) \quad \dots\dots\dots\dots\dots(1)$$

$$= u_1 - (u_2 - u_3) - (u_4 - u_5) - \dots - (u_{2n-2} - u_{2n-1}) - u_{2n} \quad (2),$$

and by condition (ii) the content of each bracket in (1) and (2) is positive. Thus (1) shews that s_{2n} is positive, and (2) shews that

$$s_{2n} < u_1,$$

and so s_{2n} tends to a limit s which does not exceed u_1.

If we can prove that s_{2n+1} also tends to the same limit s, the theorem is proved. Now

$$s_{2n+1} = s_{2n} + u_{2n+1},$$

and by condition (i) $u_{2n+1} \to 0$ as $n \to \infty$; hence it follows that s_{2n+1} tends to the same limit as s_{2n}.

COROLLARY. *If $u_n \to \alpha \neq 0$, but condition* (ii) *is still satisfied, the alternating series oscillates finitely.*

For if $\lim s_{2n} = s$, $\lim s_{2n+1} = s + \alpha$, and the sum of the series oscillates between s and $s + \alpha$.

5·62. *Example. If x and s are real, test for convergence the series* $\sum_{n=1}^{\infty} \frac{x^{n-1}}{n^s}$.

(1) *Absolute convergence.* Examine the series $\sum \frac{|x|^{n-1}}{n^s}$.

(*a*) $\left|\frac{u_{n+1}}{u_n}\right| = \left(\frac{n}{n+1}\right)^s |x|$, and so $\lim \left|\frac{u_{n+1}}{u_n}\right| = |x|$.

Thus $\sum |u_n|$ converges if $|x| < 1$; $\sum u_n$ *converges absolutely if* $|x| < 1$, and $\sum |u_n|$ diverges if $|x| > 1$; $\sum u_n$ does not converge absolutely if $|x| > 1$.

In the latter case, since $\left|\frac{u_{n+1}}{u_n}\right| > 1$ for $n > m$, u_n does not tend to zero and so $\sum u_n$ cannot converge; since the terms *increase*, $\sum u_n$ cannot oscillate finitely. Thus *when* $|x| > 1$, $\sum u_n$ *diverges* (or *oscillates infinitely*).

(*b*) Now consider $|x| = 1$. The series of moduli is then $\sum 1/n^s$, which converges if $s > 1$, diverges if* $s \leqslant 1$.

Hence $\sum \frac{x^{n-1}}{n^s}$ *converges absolutely when* $|x| = 1$ *and* $s > 1$.

(2) *Non-absolute convergence.*

Clearly the only case where conditional convergence is possible is when $|x| = 1$ and $s \leqslant 1$, when the series of moduli diverges.

If $x = 1$ the case reduces to (*b*) above. It remains therefore to consider

* See § 5·41 (*a*) and (*b*). $\sum 1/n^s$ when $s < 1$ is a divergent series by comparison with $\sum 1/n$.

only $x=-1$. The given series then is $\Sigma(-)^{n-1}/n^s$, this is an alternating series, and we appeal to the test given above.

Now $1/n^s \to 0$ provided that $s>0$. In this case also

$$\frac{1}{m^s} > \frac{1}{(m+1)^s},$$

and so *the conditions for convergence are satisfied if* $s>0$.

(3) *Remaining cases.*

(*a*) Let $|x|=1$, $s=0$, then when

$x=1$, series is $1+1+1+1+\dots$ which *diverges*,

$x=-1$, series is $1-1+1-1+\dots$ which *oscillates finitely*.

(*b*) Let $|x|=1$, $s<0$. Write $s=-p$ so that $p>0$, then when

$x=1$, series is Σn^p which clearly *diverges*,

$x=-1$, series is $\Sigma(-)^{n-1} n^p$ which *oscillates infinitely*.

5·7. The multiplication theorem.

We now prove the fundamental theorem on the multiplication of two infinite series. There are two cases to be considered: (1) when the series are of positive terms, and (2) when the series are absolutely convergent.

THEOREM 1. *Let Σu_n and Σv_n be two series of positive terms which are known to be convergent to sums U and V respectively, then the series Σw_n converges to the sum UV, where*

$$w_n = u_1 v_n + u_2 v_{n-1} + \dots + u_n v_1.$$

The proof depends upon the two inequalities

$$W_{2n} \geqslant U_n V_n \quad \dots\dots\dots(1),$$

$$W_n \leqslant U_n V_n \quad \dots\dots\dots(2),$$

and these are most easily recognised by arranging the terms in a double array as follows:

$$\begin{array}{llll|lll}
u_1 v_1, & u_2 v_1, & \dots, & u_n v_1; & u_{n+1} v_1, & \dots, & u_{2n} v_1 \\
u_1 v_2, & u_2 v_2, & \dots, & u_n v_2; & u_{n+1} v_2, & \dots, & u_{2n} v_2 \\
 & & & & & & \\
u_1 v_n, & u_2 v_n, & \dots, & u_n v_n; & u_{n+1} v_n, & \dots, & u_{2n} v_n \\
\hline
 & & & & & & \\
u_1 v_{2n}, & u_2 v_{2n}, & \dots, & u_n v_{2n}; & u_{n+1} v_{2n}, & \dots, & u_{2n} v_{2n}
\end{array}$$

The product $U_n V_n$ clearly consists of all the terms in the top left-hand quarter square, while W_n consists only of those terms in this quarter square which lie in the right diagonal and above it, for

$$W_n = u_1v_1 + (u_1v_2 + u_2v_1) + (u_1v_3 + u_2v_2 + u_3v_1) + \ldots + w_n.$$

Hence (2) is established.

Inequality (1) is almost obvious, for w_{2n} contains all the terms in the right diagonal of the complete square, and of these terms only u_nv_n can belong to U_nV_n: hence W_{2n} certainly contains all the terms of U_nV_n and more.

Since
$$U_n \leqslant U, \quad V_n \leqslant V,$$
it follows from (2) that
$$W_n \leqslant UV,$$
and so W_n tends to a limit W which cannot exceed UV.

Again,
$$W = \lim_{n\to\infty} W_{2n} \geqslant \lim U_n V_n = (\lim U_n) . (\lim V_n),$$
and so
$$W \geqslant UV.$$
Hence
$$W = UV.$$

THEOREM 2. *If Σu_n and Σv_n are two absolutely convergent series, then* (i) *Σw_n is absolutely convergent, and* (ii) *its sum is UV.*

Let
$$\omega_n = |u_1||v_n| + \ldots + |u_n||v_1|,$$
and let
$$\Omega_n = \sum_1^n \omega_n.$$

To prove (i), observe that $|w_n| \leqslant \omega_n$ and that $\Sigma\omega_n$ is convergent by Theorem 1; hence Σw_n is absolutely convergent.

To prove (ii) write
$$S_n = \sum_1^n |u_n| \text{ and } T_n = \sum_1^n |v_n|,$$
then $S_nT_n - \Omega_n$ is the sum of the moduli of the terms in $U_nV_n - W_n$, and so
$$|U_nV_n - W_n| \leqslant |S_nT_n - \Omega_n|;$$
and since by Theorem 1 $\Omega_n \to ST$, it follows that, for $n > n_0$,
$$|U_nV_n - W_n| < \epsilon,$$
and hence
$$W_n \to UV.$$

5·71. Application to power series.

From the preceding theorems we deduce that if both the power series $\sum_0^\infty a_nx^n$, $\sum_0^\infty b_nx^n$ are absolutely convergent in the interval $(-r, r)$, then their product $\Sigma c_n x^n$ is absolutely convergent in the same interval, where

$$c_0 = a_0 b_0, \quad c_n = a_0 b_n + a_1 b_{n-1} + \ldots + a_{n-1} b_1 + a_n b_0.$$

If the region of convergence of the two given series is different, $\Sigma c_n x^n$ certainly converges absolutely in the smaller of the two intervals, and it may or may not converge outside this interval.

The reader should observe that without discussing Abel's theorem on power series*, this is the most that can be proved about the multiplication of two power series.

5·8. The Taylor and Maclaurin series.

The theorems discussed in §§ 4·45 and 4·46 provided a means of obtaining an expansion of $f(a+h)$ or of $f(h)$ in a series of ascending powers of h up to the term of any order n, and several forms of the remainder $R_n(h)$ were found. We are now in a position to consider the possibility of expanding $f(a+h)$ (or $f(h)$) in an *infinite series* of ascending powers of h.

If we assume that $f(x)$ is a function admitting continuous derivatives of any order n, however large n may be taken, *it is necessary and sufficient for the convergence of the Taylor or Maclaurin series that $R_n(h)$ should tend to zero as n tends to infinity.*

If $S_n(h)$ denote the sum to n terms of the Maclaurin series, it is a necessary and sufficient condition that

$$S_n(h) \to f(h) \text{ as } n \to \infty,$$

that, given ϵ, there is a number $\nu(\epsilon)$ such that, for $n \geqslant \nu$,

$$|S_n(h) - f(h)| < \epsilon.$$

Now $$f(h) = S_n(h) + R_n(h),$$

so that the convergence condition is that

$$|R_n(h)| < \epsilon \text{ for } n \geqslant \nu,$$

in other words $$R_n(h) \to 0 \text{ as } n \to \infty.$$

In some cases this condition is found to hold for all values of h, but in general $R_n(h)$ only tends to zero† when h is restricted to lie between certain limits, for example when $-1 < h < 1$.

* See § 13·5, Theorem 3.

† The reader should observe that R_n is a function of n, h and of an unknown variable θ which lies between 0 and 1. For *Taylor's* series R_n also depends on a.

5·81. Examples.

(1) *Let* $f(x)=e^x$, then $f^{(n)}(x)=e^x$ and $f^{(n)}(0)=1$ for all values of n. Hence Maclaurin's theorem with the Lagrange form of the remainder gives

$$e^x=1+x+\frac{x^2}{2!}+\ldots+\frac{x^{n-1}}{(n-1)!}+\frac{x^n}{n!}e^{\theta x},$$

where $0<\theta<1$.

To obtain an infinite series for e^x we must shew that $R_n(x)\to 0$ as $n\to\infty$, and this is true for all values of x. To prove this, write

$$n!=1.2.3\ldots.(n-1).n,$$
$$n!=n(n-1)(n-2)\ldots 2.1,$$

so that $(n!)^2$ is the product of n factors of the form $p(n-p+1)$, where

$$1\leqslant p\leqslant n.$$

Now, when $p>1$,

$$p(n-p+1)=p(n-p)+p>(n-p)+p=n,$$

and so $$(n!)^2>n^n,$$

that is, $$n!>(\sqrt{n})^n.$$

It follows that

$$\frac{|x|^n}{n!}<\left(\frac{|x|}{\sqrt{n}}\right)^n\to 0 \text{ as } n\to\infty,$$

and so, since $e^{\theta x}<1+e^x$, $R_n(x)\to 0$ as $n\to\infty$ *for all values of* x.

The result also follows from the fact that $|x|^n/n!$ is the general term of a convergent series, see Theorem, § 5·11.

(2) *Let* $$f(x)=\log(1+x),$$

then $$f'(x)=\frac{1}{1+x},\quad f^{(n)}(x)=(-)^{n-1}\frac{(n-1)!}{(1+x)^n}.$$

By Maclaurin's theorem,

$$\log(1+x)=x-\frac{x^2}{2}+\frac{x^3}{3}-\ldots+R_n(x),$$

where $$R_n(x)=(-)^{n-1}\frac{x^n}{n}\left(\frac{1}{1+\theta x}\right)^n \quad\ldots\ldots\ldots\ldots\text{(L)},$$

and $$R_n(x)=(-)^{n-1}\frac{x^n}{1+\theta x}\left(\frac{1-\theta}{1+\theta x}\right)^{n-1} \quad\ldots\ldots\ldots\text{(C)}.$$

We assume throughout that $x>-1$, for otherwise the derivatives of $f(x)$ do not exist in the interval $(x, 0)$.

(i) *Suppose that x is positive and $x \leqslant 1$.*

In this case remainder (L) shews that

$$|R_n(x)| < \frac{1}{n},$$

and so $R_n(x) \to 0$ as $n \to \infty$.

(ii) *Suppose that x is negative and $x > -t > -1$.*

Here the remainder (C) must be used, and we see that

$$|R_n(x)| < \frac{t^n}{1-t},$$

for $\left|\dfrac{1-\theta}{1+\theta x}\right|$ is a positive fraction; and since $0 < t < 1$, it follows that

$$R_n(x) \to 0 \text{ as } n \to \infty.$$

We deduce that

$$\log(1+x) = x - \frac{x^2}{2} + \frac{x^3}{3} - \ldots,$$

provided that* $-1 < x \leqslant 1$.

The use of the Taylor and Maclaurin theorems to find expansions of given functions in series of ascending powers of the variable is limited in its practicability by the difficulty of calculating the successive derivatives, and by the unwieldy form of the remainder $R_n(x)$. In fact it is only in a *few* cases that $R_n(x)$ assumes a manageable form.

In the next section a method of expanding a given function in a series of ascending powers of the variable is considered which may be used in some cases where the direct application of Maclaurin's theorem is impossible because the remainder cannot be easily obtained.

5·82. Example.

Find an expansion for $\frac{1}{2}(\text{arc}\sin x)^2$ in ascending powers of x.

Let $f(x) = \frac{1}{2}(\text{arc}\sin x)^2$, then

$$f'(x) = \frac{\text{arc}\sin x}{(1-x^2)^{\frac{1}{2}}}, \quad f''(x) = \frac{1}{1-x^2} + x\frac{\text{arc}\sin x}{(1-x^2)^{\frac{3}{2}}},$$

and successive derivatives become increasingly more complicated, and $R_n(x)$ would be difficult to obtain.

* For a justification that the sum of the series when $x = 1$ is log 2, see also § 13·5 (Example).

Let $y=\frac{1}{2}(\text{arc}\sin x)^2$, then it is easy to see that y satisfies the differential equation

$$(1-x^2)\,y_2-xy_1-1=0 \quad \text{..................(1)}.$$

Differentiate n times by Leibniz's theorem, and we get

$$(1-x^2)\,y_{n+2}-2nxy_{n+1}-n(n-1)\,y_n-xy_{n+1}-ny_n=0,$$

or

$$(1-x^2)\,y_{n+2}-(2n+1)\,xy_{n+1}-n^2y_n=0 \quad \text{.........(2)}.$$

If we assume that a Maclaurin series exists for this function, the coefficients are the values of $y, y_1, y_2, \ldots, y_n, \ldots$ when $x=0$.

From the values already found,

$$y_0=0, \quad (y_1)_0=0, \quad (y_2)_0=1,$$

and from equation (2), by putting $x=0$,

$$(y_{n+2})_0=n^2(y_n)_0;$$

and so

$$(y_{2m+1})_0=0, \quad (m=0, 1, 2, \ldots),$$

and

$$(y_4)_0=2^2(y_2)_0=2^2,$$

$$(y_6)_0=4^2(y_4)_0=2^2\,.\,4^2,$$

$$\ldots \quad \ldots \quad \ldots$$

$$(y_{2m})_0=2^2\,.\,4^2\,.\,6^2\ldots(2m-2)^2.$$

On the assumption that the Maclaurin series with these coefficients has the sum $\frac{1}{2}(\text{arc}\sin x)^2$, we get

$$\tfrac{1}{2}(\text{arc}\sin x)^2=\frac{x^2}{2\,!}+\frac{2^2}{4\,!}x^4+\frac{2^2\,.\,4^2}{6\,!}x^6+\ldots$$

$$+\frac{2^2\,.\,4^2\,.\ldots.\,(2m-2)^2}{(2m)\,!}x^{2m}+\ldots \quad \text{......(3)}.$$

It can be shewn that this series converges if $x^2 \leqslant 1$, and diverges if $x^2>1$.

Since we have not been able to find an expression for $R_n(x)$, we have not *proved* that the series (3) is the Maclaurin expansion for $\frac{1}{2}(\text{arc}\sin x)^2$. In general, however, when a *convergent* infinite series for $f(x)$ is found by the preceding method, it may be *expected to represent* the function $f(x)$ within its range of absolute convergence.

What we have *proved* is that the series (3), which can be shewn to be convergent when $x^2 \leqslant 1$, has a sum-function, say $F(x)$, which is such that if $f(x)=\frac{1}{2}(\text{arc}\sin x)^2$, then

$$F(0)=f(0), \quad F'(0)=f'(0), \ldots, \quad F^{(n)}(0)=f^{(n)}(0), \ldots \quad \text{...(4)}.$$

We have *not* shewn that there may not be several functions $F(x)$, which are not identical with $f(x)$ for *all* values of x when $x^2 \leqslant 1$, but which satisfy the equations (4)*.

A more satisfactory method of dealing with power series will be found in Chapter XIII.

5·9. A useful theorem in convergence theory.

If a_n is never zero and

$$\frac{a_n}{a_{n+1}} = 1 + \frac{\mu}{n} + O\left(\frac{1}{n^2}\right),$$

then $a_n \to 0$ if $\mu > 0$, $a_n \not\to 0$ if $\mu \leqslant 0$.

Case 1. *Let* $\mu > 0$. Then, for $n \geqslant m$,

$$\frac{a_n}{a_{n+1}} > 1 + \frac{\mu}{2n}.$$

Writing down these inequalities from $n = m$ to $n-1$ and multiplying, we get

$$\frac{a_m}{a_n} > \left(1 + \frac{\mu}{2m}\right)\left(1 + \frac{\mu}{2(m+1)}\right) \dots \left(1 + \frac{\mu}{2(n-1)}\right)$$

$$> 1 + \frac{\mu}{2}\left(\frac{1}{m} + \frac{1}{m+1} + \dots + \frac{1}{n-1}\right).$$

Since the right-hand side is part of the divergent series $\Sigma \frac{1}{n}$, $\frac{a_m}{a_n} \to \infty$, so $a_n \to 0$ as $n \to \infty$.

Case 2. *Let* $\mu < 0$. Then, for $n \geqslant m$, $a_n/a_{n+1} < 1$, so the terms increase; hence $a_n \not\to 0$.

Case 3. *Let* $\mu = 0$. Then

$$\frac{a_n}{a_{n+1}} = 1 + O\left(\frac{1}{n^2}\right) = 1 + \frac{\theta_n}{n^2},$$

where $|\theta_n| < k$.

Since, for $n \geqslant m$, $\left|\frac{a_n}{a_{n+1}}\right| < 1 + \frac{k}{n^2}$, proceeding as in Case 1,

$$\left|\frac{a_m}{a_{n+1}}\right| < \left(1 + \frac{k}{m^2}\right)\left(1 + \frac{k}{(m+1)^2}\right) \dots \left(1 + \frac{k}{n^2}\right).$$

Assuming, from the theory of infinite products that $\prod_m^n \left(1 + \frac{k}{n^2}\right) \to$ a limit $\lambda \neq 0$, as $n \to \infty$; $\lim_{n \to \infty} |a_{n+1}| \geqslant \frac{|a_m|}{\lambda} > 0$. Hence $a_{n+1} \not\to 0$.

* If $C(x) = e^{-1/x^2}$, $x \neq 0$, $C(0) = 0$, then, if $F(x) = f(x) + C(x)$, equations (4) are satisfied. See Examples V, 17.

5·91. Application of the above theorem.

The theorem just proved is very useful in testing a real power series Σu_n, where $u_n = a_n x^n$. In spite of this, it is rarely given in the text-books.

Example. If c and x are real, test for convergence the power series $\sum_1^\infty u_n = \sum_1^\infty a_n x^n$

where
$$a_n = \frac{c(c+1)(c+2)\dots(c+n-1)}{n!}.$$

We test the series $\Sigma |u_n|$ first. Since

$$\left|\frac{u_{n+1}}{u_n}\right| = \frac{n+c}{n+1}|x|, \quad \lim_{n\to\infty}\left|\frac{u_{n+1}}{u_n}\right| = |x|,$$

and, using the ratio test, we have the results:

(1) If $|x| < 1$, $\Sigma |u_n|$ is convergent so Σu_n is absolutely convergent.

(2) If $|x| > 1$, $\Sigma |u_n|$ is divergent, which gives no information about Σu_n. But, when $\lim |u_{n+1}|/|u_n| > 1$, for $n \geqslant m$, $|u_{n+1}|/|u_n| > 1$, so the terms increase. Hence $u_n \nrightarrow 0$, so Σu_n is divergent or oscillates infinitely.

(3) Case when $|x| = 1$.

(*a*) If $x = 1$, as soon as $n > -c$, the series, Σa_n, is of positive terms, and

$$\frac{a_n}{a_{n+1}} = \frac{n+1}{n+c} = 1 + \frac{1-c}{n} + O\left(\frac{1}{n^2}\right) \quad \dots\dots\dots(1).$$

Hence, by Gauss' test, Σa_n, which is Σu_n with $x = 1$, is convergent if $\mu = 1 - c > 1$, divergent if $1 - c \leqslant 1$. Hence when $|x| = 1$, we have absolute convergence if $c < 0$.

(*b*) If $x = -1$, the series is $\Sigma(-1)^n a_n$. We need only consider $c \geqslant 0$, because the series when $|x| = 1$ is absolutely convergent when $c < 0$. Note that if $c = 0$ the series is $\Sigma 0$, a trivial case of convergence.

Now we use the theorem of § 5·9. By this, referring to (1) above, if

$$\mu = 1 - c > 0, \quad a_n \to 0.$$

Also, when $\mu > 0$, for $n \geqslant m$,

$$\frac{a_n}{a_{n+1}} > 1,$$

so that the terms decrease. Hence the two conditions of the alternating series test for $\Sigma(-1)^n a_n$ are satisfied. Thus, when $x = -1$, the power series is conditionally convergent if $0 < c < 1$.

Also, if $\mu \leqslant 0$, $a_n \nrightarrow 0$ so there cannot be any more convergence if $1 - c \leqslant 0$, i.e. $c \geqslant 1$.

To sum up, the given series is:

(i) absolutely convergent if $|x| < 1$; and when $|x| = 1$ if $c < 0$;

(ii) conditionally convergent when $x = -1$ if $0 < c < 1$;

(iii) divergent (or infinitely oscillating) if $|x| > 1$; and when $|x| = 1$, if $c \geqslant 1$.

EXAMPLES V.

1. Examine for convergence or divergence the series Σu_n for which u_n (from some index onwards) has the values

$$\frac{n(n+1)}{(n+2)^2},\quad \frac{3(n+1)(n+2)}{n^3\sqrt{n}},\quad \frac{1}{n^s\log n}\ (s\neq 1),\quad \frac{1.3\ldots.(2n-1)}{2.4\ldots.2n}\cdot\frac{1}{\sqrt{n}},$$

$$\left\{\frac{1.5.9\ldots.(4n-3)}{4.8.12\ldots.4n}\right\}^2.\sqrt{n},\quad a^{1/n}-1\ (a>0),\quad \frac{1}{n(\log n)^t},\quad \frac{1}{n^s(\log n)^t},$$

$$\left(1-\frac{1}{n}\right)^{n^2},\quad e^{\pm\sqrt{n}}x^n\ (x>0).$$

2. If $a_n>0$, $b_n>0$ and Σb_n diverges, shew that, if $a_n/b_n\to l$, then

$$\frac{a_1+a_2+\ldots+a_n}{b_1+b_2+\ldots+b_n}\to l.$$

Deduce that if a_n is bounded and Σa_n is divergent, then so is $\Sigma\dfrac{a_n}{1+a_n{}^2}$.

3. If Σa_n diverges, discuss the series

$$\Sigma\frac{a_n}{1+na_n},\quad \Sigma\frac{a_n}{1+n^2a_n}\quad (a_n>0).$$

4. If Σa_n is a convergent series of positive terms, prove that $\Sigma(a_na_{n+1})^{\frac{1}{2}}$ is also convergent. Shew by an example that the converse is not necessarily true, unless $\{a_n\}$ is monotonic.

Prove also that

$$\Sigma\sqrt{a_n}/n,\quad \Sigma\sqrt{a_n}/n^{\frac{1}{2}+p} \text{ for every value of } p>0$$

both converge.

5. Test for convergence the series

$$\frac{2}{1^3}-\frac{3}{2^3}+\frac{4}{3^3}-\frac{5}{4^3}+\ldots,$$

$$\frac{2}{1^2}-\frac{3}{2^2}+\frac{4}{3^2}-\frac{5}{4^2}+\ldots.$$

6. (i) Shew that if $x>0$, the series

$$1+2x+3^2\frac{x^2}{2!}+4^3\frac{x^3}{3!}+5^4\frac{x^4}{4!}+\ldots$$

converges if $x<1/e$.

(ii) Test the series

$$\frac{a}{b}+\frac{a(a+c)}{b(b+c)}x+\frac{a(a+c)(a+2c)}{b(b+c)(b+2c)}x^2+\ldots.$$

7. Test for convergence, for all real values of x, the series whose general terms are

$$\text{(i)}\ \frac{x^n}{(a+n)^s}\ (a>0), \quad \text{(ii)}\ \frac{x^n}{n \log n}.$$

8. Prove that for the binomial series, when x and m are real,

$$1+mx+\frac{m(m-1)}{2!}x^2+\ldots+\frac{m(m-1)\ldots(m-n+1)}{n!}x^n+\ldots$$

there is absolute convergence if $|x|<1$. Shew also that if $x=1$ the series converges (conditionally) provided that $m>-1$: if $x=-1$ there is convergence when $m>0$.

9. If a, b, c and x are all real, examine for convergence the hypergeometric series

$$\Sigma\frac{a(a+1)\ldots(a+n-1).b(b+1)\ldots(b+n-1)}{c(c+1)\ldots(c+n-1)}\cdot\frac{x^n}{n!}.$$

Deduce the results of Question 8 from this.

10. Find the limits, as $n\to\infty$, of the functions

$$n^{1/n}, \quad (\log n)^{1/n}, \quad (n!)^{1/n}.$$

[See § 5·3: the answers are 1, 1, ∞.]

11. Prove that, when $|x|<1$,

$$-\frac{\log(1-x)}{1-x}=x+(1+\tfrac{1}{2})x^2+(1+\tfrac{1}{2}+\tfrac{1}{3})x^3+\ldots.$$

12. Expand $\sqrt{2}\sin\pi x$ in a Taylor series about the point $x=\frac{1}{4}$, and justify carefully the steps.

13. Expand $\operatorname{arc\,tan}\dfrac{a-x}{a+x}$ in a series of ascending powers of x and determine its region of convergence.

14. If

$$y=\frac{\operatorname{arc\,sin} x}{\sqrt{1-x^2}},$$

prove that

$$(1-x^2)\frac{dy}{dx}-xy=1.$$

Hence obtain an expression for y in ascending powers of x.

15. Prove that the series

$$\{(1+\tfrac{1}{2})^2-(1+\tfrac{1}{3})^2\}+\{(1+\tfrac{1}{4})^2-(1+\tfrac{1}{5})^2\}+\ldots+\left\{\left(1+\frac{1}{2n}\right)^2-\left(1+\frac{1}{2n+1}\right)^2\right\}+\ldots$$

converges, but that the series obtained by removing the brackets oscillates.

[This example shews that we may not, without consideration, *omit* brackets occurring in a series. The series $0+0+0+\ldots$ is certainly convergent to sum 0. By writing everywhere $(1-1)$ for 0 we get

$$(1-1)+(1-1)+\ldots\equiv\Sigma(1-1)=0,$$

which is correct, but by omitting the brackets we obtain the oscillating series

$$1-1+1-1+\ldots,$$

the sum of which is 0 or 1 according as the number of terms is even or odd.]

16. Shew that, if $$s=1-\tfrac{1}{2}+\tfrac{1}{3}-\tfrac{1}{4}+\dots,$$
the sum of the series, when rearranged as follows,
$$1+\tfrac{1}{3}-\tfrac{1}{2}+\tfrac{1}{5}+\tfrac{1}{7}-\tfrac{1}{4}+\dots$$
is $\tfrac{3}{2}s$.

17. Prove that Cauchy's function
$$C(x)=e^{-1/x^2} \textit{ when } x\neq 0, \quad C(0)=0$$
has a Maclaurin expansion
$$0+0\,.\,x+0\,.\,x^2+\dots+0\,.\,x^n+\dots.$$
Hence prove that the function $f(x)=C(x)+e^x$ has a Maclaurin expansion
$$1+x+\frac{x^2}{2!}+\frac{x^3}{3!}+\dots$$
which, although it is convergent for all values of x, does not converge to the sum $f(x)$.

[We have to shew that $C'(0)=C''(0)=\dots=C^{(n)}(0)=\dots=0$.
$$C'(x)=\frac{2}{x^3}C(x)\ (x\neq 0);\ C'(0)=\lim_{h\to 0}\frac{C(h)-C(0)}{h}=0;$$
and by proceeding in this way, if $x\neq 0$,
$$C^{(n)}(x)=C(x)\left\{\frac{2^n}{x^{3n}}+\text{terms of lower degree}\right\};\ C^{(n)}(0)=0.]$$

18. Prove that, if $a>1$,
$$F(x)=\sum_{n=0}^{\infty}\frac{1}{(2n)!}\frac{1}{1+a^{2n}x}$$
is a series which converges for all values of $x\geqslant 0$, and has derivatives of every order for these values of x.

Shew also that the Maclaurin series for $F(x)$ is
$$\tfrac{1}{2}\sum_{m=0}^{\infty}(-)^m\left(e^{a^m}+e^{-a^m}\right)x^m,$$
and that it is divergent for all positive values of x.

[The validity of term-by-term derivation of the infinite series may be assumed. It can be shewn that
$$F^{(m)}(0)=(-)^m m!\sum_{n=0}^{\infty}\frac{a^{2mn}}{(2n)!},$$
whence the required result easily follows.

This example shews that the Taylor series of a given function $f(x)$ need not converge, if $f(x)$ and its derivatives of every order exist and are continuous in a given interval.]

19. If $\{a_n\}$ is a positive decreasing sequence and $a_n\to 0$ as $n\to\infty$, prove that the same is true of $\{b_n\}$ where
$$b_n=(a_1+a_2+\dots+a_n)/n.$$

Hence prove that the series
$$a_1-\tfrac{1}{2}(a_1+a_2)+\tfrac{1}{3}(a_1+a_2+a_3)-\dots$$
is convergent.

CHAPTER VI

INEQUALITIES

6·1. Introduction.

Inequalities play an important part in mathematics, and the well-known elementary ones may be found in most text-books on Algebra. There are, however, several inequalities, which may be termed elementary*, of which it is not easy to find proofs except by reference to original memoirs. These inequalities have become especially important because of their frequent use in the theory of convergence. In this chapter a systematic account of the more important inequalities will be given, and, in addition, some of the simpler recent theorems on the theory of convergence of series of positive terms will be proved.

For the sake of completeness the proofs of two elementary inequalities will now be given.

6·11. The arithmetic, geometric and harmonic means.

Let $a_1, a_2, \ldots, a_n$ be n arbitrary real numbers; their *arithmetic mean* $A(a)$ is defined to be

$$\frac{a_1+a_2+\ldots+a_n}{n}.$$

If the numbers are all *positive*, the *geometric* and *harmonic means* are defined to be

$$G(a)=\sqrt[n]{(a_1 a_2 \ldots a_n)}, \quad H(a)=\frac{n}{\dfrac{1}{a_1}+\dfrac{1}{a_2}+\ldots+\dfrac{1}{a_n}}.$$

THEOREM. *If the numbers $a_1, a_2, \ldots, a_n$ are positive and not all equal, then*

$$H(a)<G(a)<A(a).$$

We prove first that† $G(a)<A(a)$, that is, that

$$a_1 a_2 \ldots a_n < \left(\frac{a_1+a_2+\ldots+a_n}{n}\right)^n \quad \ldots\ldots\ldots\ldots(1).$$

* In particular the inequalities of Hölder and Minkowski. See §§ 6·41, 6·42.

† The proof here given is due to Cauchy (1897).

Clearly, unless $a_1 = a_2$,

$$a_1 a_2 = \left(\frac{a_1+a_2}{2}\right)^2 - \left(\frac{a_1-a_2}{2}\right)^2 < \left(\frac{a_1+a_2}{2}\right)^2;$$

$$a_1 a_2 a_3 a_4 \leqslant \left(\frac{a_1+a_2}{2}\right)^2 . \left(\frac{a_3+a_4}{2}\right)^2 \leqslant \left(\frac{a_1+a_2+a_3+a_4}{4}\right)^4$$

with equality throughout only when $a_1 = a_2 = a_3 = a_4$; and by proceeding in this way we get, if $n = 2^m$,

$$a_1 a_2 \dots a_n < \left(\frac{a_1+a_2+\dots+a_n}{n}\right)^n \qquad \dots\dots\dots\dots(2).$$

If now n is not a term of the set 2, 4, 8, ..., 2^m, ..., let n be less than 2^m and write

$$k = \frac{a_1+a_2+\dots+a_n}{n}.$$

Suppose that in the left-hand side of (2) the last $2^m - n$ factors are equal to k, then

$$a_1 a_2 \dots k^{2^m - n} < \left\{\frac{a_1+a_2+\dots+a_n+(2^m-n)k}{2^m}\right\}^{2^m}$$

or
$$a_1 a_2 \dots k^{2^m - n} < k^{2^m},$$

hence
$$a_1 a_2 \dots a_n < k^n = \left(\frac{a_1+a_2+\dots+a_n}{n}\right)^n.$$

By writing $1/a_n$ for a_n in the above we easily deduce that

$$H(a) < G(a).$$

6·12. Cauchy's inequality.

Let $a_1, a_2, \dots, a_n$; $b_1, b_2, \dots, b_n$ be two sets of arbitrary real numbers, then

$$(\Sigma ab)^2 \leqslant \Sigma a^2 . \Sigma b^2 \qquad \dots\dots\dots\dots\dots\dots\dots\dots(1),$$

the equality sign only holding if the numbers a_ν, b_ν are proportional to each other, that is when $\lambda a_\nu + \mu b_\nu = 0$ $(\nu = 1, 2, \dots, n)$, $\lambda^2 + \mu^2 > 0$.

To prove this, suppose that λ and μ are variable, and consider the quadratic form

$$\Sigma(\lambda a + \mu b)^2 \equiv A\lambda^2 + 2B\lambda\mu + C\mu^2 \qquad \dots\dots\dots\dots(2).$$

The left-hand side of (2) is clearly either positive or zero, and it can only be zero if $\lambda a_\nu + \mu b_\nu = 0$ $(\nu = 1, 2, \dots, n)$.

The numbers A and C are essentially positive, for they are respectively the expressions on the right-hand side of (1). Hence the expression

$$A\lambda^2 + 2B\lambda\mu + C\mu^2$$

can only have the same sign as A (or C) for all values of λ and μ, if

$$B^2 - AC \leqslant 0,$$

the equality sign corresponding to the case when the quadratic expression vanishes. The inequality is therefore established.

6·2. A fundamental inequality.

THEOREM*. *If $p > 1$, $x > 1$, then for all real values of p,*

$$x^p - 1 > p(x-1) \quad \text{.....................(1).}$$

(i) Let s be a positive integer and let y be greater than 1, then

$$sy^s > 1 + y + y^2 + \ldots + y^{s-1} = \frac{y^s - 1}{y - 1},$$

and if we write $x = y^s$ we get

$$s(x^{1/s} - 1) > (x-1)/x \quad \text{.....................(2).}$$

(ii) We have identically

$$\frac{y^{s+1}-1}{s+1} - \frac{y^s - 1}{s} = \frac{y-1}{s(s+1)} \{(y^s - y^{s-1}) + (y^s - y^{s-2}) + \ldots + (y^s - 1)\}.$$

If the expression in the large bracket be divided by $y - 1$ there will be $1 + 2 + 3 + \ldots + s = \frac{1}{2}s(s+1)$ terms, every one of which is not less than 1; hence

$$\frac{y^{s+1}-1}{s+1} - \frac{y^s - 1}{s} \geqslant \tfrac{1}{2}(y-1)^2 \quad \text{..................(3).}$$

Next let $r = s + t$, and by writing down the t inequalities of the same type as (3) and adding we get

$$\frac{y^r - 1}{r} - \frac{y^s - 1}{s} \geqslant \tfrac{1}{2}t(y-1)^2 \quad \text{...............(4).}$$

Now replace y^s by x and use (2); it follows that

$$\frac{x^{r/s} - 1}{r/s} - (x-1) \geqslant \tfrac{1}{2}ts(x^{1/s} - 1)^2 > \tfrac{1}{2}\frac{t}{s}\left(\frac{x-1}{x}\right)^2 \quad \text{......(5).}$$

(iii) Now let $\dfrac{r}{s} = p_n$ be a rational approximation to p†; it is clear that t/s tends to the positive limit $p - 1$ when $n \to \infty$, so that

* The theorem given here is part of the well-known inequality

$$px^{p-1}(x-1) > x^p - 1 > p(x-1), \quad (p > 1, x > 1),$$

of which a proof, *for rational values of p only*, is given in Chrystal's *Algebra*, II (1906), 43. The proof here given is due to Hardy, *Journal London Math. Soc.* IV, 68.

† See § 6·3 below.

the left-hand side of (5) is greater than a positive number which is independent of n. Hence, in the limit, (5) gives (1) and *the strict sign of inequality* is preserved. The theorem is therefore proved.

Note. The reader should observe that in the above proof we have been able to make a passage to a limit and preserve the sign of inequality. In general a passage to a limit causes the sign $<$ to degenerate into $\leqslant$. This fact is one of the chief difficulties which is met with in proving inequalities. The degeneration into $\leqslant$ when a passage to a limit is made usually makes it impossible to specify the exact conditions under which the equality sign can hold.

6·3. On the definition of x^p when p is irrational and $x > 0$.

In Chapter I no definition was given of x^p when p is irrational. By presupposing a knowledge of the exponential and logarithmic functions of a real variable, x^p may be defined as $e^{p \log x}$: see § 4·33, (3).

An alternative method is to appeal to Cantor's method of defining an irrational number as the limit of a suitable infinite aggregate of rational numbers. In the preceding section this method has been used; accordingly a brief account of Cantor's theory will now be given*.

The essence of Cantor's theory lies in the postulation of the existence of the aggregate of real numbers ordered in a definite manner by prescribed rules. Any element of the aggregate is regarded as capable of symbolical representation by means of a convergent sequence whose elements are rational numbers. The aggregate of real numbers contains within itself an aggregate of objects which is similar to the ordered aggregate of rational numbers, already discussed in Chapter I, in the sense that to each rational number there corresponds a certain real number. The relative order of any two rational numbers in the ordered aggregate of rational numbers is the same as the relative order of the two corresponding real numbers in the new aggregate of real numbers. The rational numbers, although logically distinct from the real numbers to which they correspond, are usually denoted by the same symbols†.

* For further details the reader should consult Hobson, *Functions of a Real Variable*, I (1921), 27 *et seq.*

† This remark has already been made in § 1·5 in connection with the Dedekind theory. Its importance lies in the fact that it obviates logical difficulties, especially in coordinating Cantor's theory with that of Dedekind. For further details see Hobson, *loc. cit.*

The order of the real numbers in their aggregate is assigned by the following rules:

(1) Any convergent sequence $\{a_n\}$, in which the elements are rational numbers, is taken to represent a real number which we denote by a.

Two such aggregates $\{a_n\}$, $\{b_n\}$ are taken to represent the same real number if they satisfy the condition that, given ϵ, a value of n can be found such that $|a_{n+m} - b_{n+m}| < \epsilon$ for this value of n and for all positive integral values of m (zero included).

(2) The real number represented by $\{a_n\}$ is regarded as *greater than* the real number represented by $\{b_n\}$ if a value of n can be found such that, for this value of n, and for all values 0, 1, 2, 3, ... of m,

$$a_{n+m} - b_{n+m} > \delta,$$

where δ is a fixed positive rational number.

A similar interpretation is easily given for "less than."

Suppose that the number p is defined by a convergent sequence $\{p_n\}$ in which all the numbers p_n are rational. We now shew that *the aggregate $\{x^{p_n}\}$ is convergent, and the number which it defines will be denoted by x^p*

Now $$x^{p_n} - x^{p_{n+m}} = x^{p_n}\{1 - x^{p_{n+m} - p_n}\} \quad \text{...............(1).}$$

Since every member of the convergent sequence $\{p_n\}$ is less than some fixed number, $|x^{p_n}| < A$, where A is constant.

If $x > 1$,

$$|x^{p_n} - x^{p_{n+m}}| < A\,|x^{|p_n - p_{n+m}|} - 1| \quad \text{............(2).}$$

Let n be chosen so that, for all values of m,

$$|p_n - p_{n+m}| < 1/q,$$

where q is a positive integer; then

$$x^{|p_n - p_{n+m}|} - 1 < x^{1/q} - 1 < \frac{x-1}{1 + x^{1/q} + x^{2/q} + \dots + x^{(q-1)/q}}.$$

It follows from (2) that

$$|x^{p_n} - x^{p_{n+m}}| < A\,.\,(x-1)/q.$$

If q be chosen so that $\dfrac{1}{q} < \dfrac{\epsilon}{A(x-1)}$, we see that n may be so chosen that

$$|x^{p_n} - x^{p_{n+m}}| < \epsilon$$

for all values of m. Hence $\{x^{p_n}\}$ is a convergent sequence.

If $x < 1$, by what we have just proved, $\left\{\dfrac{1}{x^{p_n}}\right\}$ is a convergent sequence, and therefore $\{x^{p_n}\}$ is also convergent (§ 3·23, (3)).

If $x = 1$, $\{x^{p_n}\} = 1$.

Thus it has been proved that in every case $\{x^{p_n}\}$ is a convergent sequence if $\{p_n\}$ is convergent.

Since
$$\{x^{p_n}\} \times \{x^{q_n}\} = \{x^{p_n+q_n}\},$$
it follows that when p and q are irrational the relation
$$x^p \,.\, x^q = x^{p+q}$$
is satisfied.

6·4. Deductions from the fundamental inequality.

THEOREM 1. *If* $0 < p < 1$, $x > 1$, *then*
$$x^p - 1 < p\,(x-1) \qquad \text{.....................(1).}$$

This follows at once from the fundamental inequality of § 6·2, for if we write $\dfrac{1}{p}$ for p in equation (1) of that section, we get
$$x^{\frac{1}{p}} - 1 > \frac{1}{p}(x-1),$$
and the desired inequality follows by writing x^p for x.

THEOREM 2. *If a and b are positive and unequal, and α and β are any two positive numbers whose sum is unity, then*
$$a^{\alpha} b^{\beta} < a\alpha + b\beta \qquad \text{.......................(2).}$$

Suppose that $a > b$, and write $a = x$, $b = 1$, $\alpha = p$, and inequality (2) reads
$$x^p < px + (1-p),$$
that is
$$x^p - 1 < p\,(x-1),$$
and inequality (2) reduces to (1).

6·41. Hölder's inequality.

If $a_1, a_2, \ldots, a_n$; $b_1, b_2, \ldots, b_n$ are all positive, and $\dfrac{1}{p} + \dfrac{1}{q} = 1$, *then*
$$\sum_{\nu=1}^{n} a_\nu b_\nu \leqslant \left\{\sum_1^n a_\nu{}^p\right\}^{\frac{1}{p}} \cdot \left\{\sum_1^n b_\nu{}^q\right\}^{\frac{1}{q}} \qquad \text{..................(1),}$$
the equality sign holding only if $a_\nu{}^p = \lambda b_\nu{}^q$ $(\nu = 1, 2, \ldots, n)$, where λ is a constant.

By Theorem 2, all the summations being from $\nu=1$ to $\nu=n$, we have*

$$\frac{\Sigma a^{\alpha} b^{\beta}}{(\Sigma a)^{\alpha}(\Sigma b)^{\beta}}=\Sigma\left(\frac{a}{\Sigma a}\right)^{\alpha}.\left(\frac{b}{\Sigma b}\right)^{\beta}<\Sigma\left(\frac{\alpha a}{\Sigma a}+\frac{\beta b}{\Sigma b}\right)=\alpha+\beta=1 \quad \ldots(2)$$

unless

$$\frac{a}{\Sigma a}=\frac{b}{\Sigma b} \quad \ldots\ldots\ldots\ldots\ldots\ldots\ldots\ldots(3),$$

that is unless $a_\nu=\lambda b_\nu\,(\nu=1,2,\ldots,n)$, in which case the inequality sign is replaced by equality.

If $\frac{1}{p}$ and $\frac{1}{q}$ be written for α and β, and a^{α}, b^{β} are replaced by a and b, the inequality (1) is established, and *the condition for equality* is seen to be that

$$a_\nu^p=\lambda b_\nu^q \quad (\nu=1,2,\ldots,n).$$

COROLLARY. *If* $p^{-1}+q^{-1}=1$, *then*

$$\frac{1}{n}\Sigma ab<\left(\frac{1}{n}\,\Sigma a^p\right)^{\frac{1}{p}}.\left(\frac{1}{n}\,\Sigma b^q\right)^{\frac{1}{q}} \quad \ldots\ldots\ldots\ldots\ldots(4).$$

This is an immediate deduction from the theorem, for Hölder's inequality is homogeneous, not only in the a_ν and b_ν but also in the sign Σ. For all such inequalities *sums* may always be replaced by *means*. This special form of Hölder's inequality is very important in connection with the extension of these theorems to inequalities between integrals.

6·42. Minkowski's inequality.

If $M_p(a)\equiv(a_1^p+a_2^p+\ldots+a_n^p)^{\frac{1}{p}}$, *where the* a_ν *are all positive, and* $p>1$, *then*

$$M_p(a+b)\leqslant M_p(a)+M_p(b) \quad \ldots\ldots\ldots\ldots\ldots(1),$$

the equality sign only holding if $a_\nu=\lambda b_\nu$ $(\nu=1,2,\ldots,n)$.

The simplest proof is a deduction from Hölder's inequality†. We have identically

$$S\equiv\Sigma(a+b)^p\equiv\Sigma a(a+b)^{p-1}+\Sigma b(a+b)^{p-1}\ldots\ldots(2).$$

* For convenience we shall write Σa for $\sum_{\nu=1}^{n} a_\nu$. Unless the contrary is stated, in all the theorems on inequalities Σ will stand for $\sum_{1}^{n}$.

† The idea of this deduction is due to F. Riesz, *Équations Linéaires* (Borel Tract), 45.

If we apply Hölder's inequality to each sum on the right-hand side, and observe that $(p-1)q=p$, we get

$$S \leqslant \{(\Sigma a^p)^{\frac{1}{p}} + (\Sigma b^p)^{\frac{1}{p}}\} \{\Sigma (a+b)^p\}^{\frac{1}{q}}$$

$$= \{(\Sigma a^p)^{\frac{1}{p}} + (\Sigma b^p)^{\frac{1}{p}}\} S^{\frac{1}{q}},$$

that is
$$S^{\frac{1}{p}} \leqslant (\Sigma a^p)^{\frac{1}{p}} + (\Sigma b^p)^{\frac{1}{p}},$$
which is the same as (1).

The conditions for equality in each application of Hölder's inequality to the summations in (2) are that

$$a_\nu{}^p = \lambda_1 (a_\nu + b_\nu)^{(p-1)q} = \lambda_1 (a_\nu + b_\nu)^p,$$

and
$$b_\nu{}^p = \lambda_2 (a_\nu + b_\nu)^p,$$

from which we see that the equality sign in (1) only holds if a_ν is proportional to b_ν $(\nu = 1, 2, \ldots, n)$.

6·43. Methods of proving inequalities.

In the preceding sections the inequalities of Minkowski and of Hölder have been proved by a method which makes them depend upon the fundamental inequality of § 6·2. The proofs have been strictly *elementary* proofs, that is to say they have only appealed to Algebra, save where the limiting process was introduced to include the case of x^p when p is irrational. Since x^p is not an algebraic function when p is irrational, it is clear that no *strictly algebraical* proof could be given to include this case. The reader should observe however that the limiting process which was used to extend the inequalities to include the case when p is irrational is the *only* limiting process involved.

The inequality of Hölder is a special case of a very general inequality proved by Jensen in his classical paper* in which he introduced the concept of "convex functions." In fact most of the elementary inequalities are only particular cases of this very general result. The deduction of inequalities from Jensen's theorem is a simple and elegant method of proof, and this method is only inferior to the one given above in that Jensen's result involves an appeal to the elementary theorems of the differential calculus, and therefore to the theory of limits and continuity of functions of a single real variable.

A third type of proof is one which appeals to the theory of maxima and minima of functions of several variables. Proofs of the inequalities of Hölder and Minkowski have been given by F. Riesz† which depend upon the investigation of the extreme values of Σax subject to the condition that Σx^p is constant. The proofs are elegant and easy to remember, but they suffer from the defect of dependence upon results which are among the later developments of the theory of functions of a real variable.

* *Acta Mathematica*, xxx (1906), 175 *et seq.* See § 6·5. † *Loc. cit. ante.*

6·5. Convex and concave functions.

Let $\phi(t)$ be a function defined for all values of t in the interval $\alpha \leqslant t \leqslant \beta$; then the function $\phi(t)$ is said to be *convex* if for every pair of unequal values t_1 and t_2 in (α, β)

$$\phi\left(\frac{t_1+t_2}{2}\right) \leqslant \frac{\phi(t_1)+\phi(t_2)}{2} \quad \text{.................(1).}$$

If the sign $\leqslant$ is replaced by $<$, $\phi(t)$ is said to be *strictly convex*.

When the sign $<$ is replaced by $>$, the function $\phi(t)$ is *concave* or *strictly concave*.

THEOREM. *If $\phi(t)$ be a function defined in (α, β), and if $\phi''(t)$ exists and is not zero, then $\phi(t)$ is strictly concave or convex according as $\phi''(t)$ is less than or greater than zero.*

Let t_1 and t_2 be two arbitrary points in (α, β), then by Taylor's theorem with remainder when $n=2$, $(t_2 > t_1)$,

$$\phi(t_1) = \phi\left(\frac{t_1+t_2}{2}\right) + \frac{t_1-t_2}{2}\phi'\left(\frac{t_1+t_2}{2}\right) + \frac{(t_1-t_2)^2}{8}\phi''(\tau_1),$$

$$\phi(t_2) = \phi\left(\frac{t_1+t_2}{2}\right) + \frac{t_2-t_1}{2}\phi'\left(\frac{t_1+t_2}{2}\right) + \frac{(t_1-t_2)^2}{8}\phi''(\tau_2),$$

where $t_1 < \tau_1 < \frac{1}{2}(t_1+t_2) < \tau_2 < t_2$.

By addition it follows that

$$\phi(t_1)+\phi(t_2) \lesseqgtr 2\phi\left(\frac{t_1+t_2}{2}\right),$$

according as $\phi''(t) \gtrless 0$. This proves the theorem.

THEOREM. *If $\phi(t)$ is strictly concave in (α, β), and if $\phi''(t)$ exists, then $\phi''(t) \leqslant 0$.*

If $h > 0$ and the points $t \pm h$ are in (α, β), then, since $\phi(t)$ is strictly concave,

$$2\phi(t) > \phi(t-h) + \phi(t+h) \quad \text{...............(1).}$$

Now, by Taylor's theorem, § 4·6,

$$\phi(t+h) = \phi(t) + h\phi'(t) + \frac{h^2}{2!}(\phi''(t)+\epsilon_1) \quad \text{......(2),}$$

$$\phi(t-h) = \phi(t) - h\phi'(t) + \frac{h^2}{2!}(\phi''(t)+\epsilon_2) \quad \text{......(3),}$$

where ϵ_1 and ϵ_2 tend to zero as $h \to 0$.

By adding (2) and (3) and taking account of (1), it follows that

$$2\phi''(t)+\epsilon_1+\epsilon_2<0,$$

and on making $h \to 0$ that

$$\phi''(t) \leqslant 0.$$

Similarly, if $\phi(t)$ is strictly convex, $\phi''(t) \geqslant 0$.

6·51. Generalisation of the convex property.

The property of a convex function can be generalised as in the following theorem.

THEOREM. *If $t_1, t_2, \ldots, t_n$ are n arbitrary values of t in the interval (α, β), and if $\phi(t)$ is convex in (α, β), then*

$$\phi\left(\frac{t_1+t_2+\ldots+t_n}{n}\right) \leqslant \frac{\phi(t_1)+\phi(t_2)+\ldots+\phi(t_n)}{n}.$$

The proof follows the same lines as the one given in § 6·11 for the theorem on arithmetic and geometric means.

The reader is advised to write out the formal proof as an exercise.

6·52. Jensen's theorem.

Let $a_1, a_2, \ldots, a_n$ be arbitrary positive numbers; then, if $\phi''(t)$ exists, and $\phi(t)$ is convex in $\alpha \leqslant t \leqslant \beta$,

$$\phi\left(\frac{a_1t_1+a_2t_2+\ldots+a_nt_n}{a_1+a_2+\ldots+a_n}\right) \leqslant \frac{a_1\phi(t_1)+a_2\phi(t_2)+\ldots+a_n\phi(t_n)}{a_1+a_2+\ldots+a_n} \quad \ldots\ldots(1).$$

Since $\phi''(t)$ exists and $\phi(t)$ is convex, $\phi''(t) \geqslant 0$. Write

$$\frac{a_1t_1+a_2t_2+\ldots+a_nt_n}{a_1+a_2+\ldots+a_n}=\beta \quad \ldots\ldots\ldots\ldots\ldots(2),$$

then, by Taylor's theorem with remainder when $n=2$,

$$\phi(t_\nu)=\phi(\beta)+(t_\nu-\beta)\phi'(\beta)+\frac{(t_\nu-\beta)^2}{2}\phi''(\tau_\nu) \ldots(3),$$

where $t_\nu<\tau_\nu<\beta$.

Let $\nu=1, 2, \ldots, n$, multiply each equation from (3) by $a_1, a_2, \ldots, a_n$ respectively and add, and we get

$$\frac{a_1\phi(t_1)+a_2\phi(t_2)+\ldots+a_n\phi(t_n)}{a_1+a_2+\ldots+a_n}=\phi(\beta)+\frac{\sum_1^n a_\nu \frac{1}{2}(t_\nu-\beta)^2\phi''(\tau_\nu)}{\sum_1^n a_\nu} \quad \ldots\ldots(4),$$

$$\geqslant \phi(\beta).$$

If $\phi(t)$ is continuous and convex, then the sign of equality can only hold if either (i) all the t are equal or (ii) $\phi(t)$ is linear in an interval containing all the t, in other words, if $\phi''(t)=0$.

COROLLARY. *If $\phi(t)$ is concave, the inequality sign in* (1) *is reversed.*

Note. The reader should observe that the above theorem is extremely general. Besides the arbitrary nature of the a_ν, the class of functions $\phi(t)$ for which $\phi''(t)$ exists in (α, β) is a very wide one. Jensen himself claims that this theorem implicitly contains almost every known inequality.

6·53. Second proof of Minkowski's inequality.

The complete statement of Minkowski's inequality is as follows:

$$M_p(a+b) \leqslant M_p(a)+M_p(b) \quad \textit{if } p \geqslant 1 \text{ ....(1)},$$

$$M_p(a+b) \geqslant M_p(a)+M_p(b) \quad \textit{if } p \leqslant 1 \text{ ....(2)},$$

the equality signs only holding if $a_\nu=\lambda b_\nu$ $(\nu=1, 2, \ldots, n)$, *or if* $p=1$.

Suppose that $p \neq 0$, and write

$$A_\nu = ta_\nu + (1-t)b_\nu \quad (\nu=1, 2, \ldots, n), \quad 0 \leqslant t \leqslant 1,$$

and let
$$\phi(t) = M_p(A) \equiv (A_1{}^p + A_2{}^p + \ldots + A_n{}^p)^{\frac{1}{p}} \text{ ....(3)}.$$

Since $\phi''(t)$ exists we can use the simple criterion of § 6·5, by which the sign of $\phi''(t)$ determines whether the function $\phi(t)$ is convex or concave. Consider the interval $0 \leqslant t \leqslant 1$ and let $t_1=0$, $t_2=1$, then

$$\text{if } \phi''(t) \geqslant 0, \quad \phi(0)+\phi(1) \geqslant 2\phi(\tfrac{1}{2}) \text{ ....(4)},$$

$$\text{if } \phi''(t) \leqslant 0, \quad \phi(0)+\phi(1) \leqslant 2\phi(\tfrac{1}{2}) \text{ ....(5)}.$$

By forming the second derivative of $\phi(t)$ we get

$$\begin{aligned}\phi''(t) = (p-1)\{A_1{}^p + \ldots + A_n{}^p\}^{\frac{1}{p}-2} \\ \times [(A_1{}^p + \ldots + A_n{}^p)\{A_1{}^{p-2}(a_1-b_1)^2 + \ldots + A_n{}^{p-2}(a_n-b_n)^2\} \\ -\{A_1{}^{p-1}(a_1-b_1) + \ldots + A_n{}^{p-1}(a_n-b_n)\}^2].\end{aligned}$$

It follows from Cauchy's inequality, when suitable changes in the notation have been made, that the content of the square brackets is certainly positive so long as $a_\nu \neq \lambda b_\nu$. Hence we see that the sign of $\phi''(t)$ is the same as that of $p-1$.

From (4) and (5) we deduce that

$$2\phi(\tfrac{1}{2})\begin{cases}\leqslant \phi(0)+\phi(1) & \text{if } p\geqslant 1,\\ \geqslant \phi(0)+\phi(1) & \text{if } p\leqslant 1.\end{cases}$$

The latter inequalities are equivalent to (1) and (2) above; the theorem is therefore proved.

6·54. Further deductions from Jensen's theorem.

To illustrate the generality of Jensen's theorem (§ 6·52) two other inequalities will be deduced from it.

(1) *The generalisation of the theorem of the arithmetic, geometric and harmonic means.*

Let $\phi(t)=-\log t$; since $\phi''(t)=1/t^2>0$ if $t>0$, then $\phi(t)$ is strictly convex in any interval in which t remains positive.

Hence from Jensen's theorem it follows that if the numbers $a_1, a_2, \ldots, a_n$; $t_1, t_2, \ldots, t_n$ are all positive, and the numbers t_ν are not all equal,

$$\frac{a_1+a_2+\ldots+a_n}{\dfrac{a_1}{t_1}+\dfrac{a_2}{t_2}+\ldots+\dfrac{a_n}{t_n}}<\exp\left(\frac{a_1\log t_1+\ldots+a_n\log t_n}{a_1+\ldots+a_n}\right)$$

$$<\frac{a_1t_1+a_2t_2+\ldots+a_nt_n}{a_1+a_2+\ldots+a_n}.$$

The second half of this inequality is direct, and the first half easily follows by writing $1/t_\nu$ for t_ν.

(2) *Jensen's inequality.*

If $a_1, a_2, \ldots, a_n$ are arbitrary positive numbers,

$$(\Sigma a^r)^{\frac{1}{r}}>(\Sigma a^s)^{\frac{1}{s}}, \text{ if } 0<r<s.$$

Let $\psi(t)=(a_1^t+a_2^t+\ldots+a_n^t)^{\frac{1}{t}}=(A_1+A_2+\ldots+A_n)^{\frac{1}{t}}$ ...(1), where $A_\nu=a_\nu^t$.

The inequality is proved if we shew that $\psi(t)$ is a *decreasing* function of t; since $\psi(t)$ is positive, we must prove that $\psi'(t)<0$.

Take the logarithmic derivative of (1), and we get

$$t^2\frac{\psi'(t)}{\psi(t)}=\frac{A_1\log A_1+\ldots+A_n\log A_n}{A_1+\ldots+A_n}-\log(A_1+\ldots+A_n)$$

......(2).

By the generalised theorem of the means proved above, we have

$$\frac{A_1 \log A_1 + \ldots + A_n \log A_n}{A_1 + \ldots + A_n} < \log\left(\frac{A_1^2 + \ldots + A_n^2}{A_1 + \ldots + A_n}\right)$$
$$< \log(A_1 + \ldots + A_n) \quad \ldots(3),$$

for since all the numbers A_ν are positive,

$$(A_1^2 + \ldots + A_n^2) < (A_1 + \ldots + A_n)^2.$$

From (2) and (3) it follows that

$$\psi'(t) < 0,$$

and the inequality has been established.

6·6. Some elementary convergence theorems.

In the preceding sections all our summations have been finite, and the question of convergence could not arise. The reader will easily see that by making $n \to \infty$ the inequalities of Hölder and Minkowski at once give rise to some simple criteria for convergence of series of positive terms. These criteria are not of much practical utility, but as immediate extensions of the inequalities they are now stated.

(1) *If* $p^{-1} + q^{-1} = 1$, *and the series* $\sum_1^\infty a_\nu^p$ *and* $\sum_1^\infty b_\nu^q$ *both converge, then so does the series* $\sum_1^\infty a_\nu b_\nu$.

(2) *If* $p > 1$, *and* $\sum_1^\infty a_\nu^p$, $\sum_1^\infty b_\nu^p$ *both converge, then so does the series* $\sum_1^\infty (a_\nu + b_\nu)^p$.

The deduction of (1) and (2) from the inequalities of Hölder and Minkowski is left as an exercise for the reader. There are, of course, corresponding criteria for divergence.

6·7. Hardy's theorem*.

If $\kappa > 1$ *and* $A_n = a_1 + a_2 + \ldots + a_n$ *is the sum of the first n terms of a positive series, then, if the series* Σa_ν^κ *converges to the sum S, the series*

$$\Sigma\left(\frac{A_\nu}{\nu}\right)^\kappa \quad \textit{and} \quad \Sigma a_\nu \left(\frac{A_\nu}{\nu}\right)^{\kappa-1}$$

* This theorem was first proved by G. H. Hardy, *Math. Zeitschrift*, VI (1920), 314–317. The proof given here is on the lines of the proof by E. B. Elliott, *Journal London Math. Soc.* I (1926), 93.

will converge to sums V and T, such that

$$V \leqslant \frac{\kappa}{\kappa-1} T \leqslant \left(\frac{\kappa}{\kappa-1}\right)^{\kappa} S.$$

Write $u_n = A_n/n$, so that

$$a_n = nu_n - (n-1) u_{n-1},$$

it being understood that any number with suffix 0 is zero.

By the generalisation of the theorem of the means, § 6·54 (1),

$$(u_n^{\kappa(\kappa-1)} u^{\kappa}_{n-1})^{\frac{1}{\kappa}} \leqslant \frac{(\kappa-1) u_n^{\kappa} + u^{\kappa}_{n-1}}{\kappa};$$

hence $$\kappa u_n^{\kappa-1} u_{n-1} \leqslant (\kappa-1) u_n^{\kappa} + u^{\kappa}_{n-1} \quad \text{............(1).}$$

For any value of n, we have the identity

$$\begin{aligned} u_n^{\kappa} - \frac{\kappa}{\kappa-1} \{nu_n - (n-1) u_{n-1}\} u_n^{\kappa-1} &= u_n^{\kappa}\left(1 - \frac{n\kappa}{\kappa-1}\right) + \frac{n-1}{\kappa-1} \kappa u_n^{\kappa-1} u_{n-1} \\ &\leqslant u_n^{\kappa}\left(1 - \frac{n\kappa}{\kappa-1}\right) + \frac{n-1}{\kappa-1} \{(\kappa-1) u_n^{\kappa} + u^{\kappa}_{n-1}\} \\ &\leqslant \frac{1}{\kappa-1} \{(n-1) u^{\kappa}_{n-1} - nu_n^{\kappa}\} \quad \text{......................(2),} \end{aligned}$$

by (1).

Now let S_n, V_n, T_n denote the sum to n terms of the three series. From (2), by taking the values 1, 2, ..., n in succession for n and adding, we get

$$V_n - \frac{\kappa}{\kappa-1} T_n \leqslant -\frac{nu_n^{\kappa}}{\kappa-1} \leqslant 0 \quad \text{..................(3).}$$

Again, by Hölder's inequality,

$$\left(\sum_1^n a_\nu b_\nu^{\kappa-1}\right)^{\kappa} \leqslant \left(\sum_1^n a_\nu^{\kappa}\right) \left(\sum_1^n b_\nu^{\kappa}\right)^{\kappa-1} \quad \text{..............(4),}$$

and if we write u_ν for b_ν in (4), it becomes

$$T_n^{\kappa} \leqslant S_n V_n^{\kappa-1} \quad \text{........................(5).}$$

From (3) and (5) we deduce that

$$V_n \leqslant \frac{\kappa}{\kappa-1} T_n \leqslant \left(\frac{\kappa}{\kappa-1}\right)^{\kappa} S_n \quad \text{..............(6).}$$

By the data, the sequences $\{S_n\}$, $\{V_n\}$ and $\{T_n\}$ are positive monotonic increasing sequences, and the first is known to have a finite limit S as $n\to\infty$. Hence by making $n\to\infty$ we get

$$V \leqslant \frac{\kappa}{\kappa-1}T \leqslant \left(\frac{\kappa}{\kappa-1}\right)^{\kappa} S,$$

which was to be proved.

6·8. Carleman's theorem*.

If the numbers $a_1, a_2, \ldots, a_n, \ldots$ are all positive, and if $\sum_1^\infty a_n$ converges to the sum S, then, if $g_n = (a_1a_2\ldots a_n)^{\frac{1}{n}}$, the series $\sum_1^\infty g_n$ converges to the sum U which cannot exceed eS.

LEMMA. *Let*

$$b_n = \frac{a_1+2a_2+\ldots+na_n}{n(n+1)},$$

then Σb_n converges to the sum S.

In § 2·6 it was proved that when $n\to\infty$ the sequence

$$\frac{s_1+s_2+\ldots+s_n}{n}$$

tends to the same limit as the sequence $\{s_n\}$. Let $s_n \equiv \sum_1^n a_n$, then

$$\sum_1^n b_n = \frac{a_1}{1.2}+\frac{a_1+2a_2}{2.3}+\frac{a_1+2a_2+3a_3}{3.4}+\ldots+\frac{a_1+2a_2+\ldots+na_n}{n(n+1)}$$

$$=a_1\left\{1-\frac{1}{n+1}\right\}+2a_2\left\{\frac{1}{2}-\frac{1}{n+1}\right\}+\ldots+na_n\left\{\frac{1}{n}-\frac{1}{n+1}\right\}$$

$$=\frac{na_1}{n+1}+\frac{(n-1)a_2}{n+1}+\ldots+\frac{a_n}{n+1}$$

$$=\frac{s_1+s_2+\ldots+s_n}{n+1}=\frac{s_1+s_2+\ldots+s_n}{n}\cdot\frac{1}{1+\frac{1}{n}}.$$

Therefore $$\sum_1^n b_n \to S \text{ as } n\to\infty.$$

* The proof is on the lines of that given by K. Knopp, *Journal London Math. Soc.* III (1928), 205. Knopp also proves two other convergence theorems, one of which (see Examples VI, 10) is easily deduced from Hardy's theorem, § 6·7.

Proof of theorem.

If b_n is as defined in the lemma, we have

$$g_n = (a_1 a_2 \dots a_n)^{\frac{1}{n}} = \frac{1}{(n!)^{\frac{1}{n}}} (a_1 . 2a_2 . \dots . na_n)^{\frac{1}{n}}$$

$$\leqslant \frac{n+1}{(n!)^{\frac{1}{n}}} b_n$$

by § 6·11.

Now $$\left(1+\frac{1}{1}\right)^1 \left(1+\frac{1}{2}\right)^2 \dots \left(1+\frac{1}{n}\right)^n = \frac{(n+1)^n}{n!} < e^n,$$

so that $$\frac{n+1}{(n!)^{\frac{1}{n}}} < e.$$

Thus we have

$$\sum_1^n g_n < e \sum_1^n b_n,$$

and by making $n \rightarrow \infty$ we deduce that $\sum_1^\infty g_n$ converges, and that its sum U cannot exceed eS.

6·81. The best possible constant.

It is not difficult to prove that e is the ***best possible constant*** in the preceding theorem. The method which will be used here is one which can be employed in other cases where it is required to shew that a certain constant is the best possible. The proof in this case is sufficiently simple to be given here as an illustration of the general method of procedure.

Write $$a_n = \frac{1}{n} \text{ for } n = 1, 2, \dots, m,$$

$$a_n = 0 \text{ for } n > m,$$

we then have to prove that

$$\lim_{m \rightarrow \infty} \left\{ \frac{\sum\limits_{n=1}^\infty g_n}{\sum\limits_{n=1}^\infty a_n} \right\} = e.$$

Now $$\frac{\sum\limits_1^m g_n}{\sum\limits_1^m a_n} = \frac{\sum\limits_1^m \left(\frac{1}{n!}\right)^{\frac{1}{n}}}{\sum\limits_1^m \frac{1}{n}},$$

and since $\Sigma\frac{1}{n}$ diverges, by Examples V, 2, if $m/(m!)^{\frac{1}{m}}$ tends to a limit as $m\to\infty$, the right-hand side tends to the same limit. It remains therefore to prove that $m/(m!)^{\frac{1}{m}}$ tends to the limit e as $m\to\infty$.

If $u_m = m^m/m!$ we see that

$$\frac{u_{m+1}}{u_m} = \left(\frac{m+1}{m}\right)^m,$$

which tends to e as $m\to\infty$. Hence, by a theorem given in § 5·3, $(u_m)^{\frac{1}{m}}$ tends to the same limit e.

This proves that e is the best possible constant in Carleman's theorem.

6·9. Hilbert's theorem.

*If $a_m \geqslant 0$ and Σa_m^2 converges, then the double series**

$$\sum_{m\leqslant n}^{\infty} \frac{a_m a_n}{m+n}$$

also converges.

From Hardy's theorem when $\kappa = 2$ we deduce that if Σa_n^2 converges to the sum S, then $\Sigma\frac{a_n A_n}{n}$ converges to a sum T, and $T \leqslant 2S$.

Now

$$\frac{a_1}{n+1} + \frac{a_2}{n+2} + \ldots + \frac{a_n}{2n} < \frac{a_1 + a_2 + \ldots + a_n}{n} = \frac{A_n}{n} \quad \ldots(1);$$

but the left-hand side of (1) is $\sum_{m=1}^{n} \frac{a_m}{m+n}$, and so

$$a_n \sum_{m=1}^{n} \frac{a_m}{m+n} < \frac{a_n A_n}{n},$$

hence

$$\sum_n a_n \sum_{m=1}^{\infty} \frac{a_m}{m+n} < \sum_n \frac{a_n A_n}{n} = T \quad \ldots\ldots\ldots\ldots\ldots(2).$$

Hence, since the left-hand side of (2) is the sum of a double series of positive terms $\sum_{m\leqslant n}^{\infty} \frac{a_m a_n}{m+n}$ in a way which certainly includes all the terms, this double series converges.

* For the general theory of double series, see Bromwich's *Infinite Series* (1926), Ch. v. The fundamental result about double series *of positive terms* is that if such a series converges to a sum S in any way, it will have the same sum if summed in any other way so as to include all the terms.

It can be shewn that

$$\Sigma\Sigma \frac{a_m a_n}{m+n} \leqslant \pi \Sigma a_n^2,$$

and that π is the best constant possible. Since most of the simpler proofs of this result involve the integral calculus no further discussion of the theorem will be given here. The reader is referred to Examples VIII, 16.

EXAMPLES VI.

1. Prove that if n positive numbers have a definite sum nc, their product cannot exceed c^n and the sum of their squares lies between nc^2 and n^2c^2.

If there be four positive numbers whose sum is $4c$ and the sum of whose squares is $8c^2$, prove that no one of them can exceed $(\sqrt{3}+1)\,c$.

2. Examine whether the following functions are convex or concave,

(i) $t^m\,(0<m<1)$, (ii) $\log t$, (iii) $t^m\,(m>1$, or $m<0)$, (iv) $t \log t$, in any positive interval for t;

$$\text{(v)}\ \log(1+e^t), \quad \text{(vi)}\ (c^2+t^2)^{\frac{1}{2}}\ (c>0), \quad \text{(vii)}\ t^4,$$

in any interval for t.

3. By writing $\phi(t)=t^q$ $(q>1,\ t>0)$, deduce Hölder's inequality from Jensen's theorem.

4. If
$$f(t)=\left(\frac{a_1^t+a_2^t+\ldots+a_n^t}{n}\right)^{\frac{1}{t}} \quad (a_\nu>0,\ \nu=1, 2, \ldots, n),$$
prove that $f(t)$ is monotonic increasing. Deduce that

$$\left(\frac{1}{n}\,\Sigma a^r\right)^{\frac{1}{r}} < \left(\frac{1}{n}\,\Sigma a^s\right)^{\frac{1}{s}} \quad \text{if } 0<r<s.$$

Shew that the latter inequality is also deducible from Hölder's.

[Write $A_\nu=a_\nu^t$, calculate the logarithmic derivative of $f(t)$, and shew that $f'(t)>0$ by applying the generalised theorem of the means, see § 6·54.]

5. Evaluate $f(-1)$, $f(0)$ and $f(1)$ from $f(t)$ as defined in the last question; and hence deduce the theorem of § 6·11.

[$f(0)$ is taken to be $\lim\limits_{t\to 0} f(t)$.]

Shew also that the theorem of § 6·11 may be deduced from the definition of a generalised convex function (§ 6·51), by writing $\phi(t)=\log t$.

6. Prove that if $\phi(t)$ be a bounded function which is convex (or concave) in $\alpha \leqslant t \leqslant \beta$, then $\phi(t)$ is a continuous function of t in (α, β).

7. Prove that if $p+q+r=1$, and all the numbers involved are positive,

$$\sum_1^n a_\nu^p b_\nu^q c_\nu^r \leqslant (\Sigma a_\nu)^p (\Sigma b_\nu)^q (\Sigma c_\nu)^r.$$

8. From Jensen's theorem, by writing $\phi(t)=t\log t$, shew that

$$\frac{a_1 t_1 + a_2 t_2 + \ldots + a_n t_n}{t_1+t_2+\ldots+t_n} < \exp\left(\frac{a_1 t_1 \log a_1 + a_2 t_2 \log a_2 + \ldots + a_n t_n \log a_n}{a_1 t_1 + a_2 t_2 + \ldots + a_n t_n}\right).$$

9. Examine the convergence of the series

$$\text{(i)}\ u_1=1, \ldots, u_n=\left\{\frac{1}{n}+\frac{1}{n^{1/p}\log n}\right\}^p \quad (n\geqslant 2),$$

$$\text{(ii)}\ u_1=1, \ldots, u_n=\left\{\frac{1}{n!}+\frac{1}{n(\log n)^a}\right\}^p \quad (n\geqslant 2),$$

where a is any constant.

10. Deduce from Hardy's theorem (§ 6·7) that if $0<p<1$, $a_n\geqslant 0$, and Σa_n converges, then

$$\sum_{n=1}^\infty \left(\frac{a_1^p+a_2^p+\ldots+a_n^p}{n}\right)^{\frac{1}{p}} < (1-p)^{-1/p}\sum_1^\infty a_n.$$

11. If all the numbers are positive, prove that, unless $x/a=y/b$,

$$x\log\frac{x}{a}+y\log\frac{y}{b} > (x+y)\log\frac{x+y}{a+b}.$$

12. If $\phi(x)$ is positive and twice differentiable, shew that a necessary and sufficient condition that $\log\phi(x)$ should be convex is $\phi\phi''-\phi'^2\geqslant 0$.

13. If $\phi(x)$ is positive, twice differentiable and convex, prove that the same is true of

$$\text{(i)}\ x^{\frac{1}{2}(s+1)}\phi(x^{-s}) \quad (s\geqslant 1,\ x>0), \qquad \text{(ii)}\ e^{\frac{1}{2}x}\phi(e^{-x}).$$

14. The set of numbers $a_1, a_2, \ldots a_{2n+1}$ is such that

$$a_\nu - 2a_{\nu+1} + a_{\nu+2} \geqslant 0 \quad (\nu=1, 2, \ldots, 2n-1);$$

prove that

$$\frac{a_1+a_3+\ldots+a_{2n+1}}{n+1} \geqslant \frac{a_2+a_4+\ldots+a_{2n}}{n},$$

the equality sign holding if the numbers are in arithmetical progression.

Deduce that, if $0<x<1$,

$$\frac{1-x^{n+1}}{n+1} > \frac{1-x^n}{n}\sqrt{x}.$$

CHAPTER VII

INTEGRAL CALCULUS

7·1. Introduction.

At the outset we shall distinguish carefully between *definite* and *indefinite integration*, which, although there is a connection between them, are processes which differ fundamentally from each other.

Indefinite integration is the process which is the inverse of derivation, and the two problems can be stated as follows: given a function $y=f(x)$ of a single real variable x, there is a definite process whereby we can find (if it exists) the function $\phi(x)$ such that

$$\frac{dy}{dx}=\phi(x) \qquad\qquad (1).$$

We have already seen that

$$\phi(x)=\lim_{h\to 0}\frac{f(x+h)-f(x)}{h}$$

whenever the above limit exists. The inverse process is not so simple: given the differential equation (1), we have to find the function $f(x)$. Thus the problem of indefinite integration is to find the function $f(x)$ whose derivative is the given function $\phi(x)$. Although rules may be given which cover this operation with various types of simple functions, indefinite integration is a tentative process, and indefinite integrals are found by trial.

It is also a problem which possesses no unique solution, for if $f(x)$ is any one solution of the differential equation (1), then so is $f(x)+c$, where c is any arbitrary constant, since the derivative of a constant is zero*.

The terminology "indefinite integration" is sanctioned by usage, but from the point of view of this treatment the process would be better described as the "calculus of primitives," retaining the word "integration" for the process known as definite integration. The use of the term "primitive" would help to emphasise that indefinite

* Since the solution is not unique only because of an arbitrary additive *constant*, the non-uniqueness is really trivial. It is, however, highly important to realise that there is an arbitrary constant in the solution of the differential equation.

integration is really the process of solving a differential equation of the first order, such as (1), for in the theory of differential equations the solution $y = f(x) + c$, where c is an arbitrary constant, is called the *complete primitive.*

A definite integral may be described as an analytical substitute for an area. In the usual elementary treatment of the definite integral, defined as the limit of a sum, it is assumed that the function of x considered may be represented graphically, and the limit in question is the area between the curve, the axis of x, and the two bounding ordinates, say at $x = a$ and $x = b$.

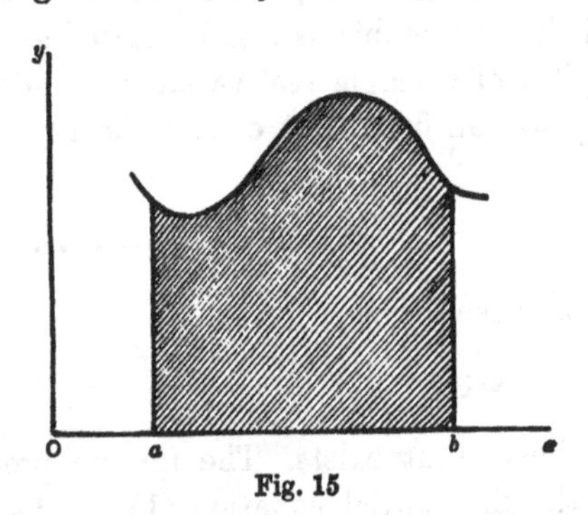

Fig. 15

If $y = f(x)$ is the function considered, the geometrical notions point to the existence of an area A, and the analytical substitute for its value is the definite integral

$$\int_a^b f(x)\,dx.$$

The formal definition of this integral is given in § 7·2, on the lines of the treatment of Riemann*, who was the first to give a rigorous arithmetical treatment independent of all appeal to geometrical intuition as an element of proof. The reader should observe that even among continuous functions there are many which cannot be represented graphically†, and for this reason alone it is necessary to replace the traditional discussion by a method which is purely arithmetical and independent of all geometrical intuition.

* "Über die Darstellbarkeit einer Funktion durch eine trigonometrische Reihe," *Göttingen Abh. Ges. Wiss.* XIII (1868).

† See, for example, Weierstrass's function mentioned in §4·34; also Examples IV, 5.

7·11. Primitives and integrals. Notation.

The class of functions which possess an integral is a wider class than that of the functions which possess a primitive.

The function $\phi(x)=0$ when $x \neq 0$, $\phi(0)=1$, is a function which in any given range of values of x (including or not including the point $x=0$) possesses a definite integral* whose value is 0. This function has no primitive, for it is impossible to find a function $f(x)$ such that $f'(x)=\phi(x)$.

The important point is that a definite integral can be defined, and all its properties can be deduced from the definition without any knowledge of the calculus of primitives at all. It is only when we require to *evaluate* an integral that we make use of primitives. To justify this we require the fundamental theorem of the integral calculus† which is not easy to prove.

Notation for indefinite integrals.

If we suppose that $y=f(x)+c$ is the complete primitive of the differential equation

$$\frac{dy}{dx}=\phi(x),$$

then the generally accepted notation‡ is to write

$$y=f(x)+c=\int \phi(x)\,dx.$$

Example. Suppose that we wish to calculate the area between the curve $y=x^2$, the axis of x and the ordinate at $x=1$.

The area in question is

$$\int_0^1 x^2\,dx,$$

and the ordinary process of elementary integral calculus involves the use of the calculus of primitives. Since

$$\frac{d}{dx}\left(\frac{x^3}{3}\right)=x^2$$

we conclude that

$$\int_0^1 x^2\,dx=\left[\frac{x^3}{3}\right]_0^1=\tfrac{1}{3}-0=\tfrac{1}{3}.$$

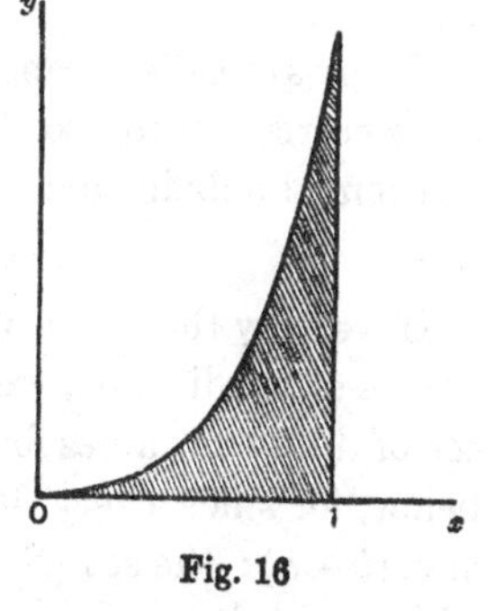

Fig. 16

* See § 7·31, Theorem 3. The single point of discontinuity can be enclosed in an interval whose length is less than ϵ, and the value of the integral is unaffected by it.

† See § 7·5 below.

‡ The use of the sign $\int$ for an indefinite integral as well as for a definite integral does not help the beginner to keep the two ideas distinct from each other. From this

The reader has probably carried out this process a great many times without realising that it needs a good deal of justification. It uses the result that if $f(x)$ be the function which satisfies the differential equation

$$\frac{dy}{dx} = \phi(x),$$

then the definite integral

$$\int_a^b \phi(x)\,dx = f(b) - f(a).$$

This result is called the fundamental theorem of the integral calculus, the proof and discussion of which are given in § 7·5.

7·2. Definition of the Riemann integral.

Let $f(x)$ be a function which exists and is bounded in the interval $a \leqslant x \leqslant b$, and suppose that m and M are the lower and upper bounds of $f(x)$ in (a, b). Take a set of points

$$x_0 = a, \quad x_1, x_2, \ldots, x_{r-1}, x_r, \ldots, x_n = b \ldots\ldots\ldots\ldots(1),$$

and write $\delta_r = x_r - x_{r-1}$.

Let M_r, m_r be the bounds of $f(x)$ in (x_{r-1}, x_r), and form the sums

$$S = \sum_{r=1}^{n} M_r \delta_r,$$

$$s = \sum_{r=1}^{n} m_r \delta_r.$$

These are called respectively *the upper and lower approximative sums* corresponding to the mode of subdivision by the points (1).

From the definition of S and s it is clear that

$$s \leqslant S.$$

If we vary the mode of subdivision of (a, b) by choosing different points of subdivision, we get a set of different values for S, and a set of different values for s. All possible upper sums S are bounded below, for since every $M_r \geqslant m$, a rough lower bound of the set $\{S\}$ is $m(b-a)$: the set $\{S\}$ therefore possesses an (exact) lower bound which will be denoted by J. Similarly the set $\{s\}$ possesses a definite (exact) upper bound which will be denoted by I.

point of view it is perhaps not the best notation to use. The notation $y = D^{-1}\phi(x)$ suggested by the theory of differential operators could be used, but it is not in accordance with custom.

The numbers J and I are of fundamental importance; they are called respectively *the upper and lower integrals of $f(x)$ in (a, b)*, and written

$$J = \overline{\int_a^b} f(x)\,dx, \quad I = \underline{\int_a^b} f(x)\,dx.$$

If $I = J$, $f(x)$ is said to be Riemann integrable (or integrable-R) in (a, b), and the common value of I and J is the integral of $f(x)$ between the limits a and b; and the integral is denoted by*

$$\int_a^b f(x)\,dx.$$

If $J \neq I$ the function $f(x)$ is not integrable-R in (a, b).

The only assumption which has been made about $f(x)$ is that it is *bounded* in (a, b). It does not follow that every bounded function is integrable, for there are bounded functions for which $J \neq I$. We subsequently prove that every *continuous* function is integrable, but integrability-R extends to a class of functions wider than the class of continuous functions.

Geometrical interpretation.

If we suppose that $f(x)$ possesses a graph, let the graph of $y = f(x)$ from $x = a$ to $x = b$ be as shewn in the figure.

In the case illustrated, $y = f(x)$ is taken to be a *continuous* function in (a, b), but that is not essential. For the particular mode of subdivision shewn in the figure, the sum S is the sum of the areas of the larger rectangles such as $AL_0P_1N_1$, and the sum s is the sum of the areas of the smaller rectangles such as $AP_0L_1N_1$. The figure suggests easily that by varying the mode of subdivision of (a, b) there will be different sums S and s and that the area required (geometrically suggested), namely, the area between the curve, the axis of x and the ordinates at $x = a$ and $x = b$, must lie between any value of S and any value of s.

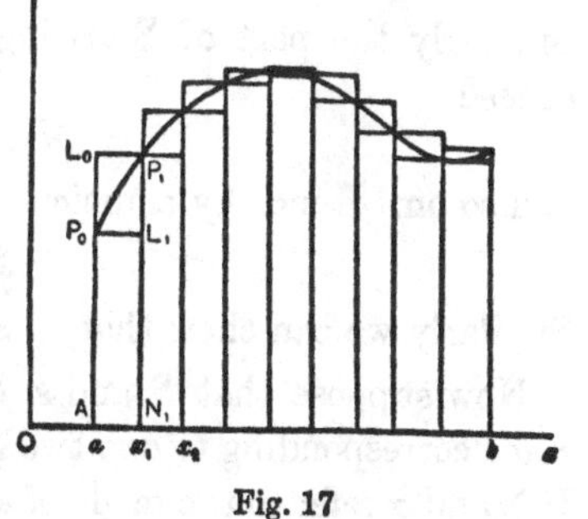

Fig. 17

The above definition is not satisfactory until we have proved analytically that *any* value of S is not exceeded by *any* value of s.

* The use of the word "limits" in this sense has the sanction of custom, but a more appropriate term would be "extreme values" or "termini," as suggested by Lamb, *Infinitesimal Calculus* (1897), 209.

We have to give the analytical demonstration of the fact, so readily suggested by the above geometrical consideration, that the area required always lies between any two selected values of S and s. We now, therefore, prove the theorem that the number J is greater than or equal to the number I.

THEOREM. *The number I cannot exceed the number J.*

Divide each of the intervals (a, x_1), (x_1, x_2), ..., (x_{n-1}, b) into smaller sub-intervals so that the new set of points of subdivision is

$$a, y_1, y_2, \ldots, y_{k-1}, x_1, y_{k+1}, y_{k+2}, \ldots, y_{l-1}, x_2, y_{l+1}, \ldots, b.$$

Let Σ and σ be the upper and lower sums for this new mode of subdivision. This new mode is said to be *consecutive* to the first mode of subdivision of (a, b).

Let us compare the parts of the two sums S and Σ arising from the interval (a, x_1). If M_1', m_1' are the bounds of $f(x)$ in (a, y_1), ..., M_k', m_k' the bounds of $f(x)$ in (y_{k-1}, x_1), then the part of Σ in question is

$$M_1'(y_1 - a) + M_2'(y_2 - y_1) + \ldots + M_k'(x_1 - y_{k-1}) \quad \ldots(1),$$

and since M_1', M_2', ..., M_k' cannot exceed M_1, the expression (1) clearly cannot exceed

$$M_1(x_1 - a).$$

Similarly the part of Σ arising from the interval (x_1, x_2) cannot exceed

$$M_2(x_2 - x_1),$$

and so on. Hence by addition

$$\Sigma \leqslant S \quad \ldots\ldots\ldots\ldots\ldots\ldots\ldots\ldots(2).$$

Similarly we can shew that $\quad \sigma \geqslant s \quad \ldots\ldots\ldots\ldots\ldots\ldots\ldots\ldots(3).$

Now suppose that S and s, S' and s' are the upper and lower sums corresponding to *any* two given modes of subdivision of (a, b). If Σ and σ refer to the mode of subdivision obtained by superposing these two, from (2) and (3) we deduce that

$$\Sigma \leqslant S, \quad \Sigma \leqslant S'; \quad \sigma \geqslant s, \quad \sigma \geqslant s';$$

and since $\Sigma \geqslant \sigma$, it follows that

$$s' \leqslant S \text{ and } s \leqslant S'.$$

Thus, since *any* upper sum is not exceeded by *any* lower sum, we must have

$$J \geqslant I.$$

COROLLARY. *Adding new points of subdivision cannot increase S and cannot decrease s.*

This follows at once from the theorem, and is contained in the inequalities (2) and (3).

We deduce at once from this corollary the important result that if by continued subdivision we obtain a sequence of upper sums

$$S_1, S_2, S_3, \ldots \qquad (4),$$

and a corresponding sequence of lower sums

$$s_1, s_2, s_3, \ldots \qquad (5),$$

it follows that

$$M(b-a) \geqslant S_1 \geqslant S_2 \geqslant \ldots$$

and

$$m(b-a) \leqslant s_1 \leqslant s_2 \leqslant \ldots.$$

Thus the sequence (4) is monotonic decreasing and the sequence (5) is monotonic increasing; and, by the theorem, no s exceeds any S.

The sums S and s certainly depend upon the way in which the subdivision is carried out. In fact, for any given mode of subdivision, the sums S and s are both functions of δ_r and of n (the number of points of subdivision in the particular mode selected). If we can find a *definite* limiting process, then the two monotonic sequences (4) and (5) will tend respectively to their lower and upper bounds. The limiting process chosen is that of making n tend to infinity in such a way that the length of the greatest of the sub-intervals δ_r shall tend to zero.

The reader should observe that the definition given above of a definite integral depends only upon the concept of a bound, and does not involve the concept of a limit at all. Bounds, though theoretically more simple, are not so convenient for formal work as limits; we therefore give a definition which depends upon a limiting process of the kind to which we have just referred.

7·21. Definition of an integral as a limit.

Subdivide (a, b) into n sub-intervals by the arbitrary set of points

$$x_0 = a, \quad x_1, x_2, \ldots, x_{n-1}, x_n = b,$$

a subdivision of (a, b) which we denote by D.

As before, write $\delta_r = x_r - x_{r-1}$, and let the length of the greatest of the sub-intervals δ_r be denoted by δ. We speak of δ as the *norm* of the subdivision D. In each sub-interval δ_r choose any point ξ_r (which may coincide with either of the end-points of δ_r) and form the sum

$$\mathscr{I}_\delta = \sum_{r=1}^{n} f(\xi_r)(x_r - x_{r-1}) = \sum_D f(\xi_r) \, . \, \delta_r .$$

Now make $\delta \to 0$. If $\mathscr{I}_\delta$ tends to a limit $\mathscr{I}$ which is independent of the choice of the points x_r, ξ_r we write

$$\mathscr{I} = \lim_{\delta \to 0} \Sigma f(\xi_r) \, \delta_r = \int_a^b f(x) \, dx.$$

This definition of a definite integral is the one usually given in elementary treatises on the Calculus. It remains to shew that this definition is equivalent to the definition which has been given in § 7·2 above. The equivalence of the two definitions is a consequence of the theorem which follows.

7·22. Darboux's theorem.

Given ϵ, there is a positive number θ such that, for all modes of subdivision whose norm does not exceed θ, the sums S are greater than J by less than ϵ, and the sums s are smaller than I by less than ϵ. In other words the sums S and s tend respectively to J and I as $n \to \infty$ in such a way that the length of the greatest sub-interval tends to zero.

To fix the ideas, consider the upper sums S. Since the upper sums have a lower bound J, there must be at least one mode of subdivision with an upper sum S_1 (say) such that

$$S_1 < J + \epsilon \quad \text{.........................(1)}.$$

Let this mode of subdivision, consisting of the points

$$a, a_1, a_2, \ldots, a_{p-1}, b \quad \text{.....................(2)},$$

be D_1. Let θ be a positive number so chosen that the lengths of all the sub-intervals of D_1 are greater than θ.

Let D_2 be the mode of subdivision by the points

$$x_0 = a, \quad x_1, x_2, \ldots, x_{n-1}, x_n = b \text{..............(3)};$$

let S_2 be its upper sum, and suppose that the norm of D_2 does not exceed θ.

Let D_3 be the mode of subdivision obtained by superposing D_1 and D_2. If the upper sum for this mode be S_3, we have

$$S_1 \geqslant S_3,$$

and so from (1) it follows that

$$S_3 < J + \epsilon \quad \text{..............................(4).}$$

The theorem is proved if we shew that θ may be so chosen that

$$S_2 < J + 2\epsilon,$$

for S_2 is the upper sum for *any* mode of subdivision D_2 whose norm does not exceed θ.

Since each sub-interval of D_1 has a length greater than θ, and the length of each sub-interval of D_2 does not exceed θ, no two of the points $a_1, a_2, \ldots, a_{p-1}$ can lie in the same sub-interval of D_2 whose points of subdivision are given by (3).

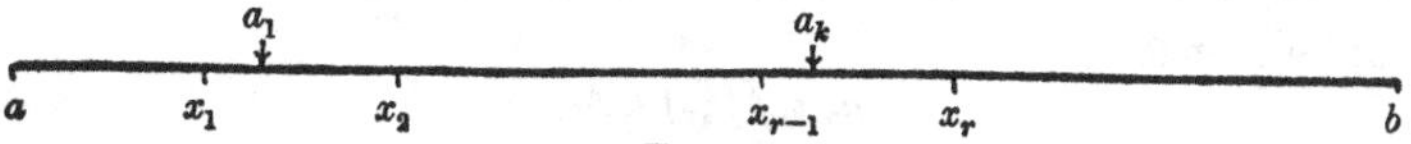

Fig. 18

The contribution of those sub-intervals of D_2 which contain no point of the set $a_1, a_2, \ldots, a_{p-1}$ is the same to S_2 as to S_3, and so

$$S_2 - S_3 = \Sigma \{M(x_{r-1}, x_r)(x_r - x_{r-1}) - M(x_{r-1}, a_k)(a_k - x_{r-1}) \\ - M(a_k, x_r)(x_r - a_k)\} \quad \text{............(5),}$$

where the summation extends only to such intervals (x_{r-1}, x_r) of D_2 as contain a point a_k of D_1, and $M(x_\alpha, x_\beta)$ denotes the upper bound of $f(x)$ in (x_α, x_β).

Now there are at most $p-1$ terms in the summation (5), and if we rewrite (5) in the form

$$S_2 - S_3 = \Sigma [\{M(x_{r-1}, x_r) - M(x_{r-1}, a_k)\}(a_k - x_{r-1}) \\ + \{M(x_{r-1}, x_r) - M(a_k, x_r)\}(x_r - a_k)],$$

the differences $M(x_{r-1}, x_r) - M(x_{r-1}, a_k)$ and $M(x_{r-1}, x_r) - M(a_k, x_r)$ are both non-negative, and certainly cannot exceed $2H$, where H denotes the upper bound of $|f(x)|$ in (a, b).

Hence $$S_2 - S_3 \leqslant 2H\,\Sigma(x_r - x_{r-1}),$$

the summation having at most $p-1$ terms, and $x_r - x_{r-1} \leqslant \theta$, thus

$$S_2 - S_3 \leqslant 2H(p-1)\theta$$

and so, by (4)

$$S_2 \leqslant J + \epsilon + 2H(p-1)\theta.$$

If now θ be chosen so that $\theta < \dfrac{\epsilon}{2H(p-1)}$, we get

$$S_2 < J + 2\epsilon.$$

A similar argument holds for s and I, and by taking the smaller of the two numbers θ involved, the same θ can be made to satisfy the conditions for both S and s, and Darboux's theorem is established*.

7·23. The importance of Darboux's theorem.

It is now easy to shew that the definition of a definite integral as the limit of a certain sum, given in § 7·21 above, is equivalent to the first definition in terms of bounds, for the sum

$$\mathscr{I}_\delta = \sum_{r=1}^{n} f(\xi_r)(x_r - x_{r-1}) \quad \text{.................(1)}$$

certainly lies between S and s for the particular mode of subdivision adopted, since

$$m_r \leqslant f(\xi_r) \leqslant M_r.$$

By Darboux's theorem we see that when the number of points of subdivision x_r increases indefinitely in such a way that the norm δ tends to zero, then

$$S \to J \text{ and } s \to I.$$

Thus when $f(x)$ is integrable, so that $J = I$, the sums S and s have a common limit which must coincide with the limit of the summation (1) as $\delta \to 0$, for $S \geqslant \mathscr{I}_\delta \geqslant s$.

Also, if $\mathscr{I}_\delta$ tends to a limit as $\delta \to 0$, both S and s tend to the same limit: for given the points x_r we can, by choice of ξ_r, make $\mathscr{I}_\delta$ as near as we please to S or as near as we please to s. Thus

$$\overline{\lim_{\delta \to 0}}\, \mathscr{I}_\delta = J, \quad \underline{\lim_{\delta \to 0}}\, \mathscr{I}_\delta = I.$$

In particular, in the definition of the integral as a limit, ξ_r may be taken to coincide either with x_{r-1} or with x_r.

7·3. A necessary and sufficient condition for integrability.

Since $J \geqslant I$, it is clearly *sufficient* for the integrability of any bounded function $f(x)$ in $a \leqslant x \leqslant b$, that

$$J - I < \epsilon.$$

THEOREM. *It is necessary and sufficient for the integrability of the bounded function $f(x)$ in (a, b) that for at least one mode of subdivision the sums S and s shall differ by less than ϵ.*

* Although in the enunciation the sums S are to be greater than J by less than ϵ, we have proved that $S_2 < J + 2\epsilon$, but by § 2·56, Lemma 1, this is trivial. If equation (1) had been written $S_1 < J + \frac{1}{2}\epsilon$, the final result could have been $S_2 < J + \epsilon$.

(i) Suppose that $$S - s < \epsilon,$$
then since $$s \leqslant I \leqslant J \leqslant S,$$
it follows that $$J - I \leqslant S - s < \epsilon.$$
The function is therefore integrable; and so the condition is *sufficient.*

(ii) It is also *necessary*, for if $f(x)$ is integrable in (a, b), $J = I$. Since I is the upper bound of the lower sums, there is at least one value of s such that
$$s > I - \tfrac{1}{2}\epsilon.$$
Similarly, there is at least one value S' of S such that $S' < J + \frac{1}{2}\epsilon$; hence, with the notation of p. 168,
$$J + \tfrac{1}{2}\epsilon > \Sigma \geqslant \sigma > I - \tfrac{1}{2}\epsilon,$$
and so $$\Sigma - \sigma < J + \tfrac{1}{2}\epsilon - (I - \tfrac{1}{2}\epsilon) = J - I + \epsilon,$$
and since $$J = I,$$
$$\Sigma - \sigma < \epsilon.$$

Other necessary and sufficient conditions for integrability may be given*, but for the purpose of this book, the one given above suffices.

7·31. Classes of bounded integrable functions.

THEOREM 1. *Every continuous function is integrable.*

Suppose that $f(x)$ is a continuous function in $a \leqslant x \leqslant b$, then by §3·44 (Corollary) the interval (a, b) can be divided into a *finite* number of sub-intervals in each of which the oscillation of $f(x)$ is less than ϵ. Suppose that
$$x_0 = a, \quad x_1, x_2, \ldots, x_n = b$$
is a set of dividing points satisfying this condition, then
$$S = \Sigma M_r (x_r - x_{r-1}),$$
$$s = \Sigma m_r (x_r - x_{r-1}),$$
so that
$$S - s = \Sigma (M_r - m_r)(x_r - x_{r-1})$$
$$< \Sigma \epsilon (x_r - x_{r-1}) = \epsilon \Sigma (x_r - x_{r-1}) = \epsilon (b - a),$$
and so $f(x)$ is integrable in (a, b).

* See, for example, Pierpont, *Theory of Functions*, I, § 498.

THEOREM 2. *Every bounded monotonic function is integrable.*

Let $f(x)$ be such a function in (a, b). If $f(x)$ is constant the theorem is obvious. To fix the ideas suppose that $f(x)$ is increasing. Divide (a, b) into n *equal* intervals of length δ, where

$$\delta < \frac{\epsilon}{f(b) - f(a)};$$

then

$$\begin{aligned} S - s &= \delta [\{f(x_1) - f(a)\} + \{f(x_2) - f(x_1)\} + \ldots + \{f(b) - f(x_{n-1})\}] \\ &= \delta \{f(b) - f(a)\} \\ &< \epsilon. \end{aligned}$$

THEOREM 3. *A bounded function whose discontinuities can be enclosed in a finite number of intervals whose total length is less than* ϵ *is integrable in* (a, b).

In the sub-intervals which do not contain discontinuities of $f(x)$, this function is continuous. Denote those sub-intervals which contain discontinuities by "d."

Now $$S - s = \Sigma (M_r - m_r)(x_r - x_{r-1}).$$

In any d-interval $M_r - m_r$ may be large, but since $f(x)$ is bounded $M_r - m_r \leqslant M - m$, where M and m are the bounds of $f(x)$ in (a, b).

Thus

$$\begin{aligned} S - s &\leqslant (M - m) \sum_d (x_r - x_{r-1}) + \epsilon \sum_{(a, b) - d} (x_r - x_{r-1}) \\ &< (M - m)\epsilon + \epsilon (b - a) = \kappa\epsilon, \end{aligned}$$

where κ is a constant.

$f(x)$ is therefore integrable in this case.

It can be proved that if $f(x)$ is bounded in (a, b) and the points of discontinuity are *infinite* in number, then $f(x)$ is integrable in (a, b) provided that the points of discontinuity can be enclosed in an infinite number of intervals of total length less than ϵ.

The proof of this result requires a more profound method than the preceding theorem, and it is outside the scope of this book*.

The reader should have no difficulty in establishing the following theorems, of which detailed proofs are not given here.

THEOREM 4. *If* $f(x)$ *is integrable in* (a, b) *it is integrable in* (α, β), *where* $a < \alpha < \beta < b$.

* For a proof of this, and of other theorems on Riemann integrability, see Hobson, *Functions of a Real Variable*, I, 433–440.

THEOREM 5. *If the function $f(x)$ is integrable in (a, b) then every other function $g(x)$ obtained by altering the value of $f(x)$ at a* FINITE *number of points of (a, b) is integrable, and*

$$\int_a^b f(x)\,dx = \int_a^b g(x)\,dx.$$

Note on the proof.

The points at which $f(x) \neq g(x)$ can be enclosed in a finite number of intervals of total length less than ϵ, and so the part of $S - s$ arising from these intervals is less than $\kappa\epsilon$, where κ is a constant. From this the integrability of $g(x)$ follows. That the integrals of $f(x)$ and $g(x)$ are identical in (a, b) can be easily proved by employing the definition of an integral as a limit (§ 7·21).

7·32. Examples.

(1) *Consider the function*

$$f(x) = 0 \quad \textit{when } x \textit{ is an integer},$$
$$f(x) = 1 \quad \textit{otherwise},$$

in the interval $(0, m)$, where m is a positive integer. This function is integrable, for its discontinuities are finite in number, being situated at the points $x = 1, 2, \ldots, m$. By Theorem 3, if each of these points be enclosed in an interval of length less than ϵ/m the total length of the "d" intervals is less than ϵ, and so $f(x)$ is integrable.

(2) *A bounded function which is not integrable* is the following:

$$f(x) = 0 \quad \text{when } x \text{ is rational},$$
$$f(x) = 1 \quad \text{when } x \text{ is irrational};$$

this function is not integrable in any finite interval (a, b).

For this function, clearly $M = 1$, $m = 0$; and suppose that

$$x_0 = a, \quad x_1, x_2, \ldots, x_{n-1}, x_n = b$$

is any mode of subdivision of (a, b). In any interval (x_{r-1}, x_r), however small, there will be values of $f(x)$ which are unity and values which are zero, and so $M_r = 1$, $m_r = 0$ $(r = 1, 2, \ldots, n)$. Thus

$$S = \Sigma M_r \delta_r = 1 . \Sigma \delta_r = b - a,$$
$$s = \Sigma m_r \delta_r = 0 . \Sigma \delta_r = 0,$$

and so $S - s = b - a$, and the necessary and sufficient condition for integrability is not satisfied.

The above example is interesting because, although it is not Riemann integrable, it is integrable in the sense of Lebesgue*, to whom a more general definition of an integral is due. The main object of Lebesgue's work was to remove the limitations on the integrand required in Riemann's treatment. The definition of a Lebesgue integral depends on the concept of the measure of a set of points, and so is outside the scope of this work.

* See Lebesgue, *Leçons sur l'Intégration* (Paris, 1904), and de la Vallée Poussin, *Intégrales de Lebesgue* (Paris, 1916).

7·4. Properties of a definite integral.

Let $f(x)$ be a bounded function which is integrable in $a \leqslant x \leqslant b$, then by definition

I. $$\int_a^b f(x)\,dx = -\int_b^a f(x)\,dx.$$

Also we have

II. $$\int_a^b dx = b - a,$$

III. $$\int_a^b kf(x)\,dx = k\int_a^b f(x)\,dx,$$

where k is a constant.

The proofs of Theorems II and III are simple, being direct deductions from the definition, and so they are left as exercises for the reader.

IV. *If c be a point in (a, b), and if $f(x)$ is integrable in (a, b), then*

$$\int_a^c f(x)\,dx + \int_c^b f(x)\,dx = \int_a^b f(x)\,dx.$$

Consider any mode of subdivision of (a, b) of which c is not one of the points of subdivision: by introducing c as an additional point, S is certainly not increased. But the sums S for the intervals (a, c) and (c, b) given by this mode of subdivision are respectively not less than

$$\int_a^c f(x)\,dx \quad \text{and} \quad \int_c^b f(x)\,dx;$$

thus every mode of subdivision of (a, b) gives a sum S such that

$$S \geqslant \int_a^c f(x)\,dx + \int_c^b f(x)\,dx \text{ ....(1)}.$$

Similarly every mode of subdivision of (a, b) gives a sum s such that

$$s \leqslant \int_a^c f(x)\,dx + \int_c^b f(x)\,dx \text{ ....(2)}.$$

Since $\int_a^b f(x)\,dx$ exists, S and s are arbitrarily near, and so it follows from (1) and (2) that

$$\int_a^b f(x)\,dx = \int_a^c f(x)\,dx + \int_c^b f(x)\,dx.$$

V. *If $f(x)$ and $g(x)$ are both integrable in (a, b) and $f(x) \geqslant g(x)$ at every point of (a, b), then*

$$\int_a^b f(x)\,dx \geqslant \int_a^b g(x)\,dx.$$

Let M_r, m_r be the bounds of $f(x)$, M_r', m_r' those of $g(x)$ in (x_{r-1}, x_r). Since $f(x) \geqslant g(x)$ in δ_r, we have

$$M_r \geqslant M_r', \quad m_r \geqslant m_r',$$

and so

$$S \geqslant S', \quad s \geqslant s'.$$

From this it also follows that

$$J \geqslant J', \quad I \geqslant I',$$

which proves the theorem.

VI. *If $f(x)$ and $g(x)$ are both integrable in (a, b), then $f(x)+g(x)$ is integrable, and*

$$\int_a^b \{f(x)+g(x)\}\,dx = \int_a^b f(x)\,dx + \int_a^b g(x)\,dx.$$

Subdivide (a, b) as usual, and let the bounds of $f(x)$ and $g(x)$ be as in V. Let $\mathbf{M}_r$ and $\mathbf{m}_r$ be the bounds of $f(x)+g(x)$ in (x_{r-1}, x_r).

From the definition of an upper bound we have, in δ_r,

$$f(x)+g(x) \leqslant M_r + M_r',$$

and so it follows that in δ_r

$$\mathbf{M}_r \leqslant M_r + M_r' \text{*}.$$

Similarly

$$\mathbf{m}_r \geqslant m_r + m_r'.$$

Hence, with an obvious notation,

$$\left.\begin{array}{l}\mathbf{S} \leqslant S + S' \\ \mathbf{s} \geqslant s + s'\end{array}\right\} \quad \text{.........................(1).}$$

Observe that we have to *prove* that $f(x)+g(x)$ is integrable: this does not necessarily follow because $f(x)$ and $g(x)$ are separately integrable.

From equations (1) we get

$$\left.\begin{array}{l}\mathbf{J} \leqslant J + J' \\ \mathbf{I} \geqslant I + I'\end{array}\right\} \quad \text{.........................(2);}$$

* If $f(x)=x$, $g(x)=1-x$ in $(0, 1)$, then clearly $M=1$, $M'=1$, $\mathbf{M}=1$, and so $\mathbf{M}=\frac{1}{2}(M+M')<M+M'$.

but $J=I$ and $J'=I'$, hence it follows that

$$\mathbf{J}=\mathbf{I},$$

and so (i) $f(x)+g(x)$ is integrable and

$$\text{(ii)} \int_a^b \{f(x)+g(x)\}\,dx = \int_a^b f(x)\,dx + \int_a^b g(x)\,dx.$$

COROLLARY. *If the functions $f_1(x), f_2(x), \ldots, f_p(x)$, where p is a fixed integer, are all integrable in (a, b), then so is their sum, and also*

$$\int_a^b \{f_1(x)+\ldots+f_p(x)\}\,dx = \int_a^b f_1(x)\,dx + \ldots + \int_a^b f_p(x)\,dx.$$

The proof easily extends step by step from the result of VI.

VII. *If $f(x)$ is integrable in (a, b), then so is $|f(x)|$, and*

$$\left|\int_a^b f(x)\,dx\right| \leqslant \int_a^b |f(x)|\,dx.$$

Define the functions $f_1(x)$ and $f_2(x)$ as follows:

$$\begin{aligned} f_1(x) &= f(x) \text{ when } f(x) \geqslant 0, \\ &= 0 \text{ otherwise}; \\ f_2(x) &= -f(x) \text{ when } f(x) \leqslant 0, \\ &= 0 \text{ otherwise}. \end{aligned}$$

Then $f(x)=f_1(x)-f_2(x)$ and $|f(x)|=f_1(x)+f_2(x)$.

Now the oscillation of $f_1(x)$ or of $f_2(x)$ in any interval cannot exceed that of $f(x)$ in the same interval, hence since $f(x)$ is integrable, so are $f_1(x)$ and $f_2(x)$, and so, by VI, $f_1(x)+f_2(x)$ is integrable; in other words $|f(x)|$ is integrable*.

Now

$$\int_a^b f(x)\,dx = \int_a^b f_1(x)\,dx - \int_a^b f_2(x)\,dx,$$

and so

$$\begin{aligned} \left|\int_a^b f(x)\,dx\right| &\leqslant \left|\int_a^b f_1(x)\,dx\right| + \left|\int_a^b f_2(x)\,dx\right| \\ &= \int_a^b f_1(x)\,dx + \int_a^b f_2(x)\,dx, \end{aligned}$$

since $f_1(x)$ and $f_2(x)$ are positive functions.

* Observe that the converse of this statement is not necessarily true; $|f(x)|$ may be integrable when $f(x)$ is not. If $f(x)=1$ when x is rational, $f(x)=-1$ when x is irrational, $f(x)$ is not integrable in any finite interval (a, b). However, $|f(x)|=1$ throughout (a, b), and is certainly integrable therein.

Hence

$$\left|\int_a^b f(x)\,dx\right| \leqslant \int_a^b \{f_1(x)+f_2(x)\}\,dx = \int_a^b |f(x)|\,dx,$$

which proves the theorem.

All the preceding theorems about definite integrals have been proved either by a direct appeal to the definition or to a result previously proved. All the main general results about definite integrals are included in the above theorems, and any other such property can be proved by a similar method. Before we can proceed to the *evaluation* of a definite integral, we must prove the fundamental theorem which is given in the next section.

7·5. The fundamental theorem of the integral calculus.

The object of this section is to discuss simply the conditions under which it may be true to assert that the operations of derivation and of integration are inverse operations. It is of the greatest possible importance in the theory, for by means of it we establish the validity of the process mentioned in the example of § 7·11, namely the employment of indefinite integration for the purpose of evaluating definite integrals. The argument will be easier to follow if the various results are proved as separate theorems. We first prove a lemma.

Lemma*. *If M and m are the bounds of an integrable function $f(x)$ in (a, b), then*

$$m(b-a) \leqslant \int_a^b f(x)\,dx \leqslant M(b-a) \qquad \text{............(1).}$$

The proof is immediate, for $S = \Sigma M_r \delta_r$ and $s = \Sigma m_r \delta_r$ both lie between $M\Sigma\delta_r$ and $m\Sigma\delta_r$, that is between $M(b-a)$ and $m(b-a)$. Since $f(x)$ is integrable, S and s have a common limit as the norm δ tends to zero. This common limit therefore also lies between $M(b-a)$ and $m(b-a)$, and since this limit is $\int_a^b f(x)\,dx$, the lemma is proved.

From (1) it follows that *if $f(x)$ is integrable in (a, b) and if $|f(x)| \leqslant N$, then*

$$\left|\int_a^b f(x)\,dx\right| \leqslant N(b-a) \qquad \text{......................(2),}$$

for N is $\max\{|m|, |M|\}$.

* This lemma is known as *the first mean-value theorem for integrals.* See § 7·8.

If μ be such that $m \leqslant \mu \leqslant M$, then (1) may be written

$$\int_a^b f(x)\,dx = \mu(b-a) \quad \text{.................(3)}.$$

THEOREM 1. *If $f(x)$ is bounded and integrable in (a, b) and*

$$F(x) = \int_a^x f(x)\,dx,$$

then $F(x)$ is a continuous function of x in (a, b).

We have $$F(x+h) - F(x) = \int_x^{x+h} f(x)\,dx,$$

and if $|f(x)| \leqslant N$ in $(x, x+h)$ we have, by (2),

$$|F(x+h) - F(x)| \leqslant N\,|h|.$$

If, given ϵ, a number $h_0 > 0$ be chosen so that $h_0 < \epsilon/N$, we have, for every h such that $x+h$ lies in (a, b) and $|h| < h_0$,

$$|F(x+h) - F(x)| < \epsilon;$$

hence $F(x)$ is continuous at the point x.

THEOREM 2. *If $f(x)$ be continuous in (a, b), then at every point in (a, b), $F(x)$ possesses a derivative which is the function $f(x)$.*

Since $f(x)$ is continuous, given ϵ, an interval $(x-\delta, x+\delta)$ can be found such that $|f(x \pm \theta\delta) - f(x)| < \epsilon$, where $0 < \theta < 1$. Hence, for $h > 0$,

$$F(x \pm h) - F(x) = \int_x^{x \pm h} f(x)\,dx$$

lies between $h\{f(x)+\epsilon\}$ and $h\{f(x)-\epsilon\}$, provided that $h < \delta(\epsilon)$. Hence, since $\{F(x \pm h) - F(x)\}/h$ lies between $f(x)+\epsilon$ and $f(x)-\epsilon$ for $h < \delta(\epsilon)$, $f(x)$ is the derivative of $F(x)$.

THEOREM 3. *If $\phi(x)$ is a function which at every point of (a, b) possesses a derivative which is a continuous function $f(x)$, then*

$$F(x) = \int_a^x f(x)\,dx = \phi(x) - \phi(a).$$

The derivative of the function $F(x)-\phi(x)$ must be zero, so that $F(x)-\phi(x)$ is constant; and by putting $x=a$, since $F(a)=0$ the constant is $-\phi(a)$, and so

$$F(x)=\phi(x)-\phi(a).$$

The importance of Theorem 2 lies in the fact that every *continuous* function has been shewn to be the derivative of a continuous function which is called its primitive (indefinite integral).

By taking all the preceding theorems together we have established the validity of the following process for evaluating $\int_a^b f(x)\,dx$, namely, to obtain, by any means we can, the function $\phi(x)$ such that

$$\frac{d}{dx}\phi(x)=f(x),$$

and then

$$\int_a^b f(x)\,dx=\phi(b)-\phi(a).$$

7·51. Note on the fundamental theorem.

By Theorem 1 of the preceding section we see that $F(x)$ is always a continuous function, though $f(x)$ may not be. In Theorem 2 it is necessary to assume the continuity of $f(x)$ in order to prove that $f(x)$ is the derivative of the continuous function $F(x)$. In Theorem 3 again the continuity of $f(x)$ is assumed. We thus obtain the important result that *whenever $f(x)$ is a continuous function it possesses a primitive, and knowledge of the primitive is equivalent to ability to evaluate $\int_a^b f(x)\,dx$, for if $\phi(x)$ is the primitive in question, the value of the integral is $\phi(b)-\phi(a)$.*

The question as to whether a primitive exists, and the question of the existence of an integral of the function $f(x)$ in (a, b) are entirely independent questions. The fundamental theorem however shews that when $f(x)$ is a continuous function in the interval (a, x), then the function

$$F(x)=\int_a^x f(x)\,dx,$$

and the function $\phi(x)$ which satisfies the differential equation

$$\frac{dy}{dx}=f(x),$$

are identical, save for an arbitrary additive constant.

Thus, whenever $f(x)$ possesses a primitive*, which it certainly does when it is continuous, the fundamental theorem enables us to evaluate

$$\int_a^b f(x)\,dx$$

if it exists.

The definite integral of $f(x)$ may however exist, if $f(x)$ possesses no primitive at all in the interval (a, b).

7·6. Change of variable in an integral.

For the purpose of evaluating definite integrals it is often useful to change the variable, and by suitable choice of the transformation the new integral is rendered easier to evaluate than the original one. We now investigate a formula for change of variable.

To change the variable in the integral $\int_a^b f(x)\,dx$ *by the transformation* $x=\phi(t)$.

We shall assume that $\phi(t)$ possesses a derivative at every point of the interval $\alpha \leqslant t \leqslant \beta$†, where $\phi(\alpha)=a$ and $\phi(\beta)=b$. Suppose also that in (α, β) the variable x is a monotonic function of t which is always increasing or always decreasing as t ranges from α to β.

Divide the range (α, β) into a finite number of sub-intervals by the points

$$t_0=\alpha, \quad t_1, t_2, \dots, t_{n-1}, t_n=\beta \quad \dots\dots\dots\dots\dots(1),$$

and let the corresponding values of x be

$$a, x_1, x_2, \dots, x_{n-1}, b \quad \dots\dots\dots\dots\dots\dots\dots\dots(2),$$

then by the mean-value theorem, if $t_{r-1} < \tau_r < t_r$,

$$x_r - x_{r-1} = (t_r - t_{r-1})\,\phi'(\tau_r).$$

Suppose that $\xi_r = \phi(\tau_r)$ is the corresponding value of x in (x_{r-1}, x_r). Since $f(x)$ is integrable in (a, b), the sum

$$f(\xi_1)(x_1-a)+f(\xi_2)(x_2-x_1)+\dots+f(\xi_n)(b-x_{n-1})\dots(3)$$

approaches the given integral as its limit as the norm of the subdivision (2) tends to zero.

* The fact that a given function (theoretically) possesses a primitive does not mean that rules for obtaining its actual value are necessarily known.

† $\phi(t)$ need only possess a right-hand derivative at α and a left-hand derivative at β, since we are not concerned with points outside the interval.

The sum (3) may also be written

$$f[\phi(\tau_1)]\phi'(\tau_1)(t_1-\alpha)+\ldots+f[\phi(\tau_r)]\phi'(\tau_r)(t_r-t_{r-1})+\ldots\ldots(4),$$

and by assuming that $f[\phi(t)]$ and $\phi'(t)$ are integrable in (α, β)* (see Examples VII, 9), the sum (4) approaches as its limit

$$\int_\alpha^\beta f[\phi(t)]\,\phi'(t)\,dt \quad\ldots\ldots\ldots\ldots\ldots\ldots(5),$$

as the norm of the subdivision (1) tends to zero. This gives the formula for change of variable required.

Extension of the theorem. If as t varies from α to β, $\phi(t)$ is not always monotonic increasing nor always monotonic decreasing we can extend the theorem to cover this case so long as the range (α, β) can be divided up into a finite number of sub-intervals in each of which $\phi(t)$ *steadily* increases or decreases as t increases.

7·61. Note on change of variable.

Not every substitution is suitable for changing the variable in an integral, and care must be taken to ensure that the above conditions hold.

Example (i).

Let $$I \equiv \int_{-1}^{1}\frac{dx}{1+x^2}=\Big[\text{arc tan}\, x\Big]_{-1}^{1}=\tfrac{1}{2}\pi.$$

By writing $x=1/u$ we get

$$I=-\int_{-1}^{1}\frac{du}{1+u^2}=-\tfrac{1}{2}\pi.$$

The reason for the discrepancy here is that the function $u=1/x$ does not possess an integrable derivative in $(-1, 1)$: in fact the function itself is undefined when $x=0$.

Example (ii). Make the transformation $x=t^{\frac{3}{2}}$ in the integral $\int_{-1}^{1} dx$. If the theorem holds, then

$$\int_{-1}^{1} dx=\int_{1}^{1}\tfrac{3}{2}t^{\frac{1}{2}}\,dt=0,$$

which is clearly untrue. To apply the theorem correctly, divide the interval $(-1, 1)$ into two sub-intervals $(-1, 0)$ and $(0, 1)$; in the former sub-interval we write $x=-t^{\frac{3}{2}}$ and make t vary from 1 to 0, and in the latter write $x=t^{\frac{3}{2}}$ and make t vary from 0 to 1. Then

$$\int_{-1}^{1} dx=3\int_{0}^{1} t^{\frac{1}{2}}\,dt=\Big[2t^{\frac{3}{2}}\Big]_{0}^{1}=2.$$

* The derivative of a differentiable function need not be integrable. Examples to illustrate this are not however very easy to construct; see Lebesgue, *Leçons sur l'Intégration* (Paris, 1904), 93–94. If we assume that $\phi'(t)$ is continuous (the usual assumption), the integral (5) certainly exists, but this condition is unnecessarily restricting.

7·7. Indefinite integration.

It is assumed that the reader is already familiar with the simpler standard forms, and with the usual elementary methods of finding indefinite integrals*. One of the most useful processes of indefinite integration is the process known as "integration by parts."

Although the reader who is familiar with the elementary theory of indefinite integration will be conversant with this method, for completeness it is included here.

Integration by parts.

Integrate with respect to x the formula†

$$\frac{d}{dx}(uv) = u\frac{dv}{dx} + v\frac{du}{dx} \quad\text{..................(1),}$$

and the required relation follows at once. It is usually written in the form

$$\int u\frac{dv}{dx}dx = uv - \int v\frac{du}{dx}dx \quad\text{..................(2).}$$

It is easily extended to operations with *definite integrals*. If $u=f(x)$, $v=g(x)$ and $f(x)g'(x)$, $f'(x)g(x)$ are both integrable in (a, b), then, from (1),

$$\frac{d}{dx}\{f(x)g(x)\} = f(x)g'(x) + f'(x)g(x),$$

and so we get

$$\int_a^b f(x)g'(x)\,dx = \Big[f(x)g(x)\Big]_a^b - \int_a^b f'(x)g(x)\,dx.$$

An elegant method of integrating rational functions, which is not usually given in elementary text-books, is the method due to Hermite which will now be described.

7·71. Hermite's method of integrating rational functions.

Every rational function $R(x)$ can be split up into an integral function $E(x)$ and a sum of rational fractions of the form A/X^n where X is prime to its derivative and to A, the degree of which is less than that of X^n. When the roots of the equation obtained by equating to zero the denominator of $R(x)$ are known, then X is

* The reader should possess a working knowledge of the contents of Gibson's *Calculus*, Ch. XIII, or of Lamb's *Calculus*, Ch. V.

† § 4·3.

either a linear or quadratic function of x corresponding to the partial fractions arising from a real root or from a pair of conjugate complex roots. The decomposition of a rational function

$$R(x) = P_1(x)/P_2(x)$$

into partial fractions is dependent upon a knowledge of the roots of the equation $P_2(x)=0$. The method which we shall now describe enables us to calculate the rational part of the integral $\int R(x)\,dx$ even when the roots of $P_2(x)=0$ are not known; and this rational part can be found by rational operations without any actual integrations being performed.

Suppose that $$R(x) = E(x) + \frac{f(x)}{F(x)},$$

then our object is to evaluate $\int \frac{f(x)}{F(x)}\,dx$ by finding the rational part of the integral without performing any integrations at all.

If $F(x) = (x-a)^\alpha (x-b)^\beta \dots (x-l)^\lambda$, then in partial fractions

$$\left.\begin{aligned}\frac{f(x)}{F(x)} = {} & \frac{A_1}{x-a} + \frac{A_2}{(x-a)^2} + \dots + \frac{A_\alpha}{(x-a)^\alpha} \\ & + \frac{B_1}{x-b} + \frac{B_2}{(x-b)^2} + \dots + \frac{B_\beta}{(x-b)^\beta} \\ & + \dots\dots\dots\dots\dots\dots \\ & + \frac{L_1}{x-l} + \frac{L_2}{(x-l)^2} + \dots + \frac{L_\lambda}{(x-l)^\lambda}\end{aligned}\right\} \quad \dots\dots\dots(1),$$

and so, in the integral there is a *logarithmic part* arising from the integration of the terms in the first column of (1), together with a *rational part*, which is the sum of the integrals of the remaining terms. If we write

$$P \equiv (x-a)(x-b)\dots(x-l),$$
$$Q \equiv (x-a)^{\alpha-1}(x-b)^{\beta-1}\dots(x-l)^{\lambda-1},$$

then $$\int \frac{f(x)}{F(x)}\,dx = \int \frac{X}{P}\,dx + \frac{Y}{Q} \quad \dots\dots\dots\dots\dots\dots(2),$$

or, by taking the derivative with respect to x,

$$\frac{f(x)}{F(x)} = \frac{X}{P} + \left(\frac{Y}{Q}\right)' \quad \dots\dots\dots\dots\dots\dots(3).$$

The polynomial Q is easily seen to be the highest common factor of $F(x)$ and $F'(x)$, and it can be obtained by the ordinary algebraic method. The polynomial P is $F(x)/Q$, and this can be found by ordinary division. If now X and Y are each chosen as polynomials with undetermined coefficients of degree one less than that of P and Q respectively, the identity (3) serves to determine the undetermined coefficients uniquely.

After multiplying through by $F(x)=PQ$, the identity (3) becomes

$$f(x)=PY'-\frac{YQ'P}{Q}+QX \qquad \text{.................(4).}$$

The right-hand side of (4) is a polynomial of degree one less than that of $F(x)$, and by equating coefficients of the same powers of x in (4), we have sufficient equations to determine the unknown coefficients in X and Y.

The important point of this method lies in the fact that the rational part of the integral, Y/Q, can be found without solving the equation $F(x)=0$. To complete the integration it remains to evaluate $\int \frac{X}{P}\,dx$, but since the roots of $P=0$ are all simple roots this gives the transcendental (logarithmic) part of the integral.

Example. Evaluate $\displaystyle\int \frac{3x^2+x-2}{(x-1)^3(x^2+1)}\,dx.$

Here $\qquad f=3x^2+x-2, \quad F=(x-1)^3(x^2+1),$

and so $\qquad F'=3(x-1)^2(x^2+1)+2x(x-1)^3, \quad Q=(x-1)^2,$

and $\qquad P=(x-1)(x^2+1).$

By writing $Y=Ax+B$, $X=Cx^2+Dx+E$, we get from (4)

$$3x^2+x-2\equiv(x-1)(x^2+1)A-(Ax+B)\,2(x^2+1)+(x-1)^2(Cx^2+Dx+E),$$

and by comparing coefficients,

$$C=0, \quad A=D=-\tfrac{5}{2}, \quad B=2, \quad E=-\tfrac{1}{2};$$

hence

$$\begin{aligned}\int\frac{3x^2+x-2}{(x-1)^3(x^2+1)}\,dx &= \frac{Y}{Q}+\int\frac{X}{P}\,dx\\ &= -\frac{5x-4}{2(x-1)^2}+\tfrac{1}{2}\int\frac{(-5x-1)}{(x-1)(x^2+1)}\,dx\\ &= -\frac{5x-4}{2(x-1)^2}-\tfrac{1}{2}\int\left(\frac{3}{x-1}-\frac{3x-2}{x^2+1}\right)dx\\ &= -\frac{5x-4}{2(x-1)^2}-\tfrac{3}{2}\log(x-1)+\tfrac{3}{4}\int\left\{\frac{2x}{x^2+1}-\frac{4}{3(x^2+1)}\right\}dx\\ &= -\frac{5x-4}{2(x-1)^2}-\tfrac{3}{2}\log(x-1)+\tfrac{3}{4}\log(x^2+1)-\text{arc tan}\,x+C.\end{aligned}$$

In the above example the factors of $F(x)$ are apparent, and it might reasonably be argued that it would have been just as easy to proceed by finding the partial fractions, and then evaluating the integral. The example however suffices to illustrate the method. If the factors of $F(x)$ had not been obvious, the process of finding the H.C.F. of $F(x)$ and $F'(x)$ would have had to be carried out in the usual way.

In every case the preceding method reduces the integration to that of a rational function X/P where all the roots of $P=0$ are *simple* roots.

7·72. Integrals reducible to rational functions.

A fairly complete account can be given of the integration of *rational* functions, but beyond this very little general theory can be given, and, since indefinite integration is a tentative process, it is frequently simpler to take a special method for a given case than to apply any general theorems, even when such exist. In a few cases there are integrals which can be reduced by simple substitutions to integrals of rational functions, although the original integrands are not themselves rational. Two types will be considered here.

$$(1) \qquad \int R\left\{x, \left(\frac{ax+b}{cx+d}\right)^{\alpha}, \left(\frac{ax+b}{cx+d}\right)^{\beta}, \ldots\right\} dx,$$

where $R(u)$ is a rational function of u; a, b, c, d are constants, and $\alpha, \beta, \ldots$ rational exponents. If m be the least common multiple of the denominators of the fractions $\alpha, \beta, \ldots$, the substitution

$$t^m = \frac{ax+b}{cx+d}$$

reduces the integrand to a rational function of t.

Example (i). $\int \frac{dx}{x^{\frac{1}{2}}+x^{\frac{1}{3}}}$. Write $x=t^6$, then

$$\int \frac{dx}{x^{\frac{1}{2}}+x^{\frac{1}{3}}} = 6\int \frac{t^5\,dt}{t^3+t^2} = 6\int \frac{t^3\,dt}{1+t},$$

which is easy to evaluate.

Example (ii). $\int \frac{x^3\,dx}{\sqrt{(x-1)}}$. Write $x-1=t^2$, and the given integral becomes

$$2\int (t^2+1)^3\,dt.$$

$$(2) \qquad \int x^m (a+bx^n)^p\,dx,$$

where a and b are constants, and m, n, p are rational exponents.

The first simplification is made by writing $x^n = t$, so that the integral becomes

$$\frac{1}{n}\int t^{\frac{m+1}{n}-1}(a+bt)^p\,dt,$$

and on writing $q = \dfrac{m+1}{n} - 1$ the integral becomes of the type

$$\phi(p, q) = \int (a+bt)^p\, t^q\, dt.$$

This type of integral is reducible to one with a rational integrand if (i) p is an integer, (ii) q is an integer, and (iii) $p+q$ is an integer.

(i) If p be an integer, $\phi(p, q) = \int R(t, t^q)\, dt$;

(ii) If q be an integer, $\phi(p, q) = \int R\{(a+bt)^p, t\}\, dt$;

(iii) If $p+q$ be an integer, $\phi(p, q) = \int R\left\{\left(\dfrac{a+bt}{t}\right)^p, t\right\} dt$;

and each of these integrals is of the type considered in (1).

7·8. Definite integration.

The mean-value theorems for integrals.

The first mean-value theorem has already been given in § 7·5. We now prove a special case of this theorem.

If $f(x)$ is CONTINUOUS *in (a, b), then there is a number ξ such that*

$$\int_a^b f(x)\,dx = (b-a) f(\xi) \quad \text{.................(1)},$$

where $a < \xi < b$.

Since $f(x)$ is continuous in (a, b) there must be points α and β of the interval $a \leqslant x \leqslant b$ such that $f(\alpha) = m$, $f(\beta) = M$, where m and M are the bounds of $f(x)$ in (a, b). Also, from results proved in § 3·45, $f(x)$ assumes every value between m and M as x ranges from α to β. Hence there is a point ξ between α and β such that $f(\xi) = \mu$, where μ is the number occurring in equation (3) of p. 180. It follows that equation (1) holds for a value ξ of x such that $a < \xi < b$.

7·81. The generalised first mean-value theorem.

Let $g(x)$ and $h(x)$ be integrable in $a \leqslant x \leqslant b$, and suppose that $h(x) > 0$ everywhere in (a, b). Then, if m and M are the bounds of $g(x)$ in (a, b),

$$m\int_a^b h(x)\,dx \leqslant \int_a^b g(x)\,h(x)\,dx \leqslant M\int_a^b h(x)\,dx \quad \text{......(2).}$$

The proof is simple, for since $m \leqslant g(x) \leqslant M$, and $h(x) > 0$,

$$mh(x) \leqslant g(x)\,h(x) \leqslant Mh(x),$$

and so (2) follows by integration* between the limits a and b.

COROLLARY. *If $g(x)$ is continuous in (a, b), then*

$$\int_a^b g(x)\,h(x)\,dx = g(\xi)\int_a^b h(x)\,dx \quad \text{............(3),}$$

where ξ lies between a and b.

The proof is immediate, and follows from the same type of argument as equation (1) above.

7·82. The second mean-value theorem.†

If $f(x)$ is monotonic and $f(x), f'(x)$ and $\phi(x)$ are all continuous in $a \leqslant x \leqslant b$, then

$$\int_a^b f(x)\,\phi(x)\,dx = f(a)\int_a^\xi \phi(x)\,dx + f(b)\int_\xi^b \phi(x)\,dx,$$

where ξ lies between a and b.

Write
$$\Phi(x) = \int_a^x \phi(t)\,dt,$$

then
$$\int_a^b f(x)\,\phi(x)\,dx = \int_a^b f(x)\,\Phi'(x)\,dx$$
$$= \Big[f(x)\,\Phi(x)\Big]_a^b - \int_a^b \Phi(x) f'(x\,dx)$$

on integrating by parts. Since $\Phi(a) = 0$, and $\Phi(x)$, being an integral, is a continuous function of x, we have, on using (3),

$$\int_a^b f(x)\,\phi(x)\,dx = f(b)\,\Phi(b) - \Phi(\xi)\int_a^b f'(x)\,dx$$
$$= f(b)\,\Phi(b) - \Phi(\xi)\,\{f(b) - f(a)\}$$
$$= f(b)\,\{\Phi(b) - \Phi(\xi)\} + f(a)\,\Phi(\xi)$$
$$= f(b)\int_\xi^b \phi(x)\,dx + f(a)\int_a^\xi \phi(x)\,dx.$$

* See Examples VII, 9.

† For proofs of this theorem under less stringent conditions, see e.g. Hobson, *Functions of a Real Variable*, I, 564 *seq*.

EXAMPLES VII.

1. Distinguish carefully between an integral and a primitive. In particular shew that for the function defined in the interval (0, 1) as follows:

$$f(x)=\sqrt{(1-x^2)} \quad \text{when } x \text{ is rational,}$$
$$=1-x \qquad \text{when } x \text{ is irrational.}$$
$$I=\underline{\int_0^1} f(x)\,dx=\tfrac{1}{2}, \quad J=\overline{\int_0^1} f(x)\,dx=\frac{\pi}{4},$$

and so $f(x)$ is not integrable in (0, 1).

2. Prove that in the interval $(-1, 1)$ the function

$$\phi(x)=x\sin\frac{1}{x^2}-\frac{1}{x}\cos\frac{1}{x^2} \quad \text{when } x\neq 0,$$
$$=0 \qquad \text{when } x=0,$$

possesses a primitive $\frac{1}{2}x^2\sin\frac{1}{x^2}$. Does $\int_{-1}^{1}\phi(x)\,dx$ exist?

3. Using the definition of an integral as the limit of a sum, by dividing the range (a, b) into n equal parts, calculate *ab initio* $\int_a^b x^2dx$.

4. Prove that, as $n\to\infty$,

$$n\sum_{p=0}^{n-1}\frac{1}{n^2+p^2}\to\frac{\pi}{4}.$$

5. Shew that the function defined as follows:

$$f(x)=\frac{1}{2^n} \quad \text{when } \frac{1}{2^{n+1}}<x\leqslant\frac{1}{2^n} \quad (n=0, 1, 2, 3, \ldots),$$
$$f(0)=0,$$

is integrable in (0, 1) although it has an infinite number of points of discontinuity.

6. Prove that if $f_n(x)$ tends *uniformly* to $f(x)$ as $n\to\infty$ in the interval $a\leqslant x\leqslant b$ (see § 3·431), then

$$\lim_{n\to\infty}\int_a^b f_n(x)\,dx=\int_a^b f(x)\,dx.$$

7. By using the identity

$$x^{2n}-1=(x^2-1)\prod_{r=1}^{n-1}\left(1-2x\cos\frac{r\pi}{n}+x^2\right),$$

shew that
$$\int_0^\pi \log(1-2x\cos t+x^2)\,dt=2\pi\log x \quad \text{if } x^2>1,$$
$$=0 \qquad \text{if } x^2<1.$$

8. In the interval (0, 1) suppose that $f(x)=\lim_{n\to\infty} f_n(x)$ where

$$f_n(x)=n \quad \text{when } x\leqslant 1/n,$$
$$=0 \quad \text{elsewhere.}$$

Explain fully why

$$\lim_{n\to\infty}\int_0^1 f_n(x)\,dx \neq \int_0^1 f(x)\,dx.$$

9. If $f(x)$ and $g(x)$ are bounded and integrable in (a, b), prove that their product $f(x)\,g(x)$ is also integrable in (a, b).

[Suppose first that both $f(x)$ and $g(x)$ are positive in (a, b). Let M_r, m_r; M_r', m_r'; $\mathbf{M}_r$, $\mathbf{m}^r$ be the bounds of f, g and fg in δ_r: shew that

$$\mathbf{M}_r-\mathbf{m}_r\leqslant M(M_r'-m_r')+M'(M_r-m_r),$$

and hence that $\mathbf{S}-\mathbf{s}$ tends to zero.

If the functions are not both positive, add constants c_1 and c_2 to make them so.]

10. Prove that if $f(x)$ is never negative and is integrable in (a, b), then, provided that $f(x)$ is continuous at c in (a, b) and $f(c)>0$,

$$\int_a^b f(x)\,dx>0.$$

11. Evaluate the following indefinite integrals:

$$\int\frac{x^2\,dx}{x^4+x^2-2},\quad \int\frac{dx}{1-x^6},\quad \int\frac{dx}{x(x^3+1)^3},\quad \int\frac{x\,dx}{(1+x)^{\frac{1}{2}}-(1+x)^{\frac{1}{3}}},$$

$$\int\frac{(x^4-x^3-3x^2-x)\,dx}{(x^3+1)^3},\quad \int x^3(1+x^2)^{\frac{1}{2}}\,dx.$$

12. Shew how to reduce the integral

$$\int x(1+x^3)^{\frac{1}{3}}\,dx$$

to that of a rational function, and hence evaluate it.

13. (i) By changing the variable by the relation $x=\tan(\frac{1}{4}\pi-t)$, or otherwise, prove that

$$\int_0^1\frac{\log(1+x)\,dx}{1+x^2}=\frac{\pi}{8}\log 2.$$

(ii) Shew that $$\int_0^{\frac{1}{2}\pi}\log\sin x\,dx=\tfrac{1}{2}\pi\log\tfrac{1}{2};$$

and evaluate $$\int_0^{\frac{1}{2}\pi}(x-\tfrac{1}{2}\pi)\tan x\,dx.$$

14. Prove Bonnet's form of the second mean-value theorem, that if $f'(x)$ is of constant sign and $f(b)$ has the same sign as $f(a)-f(b)$, then

$$\int_a^b f(x)\,\phi(x)\,dx=f(a)\int_a^\xi\phi(x)\,dx, \text{ where } \xi \text{ lies between } a \text{ and } b.$$

Shew that, if $q>p>0$, $$\left|\int_p^q\frac{\sin x\,dx}{x}\right|<\frac{2}{p}.$$

15. Let $f(x)$ be a bounded integrable function in (α, β) and $F(x)$ a single-valued function whose derivative is $f(x)$: prove that

$$\int_{\alpha}^{\beta} f(x)\,dx = F(\beta) - F(\alpha).$$

[Subdivide (α, β) into n intervals, apply the mean-value theorem (§ 4·42) for each sub-interval, and add the results.]

16. In the second mean-value theorem (§ 7·82), shew that $f(x)$ must be monotonic, by proving that the theorem does not hold in $(-\frac{1}{2}\pi, \frac{1}{2}\pi)$ if $f(x) = \cos x$, $\phi(x) = x^2$.

[Note that $\int_{-\frac{1}{2}\pi}^{\frac{1}{2}\pi} x^2 \cos x\,dx > 0$.]

17. Prove that

$$\text{(i)}\quad 0{\cdot}5 < \int_0^{\frac{1}{2}} \frac{dx}{\sqrt{(1-x^{2n})}} < 0{\cdot}524 \qquad (n>1),$$

$$\text{(ii)}\quad 0{\cdot}573 < \int_1^2 \frac{dx}{\sqrt{(4-3x+x^3)}} < 0{\cdot}595,$$

$$\text{(iii)}\quad xe^{-x^2} < \int_0^x e^{-t^2}\,dt < \arctan x \quad (x>0).$$

[In (ii) put $x = 1+u$, then $2+3u^2 < 2+3u^2+u^3 < 2+4u^2$ $(0<u<1)$.]

18. Shew that the remainder after n terms in Taylor's theorem may be given in the form

$$R_n = \frac{h^n}{(n-1)!}\int_0^1 (1-t)^{n-1} f^{(n)}(a+th)\,dt.$$

Deduce, from the above, the other forms of R_n given in § 4·451.

19. Let $\phi(x) \geqslant 0$, $\psi(y) \geqslant 0$, $\phi(0) = \psi(0) = 0$ and let $y = \phi(x)$, $x = \psi(y)$ be strictly increasing, continuous and inverses of each other in $x \geqslant 0, y \geqslant 0$. Shew that, if $\phi'(x)$ is continuous,

$$ab \leqslant \int_0^a \phi(x)\,dx + \int_0^b \psi(y)\,dy.$$

20. Let $f(x)$ and $g(x)$ be positive functions each integrable in $a \leqslant x \leqslant b$ and write

$$M_p(f) \equiv \left(\int_a^b \{f(x)\}^p\,dx\right)^{\frac{1}{p}};$$

prove Minkowski's inequality for integrals, that

$$M_p(f+g) \leqslant M_p(f) + M_p(g) \text{ if } p \geqslant 1,$$
$$M_p(f+g) \geqslant M_p(f) + M_p(g) \text{ if } p \leqslant 1.$$

[See § 6·53.]

CHAPTER VIII

EXTENSIONS AND APPLICATIONS OF THE INTEGRAL CALCULUS

8·1. Infinite integrals.

The theory of infinite integrals may be said to belong properly to the domain of the theory of infinite series; and the theory of the convergence of infinite integrals may be developed as a parallel to the theory of convergence of infinite series*. The reader should realise that an infinite integral is a double limit, and the discussion of double limit problems (except in simple cases) is not an easy matter. Simple cases of infinite integrals occur in elementary problems; for example, the problem of finding the area between a plane curve and its asymptote. The various types of infinite integral are considered below.

I. *The interval increasing without limit.*

(*a*) Suppose that $f(x)$ is bounded and integrable in a range (a, X), then we define

$$\int_a^\infty f(x)\,dx$$

to be

$$\lim_{X\to\infty}\int_a^X f(x)\,dx$$

provided that this limit exists.

When the limit exists, the infinite integral may be said to *converge* or to *exist.* On the other hand if the above limit is infinite, the infinite integral *diverges* or *does not exist.*

Example (i).

$$\int_1^\infty \frac{dx}{x^{\frac{3}{2}}}=2,$$

for

$$\int_1^X \frac{dx}{x^{\frac{3}{2}}}=\left[-2x^{-\frac{1}{2}}\right]_1^X=2\left(1-\frac{1}{X^{\frac{1}{2}}}\right).$$

$$\lim_{X\to\infty}\int_1^X \frac{dx}{x^{\frac{3}{2}}}=2.$$

* This is done by Hardy, *Pure Mathematics*, Chap. VIII.

Example (ii). $\int_1^\infty \log x\,dx$ does not exist, for

$$\int_1^X \log x\,dx = \Big[x\log x - x\Big]_1^X = X\log X - X + 1;$$

and as $X \to \infty$ the right-hand side increases indefinitely.

(*b*) Similar remarks apply to integrals having lower limits $-\infty$.

We define
$$\int_{-\infty}^b f(x)\,dx$$

to be
$$\lim_{X\to-\infty}\int_X^b f(x)\,dx$$

whenever this limit exists.

(*c*) If the infinite integrals $\int_{-\infty}^a f(x)\,dx$ and $\int_a^\infty f(x)\,dx$ are both convergent we say that
$$\int_{-\infty}^\infty f(x)\,dx$$

exists, and
$$\int_{-\infty}^\infty f(x)\,dx = \int_{-\infty}^a f(x)\,dx + \int_a^\infty f(x)\,dx.$$

It is easy to shew that the value of this infinite integral is independent of the arbitrary point a used in the definition.

Infinite integrals of the type discussed above are sometimes called infinite integrals *of the first kind.* A further extension of the definition of an integral is required to include the case in which the integrand may become infinite at one or more points in the range of integration, the points of infinite discontinuity of the integrand being finite in number. Such integrals are sometimes called *improper integrals.*

II. *Integrand becoming infinite at certain points.*

(*a*) Suppose that the lower limit a is the only point of infinite discontinuity of the function $f(x)$, which we shall suppose to be bounded and integrable in $(a+\epsilon, b)$, where $\epsilon > 0$.

If the integral
$$\int_{a+\epsilon}^b f(x)\,dx$$

tends to a finite limit as ϵ tends to zero, this limit is denoted by

$$\int_a^b f(x)\,dx.$$

If the above integral does not possess a finite limit as $\epsilon \to 0$ no meaning can be attached to

$$\int_a^b f(x)\,dx.$$

(*b*) Similarly when the upper limit b is the only point of infinite discontinuity of $f(x)$, which is bounded and integrable in $(a, b-\epsilon)$, we define

$$\int_a^b f(x)\,dx \text{ to be } \lim_{\epsilon \to 0} \int_a^{b-\epsilon} f(x)\,dx$$

whenever the limit exists finitely.

(*c*) Suppose finally that c, where $a < c < b$, is a point at which $f(x)$ becomes infinite. The improper integral

$$\int_a^b f(x)\,dx = \lim \left\{ \int_a^{c-\epsilon} f(x)\,dx + \int_{c+\epsilon'}^b f(x)\,dx \right\} \quad \text{......(1)}$$

when $\epsilon \to 0$ and $\epsilon' \to 0$ *independently.*

It sometimes happens that no definite limit exists when ϵ and ϵ' tend to zero *independently,* but that a limit does exist when $\epsilon = \epsilon'$. This leads us to introduce Cauchy's definition of the *principal value* of an integral.

When the limit on the right-hand side of (1) does not exist we may be able to define the *principal value* of the improper integral,

$$P \int_a^b f(x)\,dx = \lim_{\epsilon \to 0} \left\{ \int_a^{c-\epsilon} f(x)\,dx + \int_{c+\epsilon}^b f(x)\,dx \right\},$$

the principal value existing provided that the limit in question exists.

Example (i).
$$\int_{-1}^1 \frac{dx}{x^{\frac{2}{3}}} = 6.$$

$$\int_{-1}^{-\epsilon} \frac{dx}{x^{\frac{2}{3}}} = \Big[3x^{\frac{1}{3}} \Big]_{-1}^{-\epsilon} = 3(1 - \epsilon^{\frac{1}{3}}) \to 3 \text{ as } \epsilon \to 0,$$

$$\int_{\epsilon'}^1 \frac{dx}{x^{\frac{2}{3}}} = 3(1 - \epsilon'^{\frac{1}{3}}) \to 3 \text{ as } \epsilon' \to 0;$$

and so
$$\int_{-1}^1 \frac{dx}{x^{\frac{2}{3}}} = 6.$$

Example (ii). $$P\int_{-1}^{1}\frac{dx}{x}=0.$$

The improper integral $\int_{-1}^{1}\frac{dx}{x}$ is not definite, for

$$\int_{-1}^{-\epsilon}\frac{dx}{x}=\Big[\log(-x)\Big]_{-1}^{-\epsilon}=\log\epsilon,$$

$$\int_{\epsilon'}^{1}\frac{dx}{x}=-\log\epsilon';$$

and so $$\int_{-1}^{1}\frac{dx}{x}=\lim\log\frac{\epsilon}{\epsilon'},$$

as ϵ and ϵ' tend to zero independently; but this limit is not definite, and depends upon the ratio $\epsilon:\epsilon'$ which may be anything we please, since ϵ and ϵ' are both arbitrary positive numbers.

However, if we put $\epsilon=\epsilon'$, we get

$$P\int_{-1}^{1}\frac{dx}{x}=\lim\log 1=\lim 0=0.$$

The general discussion of conditions and criteria for the convergence of infinite integrals will not be given here. The reader will have seen by the preceding examples that whenever the integrand is one whose primitive is known, or which can be calculated by known rules, it is an easy matter to decide whether the limits in question exist or not.

For infinite integrals of the first kind, when $f(x)$ is a *positive* integrand for which $\int f(x)\,dx$ is not known, the question of convergence can be decided if we can find a function $\phi(x)$ whose primitive is known, and which satisfies the inequality

$$f(x)\leqslant k\phi(x) \quad\text{..........................(1)},$$

for all values of x in the range of integration.

From (1) we know that

$$\int_{a}^{X}f(x)\,dx\leqslant k\int_{a}^{X}\phi(x)\,dx,$$

and the limit of the left-hand side, as $X\to\infty$, exists provided that the limit of the right-hand side exists and is finite.

In particular if $a>0$, a suitable function $\phi(x)$ is $1/x^{\alpha}$, where $\alpha>1$.

8·2. The Gamma function.

An infinite integral of importance in Analysis is that which defines the function $\Gamma(a)$. If $a > 0$, then

$$\lim_{\substack{X\to\infty \\ \epsilon\to 0}} \int_\epsilon^X x^{a-1} e^{-x}\, dx$$

exists, and it is a convenient definition for the function $\Gamma(a)$.

To prove the existence of the infinite integral in question, consider the two integrals

$$\int_\epsilon^1 x^{a-1} e^{-x}\, dx \text{ and } \int_1^X x^{a-1} e^{-x}\, dx.$$

In the first integral, when x is small, the integrand behaves like x^{a-1}, and provided that $1 - a < 1$, the integral exists at the lower limit.

The second integral certainly exists if $a > 0$, for

$$e^x > \frac{x^n}{n!} > \frac{x^{a+1}}{n!},$$

so long as $n > a + 1$; and so

$$x^{a-1} e^{-x} < n!/x^2.$$

The integral therefore does not exceed a constant multiple of

$$\int_1^X \frac{dx}{x^2},$$

which converges as $X \to \infty$.

It is accordingly a valid definition for $\Gamma(a)$, if $a > 0$, to write

$$\Gamma(a) = \int_0^\infty x^{a-1} e^{-x}\, dx \quad \text{.................(1).}$$

If $a > 1$, we have, on integrating by parts,

$$\Gamma(a) = -\left[x^{a-1} e^{-x}\right]_0^\infty + (a-1)\int_0^\infty x^{a-2} e^{-x}\, dx;$$

hence $$\Gamma(a) = (a-1)\,\Gamma(a-1) \quad \text{.................(2);}$$

and $$\Gamma(1) = \int_0^\infty e^{-x}\, dx = 1 \quad \text{....................(3).}$$

If a is an integer it follows from (2) and (3) that

$$\Gamma(a) = (a-1)!;$$

if a is not an integer the equation (2) reduces the calculation of $\Gamma(a)$ to that of $\Gamma(p)$, where $0 < p < 1$.

8·21. The Beta function.

If $m > 0$, $n > 0$ we define

$$\mathrm{B}(m, n) \text{ to be } \int_0^1 x^{m-1}(1-x)^{n-1}\,dx.$$

The verification that, under the given restrictions, the integral exists is left to the reader.

By changing the variable by the substitution $x = 1 - y$ it is seen that

$$\mathrm{B}(m, n) = \mathrm{B}(n, m).$$

In § 11·8, by means of a double integral, the relation between the Beta and the Gamma function is investigated.

8·3. The Legendre polynomials.

Partly on account of the interesting methods of proof, and partly on account of their importance in Applied Mathematics, a section will be devoted to the discussion of Legendre's polynomial $P_n(x)$.

These polynomials first arose from a consideration of the expansion of $(1-2xh+h^2)^{-\frac{1}{2}}$ in a series of ascending powers of h. If $|2xh - h^2| < 1$, this function can be expanded in ascending powers of $2xh - h^2$: if, in addition, $|2xh| + |h|^2 < 1$, these powers can be multiplied out and the resulting series rearranged in any manner, since the expansion of

$$\{1 - (|2xh| + |h|^2)\}^{-\frac{1}{2}}$$

in powers of $|2xh| + |h|^2$ is then absolutely convergent. If it is arranged in powers of h we write

$$(1-2xh+h^2)^{-\frac{1}{2}} = P_0(x) + h\,P_1(x) + h^2\,P_2(x) + \ldots + h^n\,P_n(x) + \ldots \qquad \text{......(1).}$$

If $|2xh - h^2| < 1$, by the binomial theorem, the left-hand side gives

$$1 + \frac{1}{2}h(2x-h) + \frac{1\,.\,3}{2\,.\,4}h^2(2x-h)^2 + \frac{1\,.\,3\,.\,5}{2\,.\,4\,.\,6}h^3(2x-h)^3 + \ldots \qquad \text{......(2).}$$

If x and h are such that $2|xh| + |h|^2 < 1$, the terms of this series may be rearranged in any way without altering either the convergence or the sum of the series. For any value of x, $|h|$ may be chosen so small that the inequality

$$2|xh| + |h|^2 < 1$$

is satisfied.

By equating the coefficients of h^n in (1) and (2), commencing with the $(n+1)$th term of the series (2) and considering the terms of the series in their reverse order, we get

$$P_n(x)=\frac{1.3\ldots.(2n-1)}{2.4\ldots.2n}\left\{(2x)^n-\frac{2n}{2n-1}\cdot\frac{n-1}{1}(2x)^{n-2}\right.$$
$$\left.+\frac{2n(2n-2)}{(2n-1)(2n-3)}\,\frac{(n-2)(n-3)}{1.2}(2x)^{n-4}-\ldots\right\}$$
$$=\frac{(2n-1)!!}{n!}\left\{x^n-\frac{n(n-1)}{2(2n-1)}x^{n-2}\right.$$
$$\left.+\frac{n(n-1)(n-2)(n-3)}{2.4.(2n-1)(2n-3)}x^{n-4}-\ldots\right\}^*$$
$$=\frac{(2n-1)!!}{n!}x^n-\frac{(2n-3)!!}{2.(n-2)!}x^{n-2}+\frac{(2n-5)!!}{2.4.(n-4)!}x^{n-4}-\ldots \qquad \ldots\ldots(3).$$

If n is even the series contains $\frac{1}{2}n+1$ terms, while if n is odd it contains $\frac{1}{2}(n+1)$ terms.

Practical application.

Legendre polynomials have a practical application to problems on potential theory where there is symmetry about an axis.

Let P be the point whose distance from the point C on the axis of z is R. From the figure it is seen that

$$R^2=r^2+c^2-2cr\cos\theta.$$

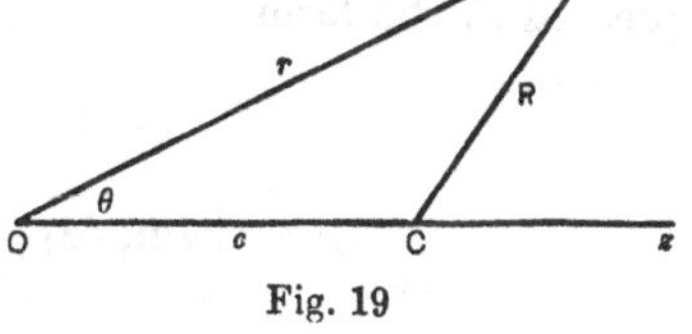

Fig. 19

By writing $x=\cos\theta$, and $h=c/r$ or $h=r/c$ according as r is greater than or less than c, we see that $1/R$ is a multiple of $(1-2xh+h^2)^{-\frac{1}{2}}$, and the potential function at P due to a charge at C is a multiple of $1/R$.

The method of development of the properties of the polynomial $P_n(x)$ adopted in the following sections has the great advantage of depending only on simple formulae for integration. It is therefore an interesting application of the integral calculus.

This method involves neither the consideration of the convergence of, nor the use of operations on, infinite series.

* The notation $(2n-1)!!$ is used to denote the product

$$(2n-1)(2n-3)(2n-5)\ldots3.1\equiv(2n)!/2^n n!$$

It will be found to be very convenient for simplifying the formulae.

8·31. Orthogonal functions.

Consider the n equations

$$\int_a^b \phi(x) \, . \, x^m \, dx = 0 \quad (m = 0, 1, 2, \ldots, n-1) \quad \ldots\ldots(1);$$

these equations express the fact that $\phi(x)$ is *orthogonal* in (a, b) to every power of x less than the nth, and therefore to every polynomial of degree less than n. Thus, if $\psi(x)$ denotes any polynomial, if the degree of $\psi(x)$ is less than n,

$$\int_a^b \phi(x) \, \psi(x) \, dx = 0.$$

Since the equations (1) are n in number, $\phi(x)$ must contain at least n independent constants, and so it must be at least of degree n. If $\phi(x)$ is taken of degree n, then the equations (1) suffice to determine it.

The equations (1) admit of solution as follows. Let

$$\Phi_1(x), \quad \Phi_2(x), \quad \ldots, \quad \Phi_m(x), \quad \ldots$$

denote the functions obtained by integrating $\phi(x)$

once, twice, ..., m times, ...;

then, by repeated integration by parts, equations (1) may be expressed in the form

$$\Big[\Phi_1(x)\Big]_a^b = 0 \quad (m=0),$$

$$\Big[x\Phi_1(x)\Big]_a^b - \Big[\Phi_2(x)\Big]_a^b = 0 \quad (m=1),$$

....

$$\Big[x^{n-1}\Phi_1(x)\Big]_a^b - \ldots + (-)^{n-1}\Big[(n-1)!\,\Phi_n(x)\Big]_a^b = 0 \quad (m = n-1).$$

These equations can be solved by inspection; they are satisfied if $\Phi_1(x), \Phi_2(x), \ldots, \Phi_n(x)$ all vanish both at a and at b, that is if $\Phi_n(x)$ contains a factor $(x-a)^n (x-b)^n$.

Since $\Phi_n(x)$ must be of degree $2n$, we have

$$\Phi_n(x) = C(x-a)^n (x-b)^n,$$

where C is constant; and so

$$\phi(x) = C \frac{d^n}{dx^n} \{(x-a)^n (x-b)^n\}.$$

This polynomial of degree n has the property of being orthogonal in (a, b) to every polynomial of lower degree.

8·32. The Legendre polynomials.

Suppose now that $a=-1$, $b=1$: if we adjust the constant C so that $\phi(x)$ begins with the term

$$\frac{(2n)!}{2^n(n!)^2}x^n=\frac{(2n-1)!!}{n!}x^n,$$

it is easily seen that $\phi(x)$ becomes Legendre's polynomial $P_n(x)$. By comparison with equation (3) of § 8·3 we see that

$$C=\frac{1}{2^n n!},$$

and then
$$P_n(x)=\frac{1}{2^n.n!}\frac{d^n}{dx^n}(x^2-1)^n.$$

This is *Rodrigues' formula for* $P_n(x)$. To shew that this formula gives the same expression for $P_n(x)$ as was obtained in § 8·3, we have, by the binomial theorem,

$$(x^2-1)^n=x^{2n}-nx^{2n-2}+\frac{n(n-1)}{2!}x^{2n-4}-\dots;$$

hence

$$\begin{aligned}\frac{d^n}{dx^n}(x^2-1)^n&=\frac{(2n)!}{n!}x^n-\frac{n(2n-2)!}{(n-2)!}x^{n-2}\\&\qquad+\frac{n(n-1)}{1.2}\frac{(2n-4)!}{(n-4)!}x^{n-4}-\dots\\&=2^n.n!\left\{\frac{(2n-1)!!}{n!}x^n-\frac{(2n-3)!!}{2.(n-2)!}x^{n-2}\right.\\&\qquad\left.+\frac{(2n-5)!!}{2.4.(n-4)!}x^{n-4}-\dots\right\};\end{aligned}$$

and so

$$P_n(x)=\frac{(2n-1)!!}{n!}x^n-\frac{(2n-3)!!}{2.(n-2)!}x^{n-2}+\frac{(2n-5)!!}{2.4.(n-4)!}x^{n-4}-\dots.$$

We can readily obtain, from the preceding formula, the results

$$P_0(x)=1,\quad P_1(x)=x,\quad P_2(x)=\tfrac{1}{2}(3x^2-1),\quad P_3(x)=\tfrac{1}{2}(5x^3-3x),$$
$$P_4(x)=\tfrac{1}{8}(35x^4-30x^2+3),\quad P_5(x)=\tfrac{1}{8}(63x^5-70x^3+15x);$$

and so on.

8·33. Special values.

The values of $P_n(1)$ and $P_n(-1)$ may be found thus. By Leibniz's theorem,

$$\frac{d^n}{dx^n}\{(x^2-1).(x^2-1)^{n-1}\}=(x^2-1)\frac{d^n}{dx^n}(x^2-1)^{n-1}$$
$$+2nx\frac{d^{n-1}}{dx^{n-1}}(x^2-1)^{n-1}+n(n-1)\frac{d^{n-2}}{dx^{n-2}}(x^2-1)^{n-1}.$$

If $x=\pm 1$ the right-hand side becomes

$$\pm 2n\left[\frac{d^{n-1}}{dx^{n-1}}(x^2-1)^{n-1}\right]_{x=\pm 1};$$

and on multiplying each side by $1/2^n n!$, we get

$$P_n(\pm 1)=\pm\frac{1}{2^{n-1}(n-1)!}\left[\frac{d^{n-1}}{dx^{n-1}}(x^2-1)^{n-1}\right]_{x=\pm 1}$$
$$=\pm P_{n-1}(\pm 1),$$

and so $$P_n(1)=1,\quad P_n(-1)=(-1)^n.$$

8·34. Two important properties.

THEOREM. *Prove that, if* $m \neq n$,

$$\text{(i)}\quad \int_{-1}^{1} P_m(x)P_n(x)\,dx=0,$$

and, if $m=n$,

$$\text{(ii)}\quad \int_{-1}^{1} P_n^2(x)\,dx=\frac{2}{2n+1}.$$

(i) is an immediate consequence of the orthogonal property, for one of the two functions $P_n(x)$, $P_m(x)$ is of degree less than the other.

To prove the second result, we write

$$2^{2n}(n!)^2\int_{-1}^{1}P_n^2(x)\,dx=\int_{-1}^{1}\frac{d^n}{dx^n}(x^2-1)^n.\frac{d^n}{dx^n}(x^2-1)^n\,dx$$

and integrate by parts; the term between the limits vanishes, and there remains the term

$$-\int_{-1}^{1}\frac{d^{n+1}}{dx^{n+1}}(x^2-1)^n.\frac{d^{n-1}}{dx^{n-1}}(x^2-1)^n\,dx.$$

Continue the process of integrating by parts until the second term in the integrand contains no differential operator; at each stage of the process the term between the limits vanishes and we get finally

$$(-)^n \int_{-1}^{1} \frac{d^{2n}}{dx^{2n}} \{(x^2-1)^n\} \cdot (x^2-1)^n \, dx = (2n)! \int_{-1}^{1} (1-x^2)^n \, dx.$$

By writing $x = \cos\theta$ the last integral becomes

$$2\,(2n)! \int_0^{\frac{1}{2}\pi} \sin^{2n+1}\theta \, d\theta = \frac{2}{2n+1} \cdot 2^{2n} (n!)^2$$

on evaluation.

Hence $$\int_{-1}^{1} P_n^2(x)\, dx = \frac{2}{2n+1}.$$

8·35. The expression of an arbitrary polynomial as a linear combination of Legendre polynomials.

Let $\psi(x)$ be any polynomial of degree n; we can find a constant c_n so that

$$\psi_1(x) = \psi(x) - c_n P_n(x)$$

is of degree $n-1$. Similarly we can find a constant c_{n-1} so that

$$\psi_2(x) = \psi_1(x) - c_{n-1} P_{n-1}(x)$$

is of degree $n-2$, and so on.

Finally $$\psi_n(x) = \psi_{n-1}(x) - c_1 P_1(x) = c_0;$$

and so, by addition,

$$\psi(x) = c_n P_n(x) + c_{n-1} P_{n-1}(x) + \dots + c_1 P_1(x) + c_0.$$

The coefficients c_r in the above expression admit of determination as follows. Multiply both sides by $P_r(x)$ and integrate, then

$$\int_{-1}^{1} \psi(x) P_r(x)\, dx = c_r \int_{-1}^{1} P_r^2(x)\, dx,$$

since all the other terms on the right-hand side vanish by the preceding theorem; and also by that theorem

$$c_r = \frac{2r+1}{2} \int_{-1}^{1} \psi(x) P_r(x)\, dx.$$

There are two important applications of this method; they are (i) *the expression of $P_n'(x)$ as a linear combination of the Legendre polynomials and* (ii) *the proof of the fundamental recurrence formula.*

(i) By the above method, clearly, since $P_n'(x)$ is of degree $n-1$,

$$P_n'(x) = \gamma_{n-1} P_{n-1}(x) + \gamma_{n-2} P_{n-2}(x) + \dots,$$

where

$$\gamma_r = \frac{2r+1}{2} \int_{-1}^{1} P_n'(x) P_r(x)\, dx,$$

and on integrating by parts,

$$\frac{2\gamma_r}{2r+1} = \Big[P_n(x) P_r(x)\Big]_{-1}^{1} - \int_{-1}^{1} P_n(x) P_r'(x)\, dx;$$

but the second integral is zero, because the degree of $P_r'(x)$ is less than n. Hence

$$\frac{2\gamma_r}{2r+1} = 1 - (-1)^{n+r},$$

and we deduce at once that

$$P_n'(x) = (2n-1) P_{n-1}(x) + (2n-5) P_{n-3}(x) + \dots.$$

(ii) Now $xP_n(x)$ is a polynomial of degree $n+1$ and therefore assumes the form

$$c_{n+1} P_{n+1}(x) + c_n P_n(x) + \dots,$$

where

$$c_r = \frac{2r+1}{2} \int_{-1}^{1} xP_n(x) P_r(x)\, dx.$$

If r is less than $n-1$, then the degree of $xP_r(x)$ is less than n, and so $c_r = 0$. Thus, with the possible exception of the first three, the coefficients c_r are zero.

Equate the coefficients of x^{n+1}, and we get

$$\frac{(2n)!}{2^n (n!)^2} = \frac{c_{n+1} (2n+2)!}{2^{n+1} \{(n+1)!\}^2},$$

that is

$$c_{n+1} = (n+1)/(2n+1).$$

Neither $xP_n(x)$ nor $P_{n+1}(x)$ contains a term in x^n, and so

$$c_n = 0.$$

If we write $x = 1$, we get

$$c_{n-1} + c_{n+1} = 1,$$

hence

$$c_{n-1} = n/(2n+1).$$

These values give us *the fundamental recurrence formula*,

$$(2n+1) xP_n(x) = (n+1) P_{n+1}(x) + nP_{n-1}(x).$$

By combining it with the formula for the derivative of $P_n(x)$, five other recurrence formulae can be established. They are

$$(1) \qquad nP_n(x) = xP_n'(x) - P'_{n-1}(x),$$

$$(2) \qquad (n+1)P_n(x) = P'_{n+1}(x) - xP_n'(x),$$

$$(3) \qquad (2n+1)P_n(x) = P'_{n+1}(x) - P'_{n-1}(x),$$

$$(4) \qquad (x^2-1)P_n'(x) = n\{xP_n(x) - P_{n-1}(x)\},$$

$$(5) \qquad (x^2-1)P_n'(x) = (n+1)\{P_{n+1}(x) - xP_n(x)\}.$$

The proofs of these formulae are left as exercises for the reader.

8·4. The concept of a plane curve.

The concept of a curve is frequently thought to be a very simple one, and it is convenient in elementary treatises to assume that the reader understands what is meant by a "curve" from the geometrical concept which enables him to draw the graph of any given simple function $y=f(x)$. We shall now investigate the question of giving a precise analytical definition of a curve, and of assigning a definite meaning to the "length" of a curve. As a preliminary, the following definition will be given.

Let $\phi(t)$ and $\psi(t)$ be continuous single-valued functions of t in the interval $\alpha \leqslant t \leqslant \beta$, and let $x=\phi(t)$, $y=\psi(t)$: then as t ranges from α to β, the points P whose coordinates are (x, y) form an aggregate of points which is called a curve.

The point (x, y) is described as the *image* of the point t. If, as t ranges from α to β, the same point (x, y) corresponds to two different values t_1 and t_2, the point is said to be a *double point* of the curve. In general, if the same point (x, y) corresponds to k different values of t, $t_1, t_2, \ldots, t_k$, the point is a *multiple point of order k*. If the point P_0 corresponding to $t=\alpha$ coincides with the point P_n corresponding to $t=\beta$, the curve is a *closed curve*.

A curve without multiple points is a *simple curve*.

Before discussing the question of assigning a meaning to the "length" of a curve, the concept of functions of bounded variation will be considered.

8·5. Functions of bounded variation.

Let $f(x)$ be a function which is defined in $a \leqslant x \leqslant b$, and suppose that (a, b) is subdivided into partial intervals by the set of dividing points

$$x_0 = a, \quad x_1, x_2, \ldots, x_r, \ldots, x_n = b \quad \ldots\ldots\ldots\ldots(1).$$

Consider the sum

$$v(a, b) = \sum_{r=1}^{n} |f(x_r) - f(x_{r-1})| \quad \dots\dots\dots\dots\dots(2);$$

if the function $f(x)$ be such that the sum (2) never exceeds some fixed positive number for all possible modes of subdivision of (a, b), then the function $f(x)$ is said to be a *function of bounded variation in* (a, b). From the definition of an upper bound, if the function $f(x)$ is of bounded variation in (a, b) there exists a number $V(a, b)$ such that for every mode of subdivision of (a, b)

$$v(a, b) = \sum_{1}^{n} |f(x_r) - f(x_{r-1})| \leqslant V(a, b),$$

and such that for at least one mode of subdivision

$$v(a, b) > V(a, b) - \epsilon.$$

Note. Functions of bounded variation are very important in Analysis: the condition of bounded variation is a sufficient condition for the existence of various types of expansion of a given function in a series. For example, Fourier series, series of harmonic functions and others, all converge when the generating function is of bounded variation.

8·51. The positive and negative variations.

Clearly $$f(b) - f(a) = \sum_{r=1}^{n} \{f(x_r) - f(x_{r-1})\};$$

suppose that this summation is divided into two parts Σ_1 and Σ_2, the first summation containing those terms for which $f(x_r) - f(x_{r-1})$ is positive, the second containing those for which it is negative. Thus

$$\Delta \equiv f(b) - f(a) = \Sigma_1 \{f(x_r) - f(x_{r-1})\} + \Sigma_2 \{f(x_r) - f(x_{r-1})\}.$$

If $f(x)$ is of bounded variation in (a, b) the value of the right-hand side cannot exceed $V(a, b)$; and so by writing

$$\Delta_r = f(x_r) - f(x_{r-1}),$$

$$\Sigma_1 \Delta_r \leqslant \tfrac{1}{2} \{V(a, b) + \Delta\},$$

$$-\Sigma_2 \Delta_r \leqslant \tfrac{1}{2} \{V(a, b) - \Delta\},$$

hence both the sums $\Sigma_1 \Delta_r$ and $-\Sigma_2 \Delta_r$ are bounded above for all possible modes of subdivision of (a, b). Further, by choosing a mode of subdivision for which

$$\Sigma\,|\Delta_r| > V(a, b) - \epsilon,$$
$$\Sigma_1 \Delta_r > \tfrac{1}{2}\{V(a, b) + \Delta - \epsilon\},$$
$$-\Sigma_2 \Delta_r > \tfrac{1}{2}\{V(a, b) - \Delta - \epsilon\}.$$

It follows that the numbers* $\frac{1}{2}(V + \Delta)$, $\frac{1}{2}(V - \Delta)$ are the upper bounds of the two summations $\Sigma_1 \Delta_r$ and $-\Sigma_2 \Delta_r$ for all possible modes of subdivision of (a, b).

If these bounds be denoted by $P(a, b)$, $N(a, b)$ respectively, we have

$$V(a, b) = P(a, b) + N(a, b),$$
$$\Delta = f(b) - f(a) = P(a, b) - N(a, b).$$

The numbers P, $-N$, and V are called respectively *the positive, negative and total variations of* $f(x)$.

8·52. Fundamental properties of functions of bounded variation.

THEOREM 1. *Every function of bounded variation can be expressed as the difference of two monotonic functions, either both increasing or both decreasing.*

To prove this theorem, observe that if x be any point of (a, b) a function $f(x)$ which is of bounded variation in (a, b) is also of bounded variation in (a, x), and consequently the positive, negative and total variations $P(a, x)$, $-N(a, x)$ and $V(a, x)$ are also bounded, and satisfy the relations

$$V(a, x) = P(a, x) + N(a, x),$$
$$f(x) - f(a) = P(a, x) - N(a, x).$$

Now if $x' > x$,

$$P(a, x') \geqslant P(a, x), \quad N(a, x') \geqslant N(a, x),$$

and so $P(a, x)$ and $N(a, x)$ are both monotonic increasing functions of x in (a, b).

Now $\quad f(x) = \{P(a, x) + f(a) + k\} - \{N(a, x) + k\} \quad \text{......(1)},$

and $\quad f(x) = \{\kappa - N(a, x)\} - \{\kappa - P(a, x) - f(a)\} \quad \text{......(2)},$

where k and κ are arbitrary constants. This proves the theorem.

* $V(a, b)$ is written to indicate in what interval the function $f(x)$ is of bounded variation. When no doubt as to the interval can exist, it is simpler to write V alone.

In virtue of this theorem, the properties of monotonic functions are all extended to the wide class of functions of bounded variation.

THEOREM 2. *If $f(x)$ is continuous and of bounded variation in (a, x), then its total variation $V(a, x)$ is continuous.*

We can find a mode of subdivision of the interval (a, x) with a point of subdivision x' as near as we please to x such that

$$v(a, x) > V(a, x) - \epsilon \text{ and } |f(x) - f(x')| < \epsilon.$$

Now $$v(a, x') = v(a, x) - |f(x) - f(x')|$$

and so $$V(a, x') \geqslant v(a, x') > V(a, x) - 2\epsilon.$$

Since $V(a, x')$ is monotonic increasing it follows that as $x' \to x - 0$

$$V(a, x') \to V(a, x).$$

Similarly $V(a, x') \to V(a, x)$ as $x' \to x + 0$.

This proves the theorem.

COROLLARY. *A continuous function of bounded variation is the difference between two continuous monotonic increasing functions.*

For, if $f(x)$ is continuous so are $P(a, x)$ and $N(a, x)$.

8·6. Rectifiable curves.

We are now in a position to consider some suitable definitions of the *length* of a curve. The reader will see that the importance of functions of bounded variation in this connection lies in the fact that, by Theorem 1 below, it is necessary and sufficient for the curve $x = \phi(t)$, $y = \psi(t)$ to be rectifiable, that the functions defining the curve should be functions of bounded variation.

Let t be a variable defined for all values of t in the interval $\alpha \leqslant t \leqslant \beta$, and let $\phi(t)$ and $\psi(t)$ be two single-valued bounded functions of t defined in (α, β); then the equations

$$x = \phi(t), \quad y = \psi(t)$$

define the arc of a plane curve.

Subdivide the interval (α, β) by the points

$$\alpha = t_0, \quad t_1, t_2, \ldots, t_r, \ldots, t_n = \beta \ldots\ldots\ldots\ldots\ldots(1),$$

then the points on the curve corresponding to these values of t may be denoted by

$$P_0, P_1, P_2, \ldots, P_r, \ldots, P_n.$$

The length of the polygonal line $P_0 P_1 \ldots P_n$, measured by

$$\sum_{r=1}^{n} P_{r-1} P_r \equiv \sum_{r=1}^{n} \{(x_r - x_{r-1})^2 + (y_r - y_{r-1})^2\}^{\frac{1}{2}} \ldots\ldots(2),$$

clearly depends upon the particular mode of subdivision of (α, β).

The summation (2) will be called the *length of an inscribed polygon.*

If the arc be such that the lengths of all the inscribed polygons have a finite upper bound L, the curve is said to be rectifiable, and the length of the arc is defined to be L. Otherwise the length of the arc is regarded as infinite (or non-existent).

If we denote the subdivision (1) of the interval (α, β) by D and write
$$\Delta x = x_r - x_{r-1}, \qquad \Delta y = y_r - y_{r-1},$$
then the summation (2) may be conveniently represented by
$$\sum_D |\Delta P| = \sum_D \sqrt{\{(\Delta x)^2 + (\Delta y)^2\}}.$$

THEOREM 1. *The necessary and sufficient condition that the arc defined by $x = \phi(t)$, $y = \psi(t)$, for $\alpha \leqslant t \leqslant \beta$, should be rectifiable is that the functions $\phi(t)$ and $\psi(t)$ are of bounded variation in (α, β).*

We have the inequalities
$$\left.\begin{array}{l}|\Delta x|\\|\Delta y|\end{array}\right\} \leqslant \sqrt{\{(\Delta x)^2 + (\Delta y)^2\}} \leqslant |\Delta x| + |\Delta y|.$$

If $\phi(t)$ and $\psi(t)$ are of bounded variation then both $\Sigma|\Delta x|$ and $\Sigma|\Delta y|$ are bounded above: hence $\Sigma|\Delta P|$ is bounded above for all modes of subdivision of (α, β).

Conversely, if $\Sigma|\Delta P|$ is bounded above, the same must be true of $\Sigma|\Delta x|$ and $\Sigma|\Delta y|$ and so $\phi(t)$ and $\psi(t)$ are of bounded variation.

This proves the theorem.

We have defined the length L of the curve above to be the upper bound of the lengths of all possible inscribed polygons. A second definition of the length of a curve (as a limit) is justified by the following theorem.

THEOREM 2. *If the functions $x = \phi(t)$, $y = \psi(t)$ are single-valued* CONTINUOUS *functions of t in $\alpha \leqslant t \leqslant \beta$, and if D be a subdivision of (α, β) of norm d, then*
$$\lim_{d \to 0} \sum_D |\Delta P| = L,$$
where L is the length of the curve.

There certainly exists a subdivision D_1 such that
$$L - \epsilon < \sum_{D_1} |\Delta P| \leqslant L \quad \text{......................(1).}$$

Suppose that the subdivision D_1 consists of p sub-intervals, the least of which has length l; and let D be a subdivision of (α, β) whose norm d satisfies the inequality $d \leqslant d_0 < l$; then no sub-interval of D contains more than one point of D_1.

Let D_2 be the subdivision consecutive to D and D_1, then

$$\sum_{D_2} |\Delta P| \geqslant \sum_{D_1} |\Delta P| \text{ and } \sum_{D_2} |\Delta P| \geqslant \sum_{D} |\Delta P| \quad \text{......(2)}.$$

Suppose that the point t_s of D_1 lies in the interval (t_{r-1}, t_r) of D, then

$$\sum_{D_2} |\Delta P| - \sum_{D} |\Delta P| = \Sigma \{P_{r-1} P_s + P_s P_r - P_{r-1} P_r\} \quad \text{...(3)},$$

the last summation containing at most p terms.

Since $\phi(t)$ and $\psi(t)$ are *continuous*, we may choose d_0 so small that each term of the summation on the right-hand side of (3) is less than ϵ/p for any value of d which does not exceed d_0; hence

$$\sum_{D_2} |\Delta P| - |\sum_{D} |\Delta P| < \epsilon \quad \text{..................(4)}.$$

Thus from (1), (2) and (4) we get

$$\sum_{D_2} |\Delta P| \geqslant \sum_{D_1} |\Delta P| > L - \epsilon,$$

and hence it follows that

$$L - \sum_{D} |\Delta P| < 2\epsilon$$

for all values of $d \leqslant d_0$. In other words

$$\lim_{d \to 0} \sum_{D} |\Delta P| = L.$$

From the preceding theorem the reader will see that if $\phi(t)$ and $\psi(t)$ are *continuous* functions of t, the length of the arc of the curve $x = \phi(t)$, $y = \psi(t)$, as t ranges from α to β, may be defined to be the limit of the length of the inscribed polygon corresponding to a subdivision D of (α, β) as the norm of D tends to zero.

THEOREM 3. *If the arc C of the curve corresponding to the interval $\alpha \leqslant t \leqslant \beta$ is rectifiable and of length s, then if s_1 and s_2 are the lengths of the arcs C_1 and C_2 which correspond to the intervals (α, γ), (γ, β), $\alpha < \gamma < \beta$,*

$$s = s_1 + s_2 \quad \text{........................(1)}.$$

Since $\phi(t)$ and $\psi(t)$ are of bounded variation in (α, β) they have *a fortiori* bounded variation in (α, γ), (γ, β), and so the arcs C_1 and C_2 are rectifiable; thus s_1 and s_2 exist.

Let D_1 and D_2 be respectively subdivisions of (α, γ), (γ, β) of norm d, then, if we assume that $\phi(t)$ and $\psi(t)$ are *continuous*,

$$s_1 = \lim_{d \to 0} \sum_{D_1} |\Delta P| \text{ and } s_2 = \lim_{d \to 0} \sum_{D_2} |\Delta P|.$$

Since

$$s = \lim_{\delta \to 0} \sum_{D} |\Delta P| \quad \text{.......................(2)},$$

when D is *any* mode of subdivision of (α, β) of norm δ, the limit in (2) is the same as when D is composed of subdivisions of the type used for D_1 and D_2, and so

$$\sum_D |\Delta P| = \sum_{D_1} |\Delta P| + \sum_{D_2} |\Delta P|,$$

and by taking the limit, equation (1) follows.

COROLLARY. *The length s of the rectifiable curve C corresponding to the interval $\alpha < t \leqslant \beta$ is a monotonic increasing function of t.*

This follows at once from the theorem.

THEOREM 4. *If $\phi(t)$ and $\psi(t)$ are continuous single-valued functions of t in (α, β), and the curve $x = \phi(t)$, $y = \psi(t)$ is rectifiable, then the length s is a continuous function $s(t)$ of t.*

Since $\phi(t)$ and $\psi(t)$ are of bounded variation, we may write

$$\phi(t) = \phi_1(t) - \phi_2(t), \quad \psi(t) = \psi_1(t) - \psi_2(t),$$

where the functions on the right-hand sides are continuous monotonic increasing functions of t (see Corollary, § 8·52).

Consider a division D of norm d of the interval $(t, t+h)$, where $h > 0$, then

$$\begin{aligned}\sum_D |\Delta P| &= \Sigma \sqrt{\{(\Delta x)^2 + (\Delta y)^2\}} \leqslant \Sigma |\Delta x| + \Sigma |\Delta y| \\ &\leqslant \Sigma \Delta\phi_1 + \Sigma\Delta\phi_2 + \Sigma\Delta\psi_1 + \Sigma\Delta\psi_2 \\ &\leqslant \delta\phi_1 + \delta\phi_2 + \delta\psi_1 + \delta\psi_2,\end{aligned}$$

where $\delta\phi_1 = \phi_1(t+h) - \phi_1(t)$, and similarly for the other functions. Since $\phi_1(t)$ is continuous, $\delta\phi_1 \to 0$ as $h \to 0$; similarly for the others, and so we can choose a number $\eta > 0$ so that

$$\sum_D |\Delta P| < \epsilon \text{ for values of } h \leqslant \eta.$$

But $s(t+h) - s(t) = \lim_{d \to 0} \sum_D |\Delta P| < \epsilon$ for values of $h \leqslant \eta$, which proves that $s(t)$ is a continuous function of t.

In virtue of the preceding theorems we see that, since s is a continuous monotonic increasing function of t, the inverse function t is a continuous monotonic increasing function of s*, so that x and y may be regarded as functions of s, say

$$x = f(s), \quad y = g(s).$$

The arc of the curve $y = F(x)$ defined in $a \leqslant x \leqslant b$ is a particular case of the above when t is identical with x.

* This follows from Theorem 2, § 10·1.

8·61. Rectification.

THEOREM. *If $\phi'(t)$ and $\psi'(t)$ are continuous functions of t in $\alpha \leqslant t \leqslant \beta$, the curve $x = \phi(t)$, $y = \psi(t)$ is rectifiable and its length s is given by*

$$s = \int_\alpha^\beta \surd[\{\phi'(t)\}^2 + \{\psi'(t)\}^2]\,dt.$$

To prove this, write

$$c_r = \surd\{(x_r - x_{r-1})^2 + (y_r - y_{r-1})^2\};$$

then, by the mean-value theorem,

$$c_r = (t_r - t_{r-1})\surd[\{\phi'(\xi_r)\}^2 + \{\psi'(\eta_r)\}^2],$$

where ξ_r and η_r lie between t_{r-1} and t_r.

We have to shew that the difference between the two expressions

$$\surd[\{\phi'(\xi_r)\}^2 + \{\psi'(\eta_r)\}^2] \quad \text{.................(1)}$$

and

$$\surd[\{\phi'(t_{r-1})\}^2 + \{\psi'(t_{r-1})\}^2] \quad \text{.................(2)}$$

tends *uniformly* to zero as $(t_r - t_{r-1})$ tends to zero.

In virtue of the identity

$$\surd(a^2 + b^2) - \surd(c^2 + d^2) = \frac{(a-c)(a+c) + (b-d)(b+d)}{\surd(a^2 + b^2) + \surd(c^2 + d^2)},$$

since

$$\left.\begin{matrix}|a| + |c| \\ |b| + |d|\end{matrix}\right\} \leqslant \surd(a^2 + b^2) + \surd(c^2 + d^2),$$

it follows that

$$|\surd(a^2 + b^2) - \surd(c^2 + d^2)| \leqslant |a - c| + |b - d| \text{.........(3)}.$$

If the left-hand side of (3) represents the difference between the expressions (1) and (2), then since $\phi'(t)$ and $\psi'(t)$ are continuous, and therefore uniformly continuous*, it follows from (3) that, if

$$a - c = \phi'(\xi_r) - \phi'(t_{r-1}) \text{ and } b - d = \psi'(\eta_r) - \psi'(t_{r-1}),$$

$$c_r = (t_r - t_{r-1})\surd[\{\phi'(t_{r-1})\}^2 + \{\psi'(t_{r-1})\}^2] + (t_r - t_{r-1})\rho_r,$$

where ρ_r tends *uniformly* to zero as $(t_r - t_{r-1}) \to 0$.

It follows that the length

$$s = \lim \Sigma c_r,$$

as the norm d of the subdivision of (α, β) tends to zero, is the integral

$$\int_\alpha^\beta \surd[\{\phi'(t)\}^2 + \{\psi'(t)\}^2]\,dt.$$

* See § 3·43.

In almost every case of practical importance the functions $\phi'(t)$ and $\psi'(t)$ are continuous functions. When $\phi'(t)$ and $\psi'(t)$ are discontinuous at a finite number of points (corresponding to the existence of cusps in the arc of the curve), all that is necessary is to divide the arc in question into several arcs, for each of which the functions $\phi'(t)$ and $\psi'(t)$ are continuous, and then add the results of integrating over the intervals for t which correspond to each of these arcs.

All the preceding analysis can be easily extended to the twisted curve

$$x = \phi(t), \quad y = \psi(t), \quad z = \chi(t).$$

For the sake of brevity the proofs have been given for plane curves only.

8·7. Curvilinear integrals.

As soon as the concept of a plane curve has been made precise, it is possible to define another type of integral which has a good deal of importance in certain branches of Applied Mathematics.

These integrals are known as *curvilinear integrals*, and we shall have occasion to employ such integrals again in Chapter XI.

Let AB be the arc of a plane curve, and suppose that $P(x, y)$ is a continuous function of the two variables x and y at every point of the curve AB, where x and y are the coordinates of a point of AB referred to a set of rectangular axes in its plane.

Subdivide the arc AB into smaller arcs by the set of points of subdivision (x_r, y_r) $(r = 0, 1, 2, \ldots, n)$, and choose any point (ξ_r, η_r) in the arc which joins the points (x_{r-1}, y_{r-1}) and (x_r, y_r).

Consider the sum *

$$P(\xi_1, \eta_1)(x_1 - a) + P(\xi_2, \eta_2)(x_2 - x_1) + \ldots + P(\xi_n, \eta_n)(b - x_{n-1}) \quad \ldots\ldots(1),$$

which is analogous to the sum used in the definition of the Riemann integral as the limit of a sum, with $P(x, y)$ replacing $f(x)$. If the sum (1) approaches a definite limit when the number of points of subdivision increases indefinitely in such a way that the length of the greatest sub-interval tends to zero, this limit may be taken as

* The notation used is an obvious one, thus the points A and B have abscissae a and b respectively, and $a = \phi(\alpha)$, $b = \phi(\beta)$.

the definition of the curvilinear integral of $P(x, y)$ with respect to x over the arc AB, and the integral is denoted by

$$\int_{AB} P(x, y)\, dx.$$

We now shew that under certain conditions the limit in question certainly exists.

Let the curve AB be defined by the equations $x = \phi(t)$, $y = \psi(t)$, so that as t varies from α to β the point (x, y) describes the curve AB.

If the functions $\phi(t)$ and $\psi(t)$ are both monotonic in (α, β), then to every value of x there corresponds one and only one value of t: similarly y is a single-valued function of t, and so, on AB, y is a single-valued function of x, say $y = \lambda(x)$.

Since $P(x, y)$ is a continuous function in both variables, the function $P\{x, \lambda(x)\}$ is a continuous function of x in $a \leqslant x \leqslant b$. Hence

$$P(\xi_r, \eta_r) = P\{\xi_r, \lambda(\xi_r)\} = \mathbf{P}(\xi_r).$$

The summation (1) may therefore be written

$$\mathbf{P}(\xi_1)(x_1 - a) + \mathbf{P}(\xi_2)(x_2 - x_1) + \ldots + \mathbf{P}(\xi_n)(b - x_{n-1}),$$

the limit of which, since $\mathbf{P}(x)$ is a continuous function of x, is the ordinary definite integral

$$\int_a^b \mathbf{P}(x)\, dx.$$

The conditions laid down above are equivalent to the geometrical restriction that a line parallel to the axis of y cannot meet the arc AB in more than one point.

The definition may be extended at once to the case where AB is a curve for which $\phi(t)$ and $\psi(t)$ are not monotonic throughout (α, β) so long as it is possible to divide up AB into a finite number of pieces on each of which the functions $\phi(t)$ and $\psi(t)$ are monotonic. Such a curve is said to be a *regular curve.*

For the curve $ACDB$, illustrated below, we have

$$\int_{ACDB} P(x, y)\, dx = \int_{AC} P(x, y)\, dx + \int_{CD} P(x, y)\, dx + \int_{DB} P(x, y)\, dx,$$

where in each of the three integrands on the right-hand side, y must be replaced by a different function of the variable x.

Curvilinear integrals of the type

$$\int Q(x, y)\,dy$$

are defined in a similar way.

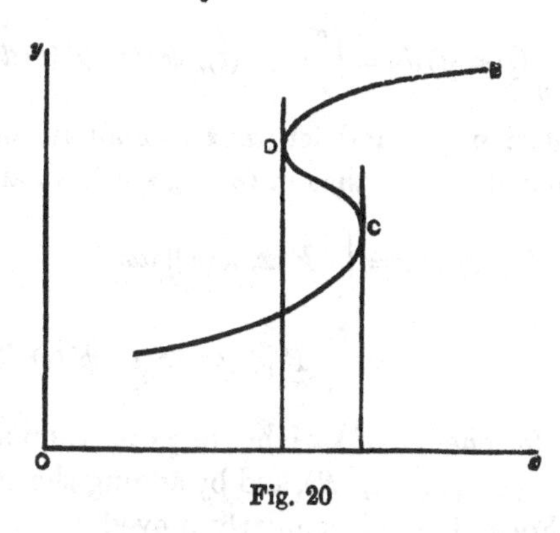

Fig. 20

8·71. Properties of curvilinear integrals.

The following elementary properties are stated; the reader is advised to prove them in detail. They are all immediate deductions from the definition.

(1) *The value of* $\int_C P(x, y)\,dx$, *when* C *is a straight line parallel to the axis of* y, *is zero.*

Similarly $\int_C Q(x, y)\,dy$, *when* C *is a straight line parallel to the axis of* x, *is zero.*

To prove these results, observe that each of the above curvilinear integrals reduces to an ordinary Riemann integral in which the limits of integration coincide.

(2) *Curvilinear integrals are additive for arcs.*

$$\int_{AC} P(x, y)\,dx = \int_{AB} P(x, y)\,dx + \int_{BC} P(x, y)\,dx.$$

(3) $\displaystyle\int_{AB} P(x, y)\,dx = -\int_{BA} P(x, y)\,dx.$

Similar results of course hold for curvilinear integrals with respect to y.

8·72. THEOREM.

If $\phi'(t)$ *and* $\psi'(t)$ *are continuous in* (α, β), *then*

$$\int_{AB} P(x, y)\, dx = \int_{\alpha}^{\beta} P\{\phi(t), \psi(t)\}\, \phi'(t)\, dt \quad \text{......(1)},$$

$$\int_{AB} Q(x, y)\, dy = \int_{\alpha}^{\beta} Q\{\phi(t), \psi(t)\}\, \psi'(t)\, dt \quad \text{......(2)}.$$

Consider equation (1), and let (α, α_1) be an arc on which $\phi(t)$ and $\psi(t)$ are monotonic, so that in (α, α_1), $y = \lambda(x)$ and

$$\int_{AA_1} P(x, y)\, dx = \int_{a}^{a_1} P\{x, \lambda(x)\}\, dx$$

$$= \int_{\alpha}^{\alpha_1} P\{\phi(t), \psi(t)\}\, \phi'(t)\, dt,$$

by the formula for change of variable in § 7·6. Similar results hold for the arcs $(\alpha_2, \alpha_3), \ldots, (\alpha_{n-1}, \beta)$, and by adding the results formula (1) is proved. Formula (2) is similarly proved.

Note. In the above theorems *sufficient* conditions for the existence of the integrals concerned have been stated, but the reader should observe that they may not all be *necessary*. For example, in equation (1) above, the integral on the right-hand side would exist if $P(x, y)$ were continuous and $\phi'(t)$ were bounded and integrable without being necessarily continuous.

The conditions given above are those which usually occur in practice, and are sufficient for all the applications of these theorems in Applied Mathematics and Physics. In general, also, the conditions stated above suffice for all the applications of curvilinear integrals which are needed in this book.

8·73. Curvilinear integrals with respect to the arc.

If s denote, as usual, the length of arc, we have seen that s is an increasing function of t, and t is a single-valued function of s, so that s may be taken as the parameter, and x and y may be expressed as functions of s: these functions will be monotonic functions of s if $\phi(t)$ and $\psi(t)$ are monotonic functions of t. Thus, if the length of the arc AB is σ, and the origin of s is chosen when $t = \alpha$, $x = a$, then clearly

$$\int_{AB} P(x, y)\, ds = \int_{0}^{\sigma} P\{x(s), y(s)\}\, ds = \int_{0}^{\sigma} \Phi(s)\, ds.$$

EXAMPLES VIII.

1. If $a > 0$, discuss whether or not the following integrals exist:

$$\int_a^\infty \frac{dx}{x^{\frac{1}{2}}},\quad \int_a^\infty \frac{dx}{x^{\frac{4}{3}}},\quad \int_a^\infty \frac{x\,dx}{b^2+x^2},\quad \int_{-\infty}^\infty \frac{dx}{1+x^2}.$$

Shew also that
$$\int_1^\infty \frac{dx}{(1+x)\sqrt{x}} = \tfrac{1}{2}\pi.$$

2. Shew that $\int_0^{\frac{1}{2}\pi} \frac{x^p}{(\sin x)^q}\,dx$ converges if $q < p+1$; and prove that $\int_0^\infty \frac{x^{a-1}}{1+x}\,dx$ converges if $0 < a < 1$.

3. Evaluate $\int_{-1}^1 (1-x^2)\,P_n'^2(x)\,dx$ and $\int_{-1}^1 (1-x^2)\,P_n^2(x)\,dx$.

4. Shew that
$$\Gamma(a) = 2\int_0^\infty e^{-t^2} t^{2a-1}\,dt = \int_0^1 \left(\log\frac{1}{u}\right)^{a-1} du,$$
and that
$$\mathrm{B}(m, n) = 2\int_0^{\frac{1}{2}\pi} \cos^{2m-1}\theta \sin^{2n-1}\theta\,d\theta = \int_0^\infty \frac{t^{n-1}}{(1+t)^{m+n}}\,dt.$$

5. Shew that $\quad n!\,P_{2n}(x) = x\frac{d^n}{d(x^2)^n}\{x^{2n-1}(x^2-1)^n\}.$

6. Prove that

(i) $u = P_n(x)$ is a solution of the differential equation
$$(1-x^2)\frac{d^2u}{dx^2} - 2x\frac{du}{dx} + n(n+1)u = 0;$$

(ii) $xP_n'(x) = nP_n(x) + (2n-3)P_{n-2}(x) + (2n-7)P_{n-4}(x) + \ldots$.

7. Prove that, if m and n are integers such that $m \leqslant n$, both being even or both odd,
$$\int_{-1}^1 \frac{dP_m}{dx}\frac{dP_n}{dx}\,dx = m(m+1).$$

8. (i) In the interval $(0, b)$, where $b > 1$, whenever x has any one of the set of values
$$1, \frac{1}{2}, \frac{1}{3}, \ldots, \frac{1}{n}, \ldots,$$
let $y = x$; and whenever x takes any one of the values
$$b, a_1, a_2, \ldots, a_m, \ldots,$$
such that $\quad 1/(m+1) < a_m < 1/m \qquad (m = 1, 2, 3, \ldots),$
let $y = 0$. Join up the sets of points by straight lines starting with $x = b$, and joining up succeeding points in the order
$$b, 1, a_1, \frac{1}{2}, a_2, \frac{1}{3}, \ldots, a_k, \frac{1}{k+1}, \ldots.$$
Prove that the function $f(x)$, whose graph is as described, is not a function of bounded variation in $(0, b)$.

(ii) If whenever x has any one of the set of values
$$1, \frac{1}{2}, \frac{1}{3}, \ldots, \frac{1}{n}, \ldots,$$

we put $y=x^2$ instead of $y=x$, prove that the function whose graph is formed in a similar way to the preceding, is a function of bounded variation in $(0, b)$.

9. Prove that a function of bounded variation in a given interval is integrable in that interval.

10. Prove that, for a circle, the length of any inscribed polygon is less than that of the circumscribed square, and deduce that the circle is a rectifiable curve.

11. (i) Find the length of the arc of the catenary $y=c\cosh x/c$ between the points $(0, c)$ and $(c, c\cosh 1)$.

(ii) Prove that the total length of the curve $x=a\cos^3 t$, $y=a\sin^3 t$ is $6a$, and sketch the curve.

12. Prove that the cardioid $r=a(1+\cos\theta)$ is divided by the line $4r=3a\sec\theta$ into two parts such that the lengths of the arcs on either side of this line are equal.

13. (i) If $x=at^2$, $y=2at$, evaluate

$$\int_C (x^2+y^2)\,dx \quad \text{and} \quad \int_C (x^2+y^2)\,dy,$$

where C is that part of the parabola lying on the left of the latus rectum.

(ii) Evaluate

$$\int_C \frac{dx}{x+y},$$

where C is the curve $x=a^2t^3$, $y=2at$, $(0\leqslant t\leqslant a)$.

14. If A is the point $(0, 1)$, B is $(0, y)$ and D is (x, y), evaluate

$$\int_C \left(\frac{x+y}{x^2+y^2}\,dx + \frac{y-x}{x^2+y^2}\,dy\right),$$

where C is the path which consists of the straight lines AB and BD.

Is the value of the integral independent of the path?

15. A chord AB of a circle which subtends an angle 2α at the centre is taken as the axis of x, and

$$u=\int_C \frac{ds}{y},$$

where C is an arc PQ of the circle lying entirely on the positive side of AB. If d is the length of the chord PQ, and y_1 and y_2 are the ordinates of P and Q, prove that

$$\sqrt{(y_1 y_2)}\sinh(\tfrac{1}{2}u\sin\alpha)=\tfrac{1}{2}d\sin\alpha.$$

16. Prove that Hilbert's double series

$$\sum_{m\leqslant n}\sum \frac{a_m a_n}{m+n}\leqslant \pi\Sigma a_n^2.$$

[Write $\quad \Sigma\Sigma\dfrac{a_m a_n}{m+n}=\Sigma\Sigma\left(\dfrac{m}{n}\right)^{\frac{1}{4}}\dfrac{a_m}{\sqrt{(m+n)}}\left(\dfrac{n}{m}\right)^{\frac{1}{4}}\dfrac{a_n}{\sqrt{(m+n)}}\leqslant\sqrt{(PQ)},$

where $\quad P\equiv\sum_m a_m^2\sum_n\sqrt{\left(\dfrac{m}{n}\right)}\dfrac{1}{m+n}\leqslant\Sigma a_m^2\displaystyle\int_0^\infty\frac{dx}{(1+x)\sqrt{x}}=\pi\Sigma a_m^2;$

and similarly for Q.]

CHAPTER IX

FUNCTIONS OF MORE THAN ONE VARIABLE

9·1. Introduction.

So far attention has been mainly directed to functions of a single real variable, and the application of the differential and integral calculus to such functions has been considered.

Functions of more than one variable were mentioned in § 3·11, and the distinction between explicit and implicit functions was made there. In this chapter we shall be mainly concerned with the application of the differential calculus to explicit functions of more than one variable. Implicit functions are considered in the next chapter.

Mainly for the sake of brevity we shall usually restrict ourselves to two or three variables only. In general, most of the theorems are easily extended to more than three variables.

9·2. Differentiability.

Consider a variable u connected with the three independent variables x, y and z by the functional relation*

$$u = u(x, y, z).$$

If arbitrary increments Δx, Δy, Δz are given to the independent variables, the corresponding increment Δu of the dependent variable of course depends upon the three increments assigned to x, y and z.

By extending the definition of differentiability for functions of one variable given in § 4·12, we say that the function $u = u(x, y, z)$ is *differentiable* at the point (x, y, z) if it possesses a determinate value in the neighbourhood of this point, and if

$$\Delta u = A\Delta x + B\Delta y + C\Delta z + \epsilon\rho \quad \text{..............(1)},$$

where $\rho = |\Delta x| + |\Delta y| + |\Delta z|$, $\epsilon \to 0$ as $\rho \to 0$, and A, B, C are independent of Δx, Δy and Δz.

In the above definition ρ may always be replaced by η, where

$$\eta = \sqrt{(\Delta x^2 + \Delta y^2 + \Delta z^2)}.$$

* The use of u for the functional symbol as well as for the dependent variable has many advantages, and is not likely to cause any ambiguity. Especially is it advantageous in the theory of differentiability, for if we write $u = f(x, y, z)$, then du and df are two different symbols for one and the same thing. This complication is avoided by replacing f by u as the functional symbol.

9·21. Partial derivatives.

By changing x to $x+\Delta x$, and allowing y and z to remain invariable, the increment of u is

$$u(x+\Delta x, y, z)-u(x, y, z).$$

By analogy with the definition of the derivative of a function of one variable, if the incrementary ratio

$$\frac{u(x+\Delta x, y, z)-u(x, y, z)}{\Delta x}$$

tends to a unique limit as Δx tends to zero, this limit is defined to be the *partial derivative* of u with respect to x and it is written

$$\frac{\partial u}{\partial x} \text{ or } u_x.$$

Similar definitions hold for $\frac{\partial u}{\partial y}$ and $\frac{\partial u}{\partial z}$.

The calculation of the partial derivatives of any given explicit function is a simple matter. If

$$u=5x^3y+3yz^2+xyz+y^3,$$

then $\frac{\partial u}{\partial x}$ is found by finding the derivative of u with respect to x and treating the variables y and z as constants. Thus

$$\frac{\partial u}{\partial x}=15x^2y+yz.$$

Similarly $$\frac{\partial u}{\partial y}=5x^3+3z^2+xz+3y^2, \quad \frac{\partial u}{\partial z}=6yz+xy.$$

9·22. The differential coefficients.

In equation (1) of § 9·2, suppose that $\Delta y=\Delta z=0$, then, on the assumption that u is differentiable at the point (x, y, z),

$$\Delta u=u(x+\Delta x, y, z)-u(x, y, z)=A\,\Delta x+\epsilon\,|\Delta x|,$$

and dividing by Δx,

$$\frac{u(x+\Delta x, y, z)-u(x, y, z)}{\Delta x}=A\pm\epsilon,$$

and by taking the limit as $\Delta x\to 0$, since $\epsilon\to 0$ as $\Delta x\to 0$,

$$\frac{\partial u}{\partial x}=A.$$

Similarly $$\frac{\partial u}{\partial y}=B, \quad \frac{\partial u}{\partial z}=C.$$

Hence when the function $u = u(x, y, z)$ is differentiable the partial derivatives $\frac{\partial u}{\partial x}, \frac{\partial u}{\partial y}, \frac{\partial u}{\partial z}$ are respectively the *differential coefficients** A, B, C, and so

$$\Delta u = \frac{\partial u}{\partial x}\Delta x + \frac{\partial u}{\partial y}\Delta y + \frac{\partial u}{\partial z}\Delta z + \epsilon\rho \quad \text{...........(2)}.$$

The differential of the dependent variable du is defined to be the principal part of Δu, so that (2) may be written

$$\Delta u = du + \epsilon\rho \quad \text{......................(3)}.$$

As we have already seen when considering functions of one variable, the differentials of the *independent* variables are identical with the arbitrary increments of these variables. If we write $u = x$, $u = y$, $u = z$ respectively, it follows that

$$dx = \Delta x, \quad dy = \Delta y, \quad dz = \Delta z \text{ ....(4)};$$

and from (4) the expression for du is seen to be

$$du = \frac{\partial u}{\partial x}dx + \frac{\partial u}{\partial y}dy + \frac{\partial u}{\partial z}dz \quad \text{..............(5)}.$$

This is a fundamental formula in the theory of differentiability.

9·23. The distinction between derivatives and differential coefficients.

In § 4·12 we saw that the necessary and sufficient condition that the function $y = f(x)$ should be differentiable at the point x is that it possesses a finite definite derivative at that point. Thus, for functions of one variable, the existence of the derivative $f'(x)$ implies the differentiability of $f(x)$ at any given point.

For functions of more than one variable this is not true. If the function $u = u(x, y, z)$ is differentiable at the point (x, y, z) the partial derivatives of u with respect to x, y and z certainly exist and are finite at this point, for then they are identical with the differential coefficients A, B and C respectively. The partial

* Historically, the original use of "differential coefficient" was in this sense, namely to express that a number such as A is the coefficient of a differential. Thus, in equation (5), $\frac{\partial u}{\partial x}$, which is the same thing as A, is the coefficient of the differential dx.

derivatives, however, may exist at a point when the function is not differentiable at that point. In other words the partial derivatives need not always be differential coefficients. This is best illustrated by means of an example.

Example. *Let*

$$f(x, y) = \frac{x^3 - y^3}{x^2 + y^2} \quad \textit{when } x \textit{ and } y \textit{ are not simultaneously zero,}$$

$$f(0, 0) = 0.$$

If this function is differentiable at the origin, then, by definition,

$$f(h, k) - f(0, 0) = Ah + Bk + \epsilon\eta \quad \text{.........................(1),}$$

where $\eta = \sqrt{(h^2 + k^2)}$, and $\epsilon \to 0$ as $\eta \to 0$.

If we write $h = \eta \cos\theta$, $k = \eta \sin\theta$, then the condition that η should tend to zero implies nothing as to the ratio $h : k$, which depends only on θ, and may be anything we please. We shew that for the given function, equation (1) does not hold. On substituting in equation (1) for h and k and dividing through by η, we get

$$\cos^3\theta - \sin^3\theta = A\cos\theta + B\sin\theta + \epsilon.$$

Since $\epsilon \to 0$ as $\eta \to 0$, we get, by taking the limit as $\eta \to 0$,

$$\cos^3\theta - \sin^3\theta = A\cos\theta + B\sin\theta,$$

which is plainly impossible, since θ is arbitrary.

The function is therefore not differentiable at (0, 0).

The partial derivatives exist however, for

$$f_x(0, 0) = \lim_{h \to 0} \frac{f(h, 0) - f(0, 0)}{h} = \lim_{h \to 0} \frac{h - 0}{h} = 1,$$

$$f_y(0, 0) = \lim_{k \to 0} \frac{f(0, k) - f(0, 0)}{k} = \lim_{k \to 0} \frac{-k}{k} = -1.$$

The above example illustrates a point of considerable importance. The explanation lies in the fact that the information given by the existence of the two first partial derivatives is limited. The values of $f_x(x, y)$ and of $f_y(x, y)$ depend only on the values of $f(x, y)$ along two lines through the point (x, y) respectively parallel to the axes of x and y: this information is incomplete, and tells us nothing at all about the behaviour of the function $f(x, y)$ as the point (x, y) is approached along a line which is inclined to the axis of x at any given angle θ which is not equal to 0 or $\frac{1}{2}\pi$.

A later example illustrates the fact that certain partial derivatives of a function may exist at a point at which the function is not even continuous.

9·24. Notation for partial derivatives of higher orders.

We have already defined the first partial derivatives of the function $u = u(x, y, z)$, and we have denoted them by u_x, u_y, u_z. Since these partial derivatives are in general also functions of x, y and z which may possess partial derivatives with respect to each of the three independent variables, we have the definitions:

$$\text{(i)} \quad \frac{\partial}{\partial x}\left(\frac{\partial u}{\partial x}\right) = \lim_{\Delta x \to 0} \frac{u_x(x+\Delta x, y, z) - u_x(x, y, z)}{\Delta x},$$

$$\text{(ii)} \quad \frac{\partial}{\partial y}\left(\frac{\partial u}{\partial x}\right) = \lim_{\Delta y \to 0} \frac{u_x(x, y+\Delta y, z) - u_x(x, y, z)}{\Delta y},$$

$$\text{(iii)} \quad \frac{\partial}{\partial z}\left(\frac{\partial u}{\partial x}\right) = \lim_{\Delta z \to 0} \frac{u_x(x, y, z+\Delta z) - u_x(x, y, z)}{\Delta z},$$

provided that each of these limits exists.

Notations for these three second-order partial derivatives differ with various writers: in this book the following symbolism is adopted: we write for the second-order partial derivatives

$$\text{(i)} \ \frac{\partial^2 u}{\partial x^2} \text{ or } u_{xx}, \quad \text{(ii)} \ \frac{\partial^2 u}{\partial y \partial x} \text{ or } u_{yx}, \quad \text{(iii)} \ \frac{\partial^2 u}{\partial z \partial x} \text{ or } u_{zx}.$$

By operating similarly on $\dfrac{\partial u}{\partial y}$ and $\dfrac{\partial u}{\partial z}$ we get six other second-order partial derivatives, making nine in all.

The following example illustrates the fact that *certain second partial derivatives of a function may exist at a point at which the function is not continuous.*

Example. Let
$$\phi(x, y) = \frac{x^3 + y^3}{x - y} \quad \textit{when } x \neq y,$$
$$\phi(x, y) = 0 \qquad \textit{when } x = y.$$

This function is discontinuous at the origin. To shew this it suffices to prove that if the origin is approached along different paths, $\phi(x, y)$ does not tend to the same definite limit. For, if $\phi(x, y)$ were continuous at (0, 0), $\phi(x, y)$ would tend to zero (the value of the function at the origin) by whatever path the origin were approached.

Let the origin be approached along the three curves

$$\text{(i)} \ y = x - x^2, \quad \text{(ii)} \ y = x - x^3, \quad \text{(iii)} \ y = x - x^4;$$

then we have
$$\text{(i)} \quad \phi(x, y) = \frac{2x^3 + O(x^4)}{x^2} \to 0 \text{ as } x \to 0,$$
$$\text{(ii)} \quad \phi(x, y) = \frac{2x^3 + O(x^4)}{x^3} \to 2 \text{ as } x \to 0,$$
$$\text{(iii)} \quad \phi(x, y) = \frac{2x^3 + O(x^4)}{x^4} \to \infty \text{ as } x \to 0.$$

Certain partial derivatives, however, exist at (0, 0), for if ϕ_{xx} denote $\dfrac{\partial}{\partial x}\left(\dfrac{\partial \phi}{\partial x}\right)$ we have, for example,

$$\phi_x(0, 0) = \lim_{h\to 0} \frac{\phi(h, 0) - \phi(0, 0)}{h} = \lim_{h\to 0} \frac{h^2}{h} = 0,$$

$$\phi_{xx}(0, 0) = \lim_{h\to 0} \frac{\phi_x(h, 0) - \phi_x(0, 0)}{h} = \lim_{h\to 0} \frac{2h}{h} = 2,$$

since $\qquad \phi(x, 0) = x^2, \quad \phi_x(x, 0) = 2x \qquad$ when $x \neq 0$.

9·3. Change in the order of partial derivation.

In almost all cases that occur in practice, the partial derivative has the same value in whatever order the different operations are performed. Thus, if $u = u(x, y, z)$, it is usually found that

$$u_{yz} = u_{zy}, \quad u_{zx} = u_{xz}, \quad u_{xy} = u_{yx}.$$

To fix the ideas, consider a function of two independent variables

$$u = u(x, y).$$

Since it is so often the case with functions which occur in practice, one is tempted to assume that *always*

$$u_{xy} = u_{yx}.$$

The following example illustrates the important fact that u_{xy} is not necessarily always equal to u_{yx}.

Example. Let

$$f(x, y) = \frac{xy(x^2 - y^2)}{x^2 + y^2} \quad \textit{when } x \textit{ and } y \textit{ are not simultaneously zero,}$$

$$f(0, 0) = 0;$$

shew that, at the origin, $\qquad \dfrac{\partial^2 f}{\partial x \partial y} \neq \dfrac{\partial^2 f}{\partial y \partial x}.$

When the point (x, y) is not the origin,

$$\frac{\partial f}{\partial x} = y\left\{\frac{x^2 - y^2}{x^2 + y^2} + \frac{4x^2y^2}{(x^2 + y^2)^2}\right\} \quad \text{.........................(1)},$$

$$\frac{\partial f}{\partial y} = x\left\{\frac{x^2 - y^2}{x^2 + y^2} - \frac{4x^2y^2}{(x^2 + y^2)^2}\right\} \quad \text{.........................(2)};$$

while at the origin, $\quad f_x(0, 0) = \lim_{h\to 0} \dfrac{f(h, 0) - f(0, 0)}{h} = 0 \quad$ ....(3);

and similarly $\qquad f_y(0, 0) = 0.$

From (1) and (2) we see that

$$f_x(0, y) = -y \qquad (y \neq 0) \text{ ....(4)},$$

$$f_y(x, 0) = x \qquad (x \neq 0) \text{ ....(5)}.$$

Now we have

$$f_{xy}(0, 0) = \lim_{h \to 0} \frac{f_y(h, 0) - f_y(0, 0)}{h} = \lim \frac{h}{h} = 1,$$

$$f_{yx}(0, 0) = \lim_{k \to 0} \frac{f_x(0, k) - f_x(0, 0)}{k} = \lim - \frac{k}{k} = -1,$$

and so

$$f_{xy}(0, 0) \neq f_{yx}(0, 0).$$

In the direct calculation of f_{xy} or of f_{yx} a double limit is involved. and the value of a double limit may depend upon the order in which the limiting operations are made. For example

$$\lim_{y \to 0} \lim_{x \to 0} \frac{x-y}{x+y} = \lim_{y \to 0} \left(\frac{-y}{y}\right) = -1,$$

$$\lim_{x \to 0} \lim_{y \to 0} \frac{x-y}{x+y} = \lim_{x \to 0} \left(\frac{x}{x}\right) = 1.$$

Now

$$f_x(a, y) = \lim_{h \to 0} \frac{f(a+h, y) - f(a, y)}{h},$$

$$f_{yx}(a, b) = \lim_{k \to 0} \frac{f_x(a, b+k) - f_x(a, b)}{k}$$

$$= \lim_{k \to 0} \frac{1}{k} \left\{ \lim_{h \to 0} \frac{f(a+h, b+k) - f(a, b+k)}{h} - \lim_{h \to 0} \frac{f(a+h, b) - f(a, b)}{h} \right\},$$

and so, if we write

$$\Delta^2 f(h, k) = f(a+h, b+k) - f(a, b+k) - f(a+h, b) + f(a, b) \qquad \text{......(1)},$$

then

$$f_{yx}(a, b) = \lim_{k \to 0} \lim_{h \to 0} \frac{\Delta^2 f(h, k)}{hk}$$

and

$$f_{xy}(a, b) = \lim_{h \to 0} \lim_{k \to 0} \frac{\Delta^2 f(h, k)}{hk}.$$

There is therefore no *a priori* reason why f_{yx} and f_{xy} should *always* be equal, in spite of the fact that in most practical cases it is found to be so. We now prove two theorems, the object of which is to set out precisely under what conditions it is allowable to assume that $f_{xy}(a, b) = f_{yx}(a, b)$.

THEOREM 1. *If* (i) f_x *and* f_y *exist in the neighbourhood of the point* (a, b) *and* (ii) f_x *and* f_y *are differentiable at* (a, b); *then*

$$f_{xy} = f_{yx}.$$

We shall prove this by taking equal increments h both for x and y, and calculating $\Delta^2 f$ in two different ways, where

$$\Delta^2 f = f(a+h, b+h) - f(a+h, b) - f(a, b+h) + f(a, b).$$

(*a*) $\Delta^2 f$ is the increment of the function

$$H(x) = f(x, b+h) - f(x, b)$$

when x changes from a to $a+h$. If we apply the mean value theorem to $H(x)$ we get, if $0 < \theta < 1$,

$$H(a+h) - H(a) = hH'(a+\theta h),$$

and so $$\Delta^2 f = h[f_x(a+\theta h, b+h) - f_x(a+\theta h, b)] \quad \ldots\ldots(1).$$

By hypothesis (ii) $f_x(x, y)$ is differentiable at (a, b), so that

$$f_x(a+\theta h, b+h) - f_x(a, b) = \theta h f_{xx}(a, b) + h f_{yx}(a, b) + \epsilon' h,$$

and $$f_x(a+\theta h, b) - f_x(a, b) = \theta h f_{xx}(a, b) + \epsilon'' h,$$

where ϵ' and ϵ'' tend to zero with h.

On subtraction and substitution of the difference in (1) we get

$$\Delta^2 f = h^2 f_{yx} + \epsilon_1 h^2 \quad \ldots\ldots\ldots\ldots\ldots\ldots\ldots(2),$$

where $\epsilon_1 = \epsilon' - \epsilon''$, so that ϵ_1 tends to zero with h.

(*b*) $\Delta^2 f$ is also the increment of the function

$$K(y) = f(a+h, y) - f(a, y)$$

as y increases from b to $b+h$; and by a similar argument

$$\Delta^2 f = h^2 f_{xy} + \epsilon_2 h^2 \quad \ldots\ldots\ldots\ldots\ldots\ldots\ldots(3).$$

If we divide equations (2) and (3) by h^2 and take the limit as h tends to zero, it follows that

$$\lim \frac{\Delta^2 f}{h^2} = f_{yx}(a, b) = f_{xy}(a, b).$$

Note. By assuming hypothesis (ii) in the above theorem we postulate the existence of *all* the second order partial derivatives of $f(x, y)$ at the point (a, b), but not necessarily their continuity.

An alternative set of hypotheses involves only the existence of one of the second order partial derivatives of $f(x, y)$ at (a, b) provided that we assume also its continuity. These hypotheses are made in the following theorem.

THEOREM 2. *If* (i) f_x, f_y, f_{yx} *all exist in the neighbourhood of the point* (a, b) *and* (ii) f_{yx} *is continuous at* (a, b), *then* (*a*) f_{xy} *also exists at* (a, b), *and* (*b*) $f_{xy} = f_{yx}$.

Consider the function

$$H(x) = f(x, b+k) - f(x, b),$$

and apply the mean value theorem to $H(x)$ in the interval $(a, a+h)$, then

$$H(a+h) - H(a) = h\,H'(a+\theta h)$$
$$= h[f_x(a+\theta h, b+k) - f_x(a+\theta h, b)].$$

Now, since $f_{yx}(x, y)$ exists in the neighbourhood of (a, b) we can apply the mean value theorem again to the right-hand side of this equation: this gives, if $0 < \theta' < 1$,

$$H(a+h) - H(a) = hkf_{yx}(a+\theta h, b+\theta' k).$$

By hypothesis (ii) f_{yx} is continuous at (a, b) and so the above equation may be written

$$H(a+h) - H(a) = hk[f_{yx}(a, b) + \epsilon] \quad\text{............(1)},$$

where ϵ tends to zero as h and k tend to zero.

Now $H(a+h) - H(a)$ is the function $\Delta^2 f(h, k)$ defined above, and so, if we divide equation (1) by hk and take the double limit we get

$$f_{xy}(a, b) = \lim_{h \to 0} \lim_{k \to 0} \frac{\Delta^2 f(h, k)}{hk} = f_{yx}(a, b)$$

from (1) above.

Thus f_{yx} exists and is also equal to f_{xy}.

Note. By examining the hypotheses in Theorems 1 and 2 above, the reader will see that *if* $\frac{\partial^2 f}{\partial x \partial y}$ *and* $\frac{\partial^2 f}{\partial y \partial x}$ *are both continuous functions of* x *and* y, *then these two derivatives are certainly equal*, for the assumption of the continuity of both these derivatives is a wider assumption than those required for proving either Theorem 1 or Theorem 2.

9·4. Differentials of higher orders.

Let u be a differentiable function of the *independent* variables x and y. The first differential of u,

$$du = \frac{\partial u}{\partial x} dx + \frac{\partial u}{\partial y} dy \quad\text{....................(1)},$$

is *differentiable* (dx and dy being regarded as constants), if $\frac{\partial u}{\partial x}$ and $\frac{\partial u}{\partial y}$ are determinate in the neighbourhood of the point considered and are both differentiable at this point. In this case the differential of du, which is called the *second differential* of u exists, and it is denoted by d^2u.

The second differential is calculated in the same way as the first; thus

$$d^2u = d\,(du) = d\left(\frac{\partial u}{\partial x}\right) dx + d\left(\frac{\partial u}{\partial y}\right) dy \quad \text{.........(2).}$$

To evaluate $d\left(\frac{\partial u}{\partial x}\right)$ we have only to replace, in the right-hand side of (1), the function u by the function $\frac{\partial u}{\partial x}$: a similar remark applies to $d\left(\frac{\partial u}{\partial y}\right)$. Since du is differentiable, we can use the theorems of the preceding section to justify our assumption that

$$\frac{\partial^2 u}{\partial x \partial y} = \frac{\partial^2 u}{\partial y \partial x};$$

hence

$$d^2u = \left(\frac{\partial^2 u}{\partial x^2} dx + \frac{\partial^2 u}{\partial y \partial x} dy\right) dx + \left(\frac{\partial^2 u}{\partial x \partial y} dx + \frac{\partial^2 u}{\partial y^2} dy\right) dy$$

$$= \frac{\partial^2 u}{\partial x^2} dx^2 + 2\frac{\partial^2 u}{\partial x \partial y} dx\,dy + \frac{\partial^2 u}{\partial y^2} dy^2 \quad \text{.................(3),}$$

where, of course, dx^2 is $dx \,.\, dx$.

Similarly we can define the successive differentials d^3u, d^4u, ... The function $u = u(x, y)$ is said to possess a differential of order n, $d^n u$, at the point (a, b) if $d^{n-1}u$ is differentiable at this point, which implies that all the partial derivatives of order $n-1$ exist in the neighbourhood of, and are differentiable at (a, b). This condition ensures the legitimacy of inverting the order of the partial derivatives with respect to x and with respect to y, and so

$$\frac{\partial^{n-1} u}{\partial x^r \partial y^s} = \frac{\partial^{n-1} u}{\partial y^s \partial x^r},$$

where $r + s = n - 1$.

The abbreviated notation for $d^n u$ is useful, and may be conveniently introduced here. We write

$$d^n u = \left(\frac{\partial}{\partial x} dx + \frac{\partial}{\partial y} dy\right)^n u,$$

and the complete expression is obtained by expanding the right-hand side as a binomial expansion and interpreting the powers of

∂ as indices denoting the order of the partial derivatives concerned: thus

$$d^n u = \frac{\partial^n u}{\partial x^n} dx^n + n \frac{\partial^n u}{\partial x^{n-1} \partial y} dx^{n-1} dy$$
$$+ \frac{n(n-1)}{2!} \frac{\partial^n u}{\partial x^{n-2} \partial y^2} dx^{n-2} dy^2 + \ldots + \frac{\partial^n u}{\partial y^n} dy^n.$$

The reader should observe that in the preceding discussion x and y are THE INDEPENDENT VARIABLES, and so dx and dy may be treated as constants*. It will be recalled that the differentials of the independent variables may be taken to be the arbitrary increments of these variables,

$$dx = \Delta x, \quad dy = \Delta y.$$

9·5. Differentiation of functions of functions.

So far attention has been directed solely to functions

$$u = u(x, y, z, \ldots)$$

where the variables $x, y, z, \ldots$ are the independent variables. We now consider a functional relation

$$u = u(x, y, z, \ldots),$$

in which the variables $x, y, z, \ldots$ are not the independent variables, but are themselves functions of other independent variables $r, s, t, \ldots$, so that, say

$$x = x(r, s, t, \ldots),$$
$$y = y(r, s, t, \ldots),$$
$$z = z(r, s, t, \ldots),$$
$$\ldots\ldots\ldots\ldots\ldots\ldots$$

We now prove the fundamental theorem on the permanence of the expression for the first differential.

THEOREM. *If $u = u(x, y, z, \ldots)$ is a differentiable function of the variables $x, y, z, \ldots$, and these variables are themselves differentiable functions of the independent variables $r, s, t, \ldots$, then u, considered*

* The statement that a set of variables $x, y, \ldots$ are "the independent variables," will always mean that they are the independent variables of the problem considered, and so the differentials $dx, dy, \ldots$ may be treated as constants. The word "independent" is used by some writers in a loose way, and sometimes when it is said that two variables x and y are "independent," in the sense that x does not depend on y, the word "independent" carries a slightly different shade of meaning. x and y may be *independent* of each other, and yet not be *the independent variables* of a particular problem.

as a function of $r, s, t, \ldots$, is differentiable, and its differential is given by

$$du = \frac{\partial u}{\partial x} dx + \frac{\partial u}{\partial y} dy + \frac{\partial u}{\partial z} dz + \ldots,$$

just as though $x, y, z, \ldots$ were the independent variables.

To fix the ideas, consider only two intermediary functions x and y and two independent variables r and s. The method of proof is evidently general.

Write $\rho = |\Delta x| + |\Delta y|$ and $\rho' = |\Delta r| + |\Delta s|$.

Since x and y are differentiable functions of r and s,

$$\left.\begin{aligned} \Delta x &= \frac{\partial x}{\partial r} \Delta r + \frac{\partial x}{\partial s} \Delta s + \epsilon' \rho' \\ \Delta y &= \frac{\partial y}{\partial r} \Delta r + \frac{\partial y}{\partial s} \Delta s + \epsilon'' \rho' \end{aligned}\right\} \quad \ldots\ldots\ldots\ldots\ldots(1),$$

where ϵ' and ϵ'' tend to zero as $\rho' \to 0$.

If ω denotes the greater of ϵ' and ϵ'', and M the greatest in absolute value of the four partial derivatives which occur in equations (1), then

$$|\Delta x| < (M + \omega)\rho', \quad |\Delta y| < (M + \omega)\rho',$$

and so $$\frac{\rho}{\rho'} = \frac{|\Delta x| + |\Delta y|}{\rho'} < 2(M + \omega):$$

thus the ratio $\rho : \rho'$ remains finite and $\rho \to 0$ when $\rho' \to 0$.

Since u is a differentiable function of x and y,

$$\Delta u = \frac{\partial u}{\partial x} \Delta x + \frac{\partial u}{\partial y} \Delta y + \epsilon\rho \quad \ldots\ldots\ldots\ldots\ldots\ldots(2),$$

where $\epsilon \to 0$ as $\rho \to 0$.

Now r and s are *the independent variables*, and so Δr and Δs may be replaced by dr and ds and equations (1) may be written

$$\Delta x = dx + \epsilon'\rho', \quad \Delta y = dy + \epsilon''\rho',$$

and, on substituting these values in (2),

$$\Delta u = \frac{\partial u}{\partial x} dx + \frac{\partial u}{\partial y} dy + \rho' \left(\epsilon \frac{\rho}{\rho'} + \epsilon' \frac{\partial u}{\partial x} + \epsilon'' \frac{\partial u}{\partial y} \right).$$

This proves the theorem, for the expression in the bracket tends to zero as $\rho' \to 0$, and so

$$du = \frac{\partial u}{\partial x} dx + \frac{\partial u}{\partial y} dy.$$

The above theorem establishes this fact of fundamental importance. *The first differential of a function is expressed always by the same formula, whether the variables concerned are independent or whether they are themselves functions of other independent variables.*

This is only true for the first differential. Differentials of higher orders are discussed in § 9·6 below.

9·51. The derivation of composite functions.

We deduce at once from the preceding theorem two important results.

(i) Let $x, y, \ldots$ be differentiable functions of a single independent variable t, then the derivative of $u = u(x, y, \ldots)$ with respect to t is obtained by dividing the differential du by dt,

$$\frac{du}{dt} = \frac{\partial u}{\partial x}\frac{dx}{dt} + \frac{\partial u}{\partial y}\frac{dy}{dt} + \ldots \qquad \text{.................(1).}$$

(ii) If $x, y, \ldots$ are differentiable functions of several variables $r, s, \ldots$ the partial derivatives of $u = u(x, y, \ldots)$ with respect to these variables are calculated by the preceding formula, except that all the derivatives concerned are now partial; thus

$$\left.\begin{aligned} \frac{\partial u}{\partial r} &= \frac{\partial u}{\partial x}\frac{\partial x}{\partial r} + \frac{\partial u}{\partial y}\frac{\partial y}{\partial r} + \ldots \\ \frac{\partial u}{\partial s} &= \frac{\partial u}{\partial x}\frac{\partial x}{\partial s} + \frac{\partial u}{\partial y}\frac{\partial y}{\partial s} + \ldots \\ &\ldots\ldots\ldots\ldots\ldots\ldots \end{aligned}\right\} \qquad \text{.................(2).}$$

This is immediate, for if we consider that of all the variables $r, s, \ldots$ concerned, r alone varies, then the conditions are the same as in (1), save that when t is the *only* independent variable

$$\frac{du}{dt}, \frac{dx}{dt}, \frac{dy}{dt}, \ldots,$$

are ordinary derivatives, but when r alone varies, the other variables being kept constant, the derivatives concerned will all be partial $\frac{\partial u}{\partial r}, \frac{\partial x}{\partial r}, \frac{\partial y}{\partial r}, \ldots$.

If the set of equations (2) are multiplied by $dr, ds, \ldots$ respectively, then on addition we reproduce the fundamental formula

$$du = \frac{\partial u}{\partial x}dx + \frac{\partial u}{\partial y}dy + \ldots \qquad \text{.................(3).}$$

Note. The above result illustrates the great advantage of the differential notation. From equation (3) we can obtain all the information that can be obtained as regards first order derivatives, whereas if all the possible sets of equations involving partial derivatives are written down, there would be a long list of equations every one of which is implicitly contained in (3).

For example, if

$$u=u(x, y, z), \quad v=v(x, y, z), \quad w=w(x, y, z),$$

and we differentiate these, we obtain *three* equations of the form

$$du=A_1 dx+B_1 dy+C_1 dz,$$
$$dv=A_2 dx+B_2 dy+C_2 dz,$$
$$dw=A_3 dx+B_3 dy+C_3 dz.$$

From the six variables u, v, w, x, y, z sets of three independent variables can be chosen in 20 different ways: corresponding to any one set of three independent variables there are nine equations involving partial derivatives. Thus the information conveyed by the above three equations involving differentials can be obtained only by the formation of no less than 180 equations involving partial derivatives.

9·6. Differentials of higher orders of a function of functions.

In the preceding section we have seen that if we write down the expression for the *first* differential of any function of several variables, the equation is always correct, whichever of the variables concerned may be the independent variables. The permanence of the expression for the first differential is a fact of very great importance in practice, and it may be stated in the following form.

Given a differentiable equation between a certain number of variables, independent or not, but differentiable, it is always allowable to differentiate the equation totally.*

Consider the function $u=u(x, y)$, when x and y are themselves functions of other independent variables, say r and s. We have seen that in any case, whether x and y are the independent variables or not,

$$du=\frac{\partial u}{\partial x}dx+\frac{\partial u}{\partial y}dy \quad \text{.....................(1).}$$

* The differential du of a function $u(x, y)$ is sometimes called a *total differential*, to distinguish it from the so-called *partial differentials*

$$d_x u=\frac{\partial u}{\partial x}dx \text{ and } d_y u=\frac{\partial u}{\partial y}dy.$$

It is then natural to speak of the process of finding du as "differentiating totally." The concept of partial differentials is not of much practical utility, and will not be used in this book. The term "total differential" may however be used occasionally.

In the formation of the second differential d^2u, however, there is an important difference when x and y are not the independent variables.

Since x and y are not the independent variables, dx and dy are no longer constants, but can be themselves differentiated and d^2x, d^2y both exist. On differentiating equation (1) we get

$$d^2u = d\left(\frac{\partial u}{\partial x}\right)dx + \frac{\partial u}{\partial x}d^2x + d\left(\frac{\partial u}{\partial y}\right)dy + \frac{\partial u}{\partial y}d^2y,$$

and on comparison with equations (2) and (3) of § 9·4, we see that

$$d^2u = \frac{\partial^2 u}{\partial x^2}dx^2 + 2\frac{\partial^2 u}{\partial x \partial y}dx\,dy + \frac{\partial^2 u}{\partial y^2}dy^2 + \frac{\partial u}{\partial x}d^2x + \frac{\partial u}{\partial y}d^2y \quad \ldots(2).$$

The formation of differentials of higher orders follows the same law, but the formulae become more lengthy and complicated. No simple general formula for d^nu can be given. The introduction of more than two intermediary variables causes no difficulty: thus when $u = u(x, y, z)$, and x, y, z are *not* the independent variables,

$$d^2u = \left(\frac{\partial}{\partial x}dx + \frac{\partial}{\partial y}dy + \frac{\partial}{\partial z}dz\right)^2 u + \frac{\partial u}{\partial x}d^2x + \frac{\partial u}{\partial y}d^2y + \frac{\partial u}{\partial z}d^2z \quad \ldots(3).$$

The reader should notice that if r and s are the independent variables, we have equations of the form

$$dx = R_1 dr + S_1 ds, \quad dy = R_2 dr + S_2 ds,$$
$$d^2x = F_1 dr^2 + 2G_1 dr\,ds + H_1 ds^2,$$
$$d^2y = F_2 dr^2 + 2G_2 dr\,ds + H_2 ds^2;$$

and, on substituting for these in equation (2), we get

$$d^2u = A\,dr^2 + 2B\,dr\,ds + C\,ds^2,$$

which expresses d^2u as a quadratic function of the differentials dr and ds of the *independent* variables.

9·61. Example.

Prove that, by the transformations $u = x - ct$, $v = x + ct$, *the partial differential equation*

$$\frac{\partial^2 y}{\partial t^2} = c^2 \frac{\partial^2 y}{\partial x^2}$$

reduces to

$$\frac{\partial^2 y}{\partial u \partial v} = 0,$$

and hence solve the equation.

In this problem x and t are the independent variables, and we wish to express y as a function of u and v, where u and v are functions of the independent variables x and t.

Now $$d^2y = \frac{\partial^2 y}{\partial u^2} du^2 + 2\frac{\partial^2 y}{\partial u \partial v} du dv + \frac{\partial^2 y}{\partial v^2} dv^2 + \frac{\partial y}{\partial u} d^2u + \frac{\partial y}{\partial v} d^2v, \quad \text{......(1)},$$

and we have to express d^2y in terms of the differentials dx and dt.

Now $$du = dx - cdt, \qquad dv = dx + cdt,$$
$$d^2u = 0, \qquad d^2v = 0^*.$$

On substitution in (1) we get

$$d^2y = \left\{\frac{\partial^2 y}{\partial u^2} + 2\frac{\partial^2 y}{\partial u \partial v} + \frac{\partial^2 y}{\partial v^2}\right\} dx^2 + B\,dx\,dt + c^2\left\{\frac{\partial^2 y}{\partial u^2} - 2\frac{\partial^2 y}{\partial u \partial v} + \frac{\partial^2 y}{\partial v^2}\right\} dt^2 \quad \text{...(2)},$$

the coefficient of $dx\,dt$ being written as B, since its actual value is not required for this problem.

Now $\frac{\partial^2 y}{\partial x^2}$ is the coefficient of dx^2 in equation (2), and so

$$\frac{\partial^2 y}{\partial x^2} = \frac{\partial^2 y}{\partial u^2} + 2\frac{\partial^2 y}{\partial u \partial v} + \frac{\partial^2 y}{\partial v^2},$$

and similarly, on taking the coefficient of dt^2 in (2),

$$\frac{\partial^2 y}{\partial t^2} = c^2\left(\frac{\partial^2 y}{\partial u^2} - 2\frac{\partial^2 y}{\partial u \partial v} + \frac{\partial^2 y}{\partial v^2}\right).$$

If these values are substituted in the given differential equation

$$\frac{\partial^2 y}{\partial t^2} = c^2\frac{\partial^2 y}{\partial x^2},$$

it follows that $$\frac{\partial^2 y}{\partial u \partial v} = 0.$$

The solution of this equation is simple, for the equation

$$\frac{\partial}{\partial u}\left(\frac{\partial y}{\partial v}\right) = 0$$

implies that $\frac{\partial y}{\partial v}$ is a function of v alone: and so the general solution of the differential equation takes the form

$$y = F(u) + f(v),$$

where F and f are arbitrary functions. The general solution of the given equation is consequently

$$y = F(x - ct) + f(x + ct).$$

The differential equation which has just been solved is the "wave equation," and is very important in Mathematical Physics.

* Note that d^2u and d^2v are zero merely because u and v are *linear* functions of x and t; and, in general, if u and v had been other than linear functions of x and t, d^2u and d^2v would not have been zero. Equation (1) above would have been incorrect if the last two terms of it had been omitted, although in this particular case they vanish.

9·7. Euler's theorem on homogeneous functions.

A function $f(x, y, z, \ldots)$ is a homogeneous function of degree n if it has the property

$$f(tx, ty, tz, \ldots) = t^n f(x, y, z, \ldots) \quad \ldots\ldots\ldots\ldots(1).$$

(1) To prove Euler's theorem, write $x' = tx, y' = ty, \ldots$, then

$$f(x', y', z', \ldots) = t^n f(x, y, z, \ldots) \quad \ldots\ldots\ldots\ldots(2),$$

and if we take the partial derivative with respect to t, we get

$$\frac{\partial f}{\partial x'}\frac{\partial x'}{\partial t} + \frac{\partial f}{\partial y'}\frac{\partial y'}{\partial t} + \ldots = nt^{n-1} f(x, y, z, \ldots),$$

that is
$$x\frac{\partial f}{\partial x'} + y\frac{\partial f}{\partial y'} + \ldots = nt^{n-1} f(x, y, z, \ldots).$$

Now put $t = 1$, so that $x' = x$, $y' = y, \ldots$ and we get

$$x\frac{\partial f}{\partial x} + y\frac{\partial f}{\partial y} + \ldots = nf(x, y, \ldots) \quad \ldots\ldots\ldots(3),$$

which is known as *Euler's theorem*.

(2) Differentiate the equation (2) m times; since t is the only independent variable, we have

$$d^m f = n(n-1)\ldots(n-m+1)\, t^{n-m} f(x, y, z, \ldots)\,.\,dt^m.$$

Now
$$d^m f = \frac{\partial^m f}{\partial x'^m} dx'^m + \frac{\partial^m f}{\partial y'^m} dy'^m + \ldots,$$

and since $x'^m = t^m x^m$, $y'^m = t^m y^m, \ldots$, when $t = 1$ we get

$$\left(x\frac{\partial}{\partial x} + y\frac{\partial}{\partial y} + \ldots\right)^m f(x,y,z,\ldots) = n(n-1)\ldots(n-m+1) f(x,y,z,\ldots)$$

which is *the generalisation of Euler's theorem*.

9·8. Taylor's theorem.

In view of Taylor's theorem for functions of one variable, it is not unnatural to expect the possibility of expanding a function of more than one variable $f(x+h,\ y+k,\ z+l, \ldots)$ in a series of ascending powers of $h, k, l, \ldots$. To fix the ideas, consider a function of two variables only; the reasoning in the general case is precisely the same.

Consider a circular domain of centre (a, b) and radius large enough for the point $(a+h, b+k)$ to be also within the domain. Suppose that $f(x, y)$ is a function such that *all the partial derivatives of order n of $f(x, y)$ are continuous in the domain.* Write

$$x = a + ht, \quad y = b + kt,$$

so that, as t ranges from 0 to 1, the point (x, y) moves along the line joining the point (a, b) to the point $(a+h, b+k)$; then

$$f(x, y) = f(a+ht, b+kt) = \phi(t).$$

Now $$\phi'(t) = \frac{\partial f}{\partial x}\frac{dx}{dt} + \frac{\partial f}{\partial y}\frac{dy}{dt} = h\frac{\partial f}{\partial x} + k\frac{\partial f}{\partial y} = df,$$

and similarly $$\phi''(t) = d^2 f, \quad \ldots, \quad \phi^{(n)}(t) = d^n f.$$

We thus see that $\phi(t)$ and its first n derivatives are continuous functions of t in the interval $0 \leqslant t \leqslant 1$, and so, by Maclaurin's theorem,

$$\phi(t) = \phi(0) + t\phi'(0) + \frac{t^2}{2!}\phi''(0) + \ldots + \frac{t^n}{n!}\phi^{(n)}(\theta t),$$

where $0 < \theta < 1$. Now put $t = 1$ and observe that

$$\phi(1) = f(a+h, b+k), \quad \phi(0) = f(a, b), \quad \phi'(0) = df(a, b),$$

$$\phi''(0) = d^2 f(a, b), \quad \ldots, \quad \phi^{(n)}(\theta t) = d^n f(a+\theta h, b+\theta k).$$

It follows immediately that

$$f(a+h, b+k) = f(a, b) + df(a, b) + \frac{1}{2!}d^2 f(a, b) + \ldots$$

$$+ \frac{1}{(n-1)!}d^{n-1}f(a, b) + R_n \quad \ldots\ldots(1),$$

where $$R_n = \frac{1}{n!}d^n f(a+\theta h, b+\theta k), \; 0 < \theta < 1.$$

Note. The reader should compare the assumptions made here about $f(x, y)$ with those made about $f(x)$ for the application of Taylor's theorem to functions of one variable in § 4·45. It was there found to be sufficient that $f^{(n)}(x)$ should exist in $a < x < a+h$, but here we have assumed that all the partial derivatives of order n are *continuous* in the domain in question. It can be shewn by examples that the Taylor expansion above does not necessarily hold if these derivatives are not continuous*.

* See Examples IX, 13.

9·81. Maclaurin's theorem.

If we put $a=b=0$, $h=x$, $k=y$, we get at once, from equation (1) above,

$$f(x, y)=f(0,0)+df(0,0)+\frac{1}{2!}d^2f(0,0)+\dots$$
$$+\frac{1}{(n-1)!}d^{n-1}f(0,0)+R_n,$$

where $$R_n=\frac{1}{n!}d^nf(\theta x, \theta y),\ 0<\theta<1.$$

The theorems easily extend to any number of variables.

EXAMPLES IX.

1. If $u=(1-2xy+y^2)^{-\frac{1}{2}}$, prove that

$$\frac{\partial}{\partial x}\left\{(1-x^2)\frac{\partial u}{\partial x}\right\}+\frac{\partial}{\partial y}\left(y^2\frac{\partial u}{\partial y}\right)=0.$$

2. Prove that $\frac{\partial^2 u}{\partial x^2}+\frac{\partial^2 u}{\partial y^2}$ is invariant for change of rectangular axes.

[If the axes be turned through an angle α, then

$$x=x'\cos\alpha-y'\sin\alpha,\quad y=x'\sin\alpha+y'\cos\alpha:$$

shew that the above expression becomes $\frac{\partial^2 u}{\partial x'^2}+\frac{\partial^2 u}{\partial y'^2}$.]

3. If $x=r\cos\theta$, $y=r\sin\theta$, prove that

(i) $\frac{\partial^2\theta}{\partial x\,\partial y}=-\frac{\cos 2\theta}{r^2}$;

(ii) $(x^2-y^2)\left(\frac{\partial^2 u}{\partial x^2}-\frac{\partial^2 u}{\partial y^2}\right)+4xy\frac{\partial^2 u}{\partial x\,\partial y}=r^2\frac{\partial^2 u}{\partial r^2}-r\frac{\partial u}{\partial r}-\frac{\partial^2 u}{\partial\theta^2}$,

where u is any twice-differentiable function of x and y.

4. Prove that the function

$$f(x, y)=(|xy|)^{\frac{1}{2}}$$

is not differentiable at the point (0, 0), but that $\frac{\partial f}{\partial x}$ and $\frac{\partial f}{\partial y}$ both exist at the origin and have the value 0.

Hence deduce that these two partial derivatives are continuous except at the origin.

5. Given that $x_1, x_2, \ldots, x_n$ are n independent variables, c is a constant $p_1, p_2, \ldots, p_n$ are n functions of u which is itself a function of the x's, and that

$$p_1^2+p_2^2+\ldots+p_n^2=0,$$
$$x_1p_1+x_2p_2+\ldots+x_np_n=cu,$$

shew that
$$\frac{\partial^2 u}{\partial x_1^2}+\frac{\partial^2 u}{\partial x_2^2}+\ldots+\frac{\partial^2 u}{\partial x_n^2}=0.$$

6. Given that z is a function of x and y and that

(i)
$$x=e^u+e^{-v},\quad y=e^v+e^{-u},$$

prove that

$$\frac{\partial^2 z}{\partial u^2}-2\frac{\partial^2 z}{\partial u\partial v}+\frac{\partial^2 z}{\partial v^2}=x\frac{\partial z}{\partial x}+y\frac{\partial z}{\partial y}+x^2\frac{\partial^2 z}{\partial x^2}-2xy\frac{\partial^2 z}{\partial x\partial y}+y^2\frac{\partial^2 z}{\partial y^2}.$$

(ii) If $x=u+v$, $y=uv$, prove that

$$\frac{\partial^2 z}{\partial u^2}-2\frac{\partial^2 z}{\partial u\partial v}+\frac{\partial^2 z}{\partial v^2}=(x^2-4y)\frac{\partial^2 z}{\partial y^2}-2\frac{\partial z}{\partial y}.$$

7. Change the independent variables from x and y to u and v in

$$\text{(i)}\quad (x^2+y^2)\left(\frac{\partial^2 z}{\partial x^2}+\frac{\partial^2 z}{\partial y^2}\right)+4xy\frac{\partial^2 z}{\partial x\partial y}+2x\frac{\partial z}{\partial x}+2y\frac{\partial z}{\partial y}=0,$$

if $2x=e^u+e^v$, $2y=e^u-e^v$; and in

$$\text{(ii)}\quad \frac{\partial^2 z}{\partial x^2}+2xy^2\frac{\partial z}{\partial x}+2(y-y^3)\frac{\partial z}{\partial y}+x^2y^2z=0,$$

if $x=uv$, $y=1/v$, deducing that z is the same function of u, v as of x, y.

8. Given that z is a function of u and v, while

$$u=x^2-y^2-2xy,\quad v=y,$$

prove that the equation

$$(x+y)\frac{\partial z}{\partial x}+(x-y)\frac{\partial z}{\partial y}=0$$

is equivalent to $\dfrac{\partial z}{\partial v}=0$.

Hence deduce that $z=f(x^2-y^2-2xy)$.

9. Prove that $f_{xy}\neq f_{yx}$ at the origin for the function

$$f(x, y)=x^2\arctan\frac{y}{x}-y^2\arctan\frac{x}{y}.$$

10. Shew that
$$\frac{d^n}{dx^n}(x^{n-1}e^{1/x})=(-)^n\frac{e^{1/x}}{x^{n+1}}.$$

11. Prove that, whatever the functions f and g,

(i) $z=xf(x+y)+yg(x+y)$

satisfies the relation* $r-2s+t=0$;

(ii) $z=xf(y/x)+g(y/x)$

satisfies the relation $rx^2+2sxy+ty^2=0$.

* p, q, r, s, t are defined in § 10·31.

12. If $f(x, y)=e^{ax}\cos by$, prove that the first four terms of the Maclaurin expansion of $f(x, y)$ are

$$1+ax+\frac{a^2x^2-b^2y^2}{2!}+\frac{a^3x^3-3ab^2xy^2}{3!}+\ldots,$$

and obtain the first four terms of the Taylor expansion of $f(x+h, y+k)$ in powers of h and k.

Discuss the validity of these expansions.

13. If $f(x, y)=(|xy|)^{\frac{1}{2}}$, prove that the Taylor expansion about the point (x, x) is not valid in any domain which includes the origin. Give reasons.

[See example 4. If a Taylor expansion were possible $(n=1)$

$$f(x+h, x+h)=f(x, x)+h\{f_x(\xi, \xi)+f_y(\xi, \xi)\},$$

where $x<\xi<x+h$. This is not valid for all x, h for it implies that

$$\begin{aligned}|x+h|&=|x|\pm h, & \xi&\neq 0,\\ &=|x|, & \xi&=0.]\end{aligned}$$

14. Shew that, if $f(x, y)$, $\dfrac{\partial f}{\partial x}$, $\dfrac{\partial f}{\partial y}$ are all continuous in a circular domain D of centre (a, b) and radius large enough for the point $(a+h, b+k)$ to be within D,

$$f(a+h, b+k)=f(a, b)+hf_x(a+\theta h, b+\theta k)+kf_y(a\quad\theta h, b+\theta k)$$

where $0<\theta<1$.

If $f(x, y)=x\sqrt{(x^2+y^2)}$, $a=b=-1$, $h=k=3$, verify that the above conditions are satisfied and find the value of θ.

15. If
$$z=x^3\phi(y/x)+y^{-3}\psi(y/x),$$
prove that
$$rx^2+2sxy+ty^2+px+qy=9z.$$

16. If $f(x, y, z)$ is a homogeneous function of x, y, z of degree n, prove that

$$\begin{vmatrix} f_{xx}, & f_{xy}, & f_{xz}\\ f_{yx}, & f_{yy}, & f_{yz}\\ f_{zx}, & f_{zy}, & f_{zz}\end{vmatrix}=\frac{(n-1)^2}{z^2}\begin{vmatrix} f_{xx}, & f_{xy}, & f_x\\ f_{yx}, & f_{yy}, & f_y\\ f_x, & f_y, & nf/(n-1)\end{vmatrix}.$$

CHAPTER X

IMPLICIT FUNCTIONS

10·1. Existence theorems for implicit functions.

Let
$$F(x_1, x_2, \ldots, x_n, u) = 0 \qquad \ldots\ldots\ldots\ldots(1)$$
be a functional relation between the $n+1$ variables $x_1, \ldots, x_n, u$, and let
$$x_1 = a_1, \ldots, x_n = a_n$$
be a set of values such that the equation
$$F(a_1, \ldots, a_n, u) = 0 \qquad \ldots\ldots\ldots\ldots(2)$$
is satisfied for at least one value of u; that is, the equation (2) in u has at least one root. We may consider u as a function of the x's, $u = \phi(x_1, x_2, \ldots, x_n)$ defined in a certain domain where $\phi(x_1, x_2, \ldots, x_n)$ has assigned to it at any point $(x_1, x_2, \ldots, x_n)$ the roots u of equation (1) at this point. We say that u is the *implicit function* defined by (1): it is, in general, a many-valued function.

More generally, consider the set of equations
$$F_p(x_1, \ldots, x_n, u_1, \ldots, u_m) = 0 \qquad (p = 1, 2, \ldots, m) \ \ldots(3)$$
between the $n+m$ variables $x_1, \ldots, x_n, u_1, \ldots, u_m$, and suppose that the set of equations (3) are such that there are points $(x_1, x_2, \ldots, x_n)$ for which these m equations are satisfied for at least one set of values $u_1, u_2, \ldots, u_m$. We may consider the u's as functions of the x's,
$$u_p = \phi_p(x_1, x_2, \ldots, x_n) \qquad (p = 1, 2, \ldots, m)$$
where the functions ϕ have assigned to them at the point $(x_1, x_2, \ldots, x_n)$ the values of the roots $u_1, u_2, \ldots, u_m$ at this point. We say that $u_1, u_2, \ldots, u_m$ constitute a *system of implicit functions* defined by the set of equations (3). These functions are, in general, many-valued.

We now prove an existence theorem for implicit functions; for simplicity we consider only three variables, but the method of proof is evidently general.

THEOREM 1. *Let $F(u, x, y)$ be a continuous function of the variables u, x, y. At the point (u_0, a, b) suppose that*

(i) $F(u_0, a, b) = 0$,

(ii) $F(u, x, y)$ *is differentiable,*

(iii) *the partial derivative* $\dfrac{\partial F}{\partial u}$ *is not zero,*

then there exists at least one function $u = u(x, y)$ reducing to u_0 at the point (a, b), and which, in the neighbourhood of this point, satisfies the equation $F(u, x, y) = 0$ identically.

Also, every function u which possesses these two properties is continuous and differentiable at the point (a, b).

Since $F = 0$, $\dfrac{\partial F}{\partial u} \neq 0$ at the point (u_0, a, b), a positive number δ can be found such that $F(u_0 - \delta, a, b)$ and $F(u_0 + \delta, a, b)$ have opposite signs, for the function F is either an increasing or a decreasing function of u when $u = u_0$. Further, since F is continuous, a positive number η can be found so that the functions

$$F(u_0 - \delta, x, y), \quad F(u_0 + \delta, x, y),$$

the values of which may be as near as we please to

$$F(u_0 - \delta, a, b), \quad F(u_0 + \delta, a, b),$$

will also have opposite signs so long as $|x - a| < \eta$, $|y - b| < \eta$.

Let x, y be any two values satisfying the above conditions; then $F(u, x, y)$ is a continuous function of u which changes sign between $u_0 - \delta$ and $u_0 + \delta$ and so vanishes somewhere in this interval.

Let $u = u(x, y)$ be the root (the *greatest* root if there is more than one); then this will be one solution of $F = 0$ which reduces to u_0 at the point (a, b).

Suppose that Δu, Δx, Δy are the increments of such a function u and of the variables x and y measured from the point (a, b). Since F is differentiable at (u_0, a, b), we have

$$\Delta F = \{F_u(u_0, a, b) + \epsilon\} \Delta u + \{F_x(u_0, a, b) + \epsilon'\} \Delta x + \{F_y(u_0, a, b) + \epsilon''\} \Delta y = 0,$$

for since $F = 0$, it follows that $\Delta F = 0$, and the numbers ϵ tend to zero with Δu, Δx and Δy, and can be made as small as we please with δ and η. Let δ and η be so small that the numbers ϵ are all less than $\frac{1}{2} |F_u(u_0, a, b)|$, which is not zero by hypothesis.

The above equation then shews that $\Delta u \to 0$ as $\Delta x \to 0$ and $\Delta y \to 0$, in other words the function $u = u(x, y)$ is *continuous* at (a, b).

It also shews that u is *differentiable* at (a, b), for

$$\Delta u = -\frac{\{F_x(u_0, a, b) + \epsilon'\}\Delta x + \{F_y(u_0, a, b) + \epsilon''\}\Delta y}{F_u(u_0, a, b) + \epsilon},$$

that is,

$$\Delta u = -\frac{F_x(u_0, a, b)}{F_u(u_0, a, b)}\Delta x - \frac{F_y(u_0, a, b)}{F_u(u_0, a, b)}\Delta y + \epsilon_1 \Delta x + \epsilon_2 \Delta y,$$

ϵ_1 and ϵ_2 tending to zero with Δx and Δy.

COROLLARY 1. *If* $\frac{\partial F}{\partial u}$ *exists and is not zero in the neighbourhood of the point* (u_0, a, b), *the solution* u *of the equation* $F = 0$ *is unique.*

For, if there were two solutions u_1 and u_2 we should have, by the mean-value theorem, if $u_1 < u' < u_2$,

$$0 = F(u_1, x, y) - F(u_2, x, y) = (u_1 - u_2) F_u(u', x, y),$$

and so $F_u(u, x, y)$ would vanish at some point in the neighbourhood of (u_0, a, b), and this is contrary to hypothesis.

COROLLARY 2. *If* $F(u, x, y)$ *is differentiable in the neighbourhood of* (u_0, a, b), *the function* $u = u(x, y)$ *is differentiable in the neighbourhood of the point* (a, b).

This is immediate, for the preceding proof is then applicable at every point (u, x, y) in that neighbourhood.

The above theorem, like most existence theorems in Analysis, is of an abstract nature, and the reader will probably appreciate the theorem better by considering the geometrical illustration, taking for simplicity the case of a function of two variables only*, say $F(u, x) = 0$.

Corollary 1 is of great importance, for by considering a function of two variables only, $F(u, x) = 0$, and taking $F(u, x) = f(u) - x$, we can enunciate *the fundamental theorem on inverse functions* as follows.

THEOREM 2. *If, in the neighbourhood of* $u = u_0$, *the function* $f(u)$ *is a continuous function of* u, *and if* (i) $f(u_0) = a$, (ii) $f'(u) \neq 0$ *in the neighbourhood of the point*† $u = u_0$, *then there exists a unique*

* See, for example, Hardy's *Pure Mathematics*, § 108.

† That is, $f(u)$ is a steadily increasing (or decreasing) function in this neighbourhood.

continuous function $u = \phi(x)$, which is equal to u_0 when $x = a$, and which satisfies identically the equation

$$f(u) - x = 0,$$

in the neighbourhood of the point $x = a$.

The function $u = \phi(x)$ thus defined is called the *inverse function* of $x = f(u)$.

Examples. If $x = u^3$, $u_0 = 0$, $a = 0$, then all the conditions of Theorem 2 are satisfied, and $u = \sqrt[3]{x}$.

Under the same conditions, if $x = u^2$ the conditions of Theorem 2 are not satisfied, for u^2 is not an always increasing (or always decreasing) function of u in the neighbourhood of the origin, since it decreases when u is negative and increases when u is positive. In fact the equation $u^2 = x$ defines two functions of x,

$$u = \sqrt{x} \text{ and } u = -\sqrt{x}.$$

These functions both vanish when $x = 0$, and each is defined for positive values of x only, so that the equation $u^2 = x$ has sometimes two solutions and sometimes none.

10·11. The logarithmic function.

So far no definition of the logarithmic function has been given, although, like the circular functions*, it has already been used to enrich our examples. To illustrate the application of the inverse function theorem, it is convenient here to define the logarithmic function, and then to shew how to deduce the properties of the exponential function from it by means of the inverse function theorem of the preceding section.

Let us adopt the method of defining the logarithm by the equation

$$\log x = \int_1^x \frac{dt}{t}, \quad x > 0.$$

Since the properties of definite integrals have been discussed in Chapter VII, most of the fundamental properties of the logarithmic function can be proved directly from these properties.

(1) By the fundamental theorem of the integral calculus, $\log x$ is a continuous function of x in the interval $(1, x)$, it increases steadily as x increases, and it has a derivative

$$\frac{d}{dx}\log x = \frac{1}{x}.$$

* See § 4·33.

Also, if $x > 2^n$,

$$\log x > \int_1^{2^n} \frac{dt}{t} = \int_1^2 \frac{dt}{t} + \int_2^4 \frac{dt}{t} + \dots + \int_{2^{n-1}}^{2^n} \frac{dt}{t},$$

but, if $t = 2^r u$,

$$\int_{2^r}^{2^{r+1}} \frac{dt}{t} = \int_1^2 \frac{du}{u},$$

hence

$$\log x > n \int_1^2 \frac{du}{u},$$

and so $\log x \to \infty$ as $x \to \infty$.

(2) *If* $0 < x < 1$, *then* $\log x < 0$, for

$$\log x = \int_1^x \frac{dt}{t} = -\int_x^1 \frac{dt}{t} < 0;$$

and, by writing $t = 1/u$, we get

$$\log x = \int_1^x \frac{dt}{t} = -\int_1^{1/x} \frac{du}{u} = -\log(1/x),$$

and so $\log x \to -\infty$ as $x \to 0+0$.

(3) *The functional equation for* $\log x$ *is*

$$\log(xy) = \log x + \log y.$$

Now

$$\log(xy) = \int_1^{xy} \frac{dt}{t} = \int_{1/y}^{x} \frac{du}{u},$$

if $t = yu$, hence

$$\log(xy) = \int_1^x \frac{du}{u} - \int_1^{1/y} \frac{du}{u} = \log x - \log(1/y)$$

$$= \log x + \log y.$$

(4) *The number e.*

This number is conveniently defined* by the equation

$$1 = \int_1^e \frac{dt}{t},$$

and the definition is unique, since $\log x$ increases steadily with x, and so it can pass only once through the value 1.

From the functional equation above it follows that

$$\log x^2 = 2\log x,\ \log x^3 = 3\log x,\ \dots,\ \log x^n = n\log x,$$

where n is a positive integer. Hence

$$\log e^n = n \log e = n.$$

* Another definition of e, as a limit, has already been given in § 2·8. The two definitions are shewn to be equivalent in § 10·13 below.

If p and q are positive integers and $e^{p/q}$ denotes the positive qth root of e^p, we have

$$p = \log e^p = \log (e^{p/q})^q = q \log e^{p/q}.$$

Thus, whatever *positive rational value* y may have,

$$\log e^y = y.$$

Also, since

$$\log e^{-y} = -\log e^y = -y,$$

the above equation extends to all *rational* values of y.

Since e^y when y is irrational may be defined by Cantor's method as described in § 6·3, if y_n is an appropriate rational approximation to y, from the equation

$$\log e^{y_n} = y_n,$$

we deduce, by the limiting process of Cantor's definition, the equation

$$\log e^y = y$$

*for all real values of y**.

10·12. The exponential function. Application of the inverse function theorem.

We now define the exponential function $\exp y$, for all real values of y, as the inverse function of $y = \log x$. That is, if $y = \log x$, we write

$$x = \exp y.$$

As x ranges from 0 to ∞, y ranges from $-\infty$ to ∞ and increases steadily throughout this range, so to each value of x there corresponds one value of y and conversely.

If $a > 0$, then, in the neighbourhood of any point $x = a$, the function $F(y, x) \equiv y - \log x = 0$ satisfies all the conditions of Theorem 2 (by taking account of differences of notation); and this justifies the assumption that a unique inverse function $x = \exp y$ exists. Since, moreover, Theorem 2 is a special case of Theorem 1 we can deduce for the exponential function (i) continuity and (ii) differentiability in the neighbourhood of the point $y = b$, where b is the value of the function $y = \log x$ when $x = a$. It is therefore unnecessary to prove for the function $\exp y$ any of the properties which have already been proved for the function $\log x$ and which can be deduced from the implicit function theorem.

* This assumes, of course, the continuity of the logarithmic function.

If e be the number defined (as above) to be the number whose logarithm is unity, the functions $\exp y$ and e^y are identical for all real values of y.

An alternative method of developing the theory of the exponential and logarithmic functions is to take as the *definition* of the exponential function the result that $\exp x$ is the sum-function of the absolutely convergent power series

$$1+x+\frac{x^2}{2!}+\frac{x^3}{3!}+\dots \quad\dots\dots\dots(1)$$

for all values of x. By using the multiplication theorem for absolutely convergent series (§ 5·7) it can be proved that

$$\exp x \,.\, \exp y = \exp(x+y) \quad\dots\dots\dots(2).$$

Also
$$\frac{\exp h - 1}{h} = 1 + \frac{h}{2!} + \frac{h^2}{3!} + \dots = 1 + \phi(h) \text{ (say)}.$$

Now
$$|\tfrac{1}{2}h| + |\tfrac{1}{2}h|^2 + |\tfrac{1}{2}h|^3 + \dots = |\tfrac{1}{2}h|/(1-|\tfrac{1}{2}h|),$$

and since $\phi(h)$ is numerically less than this series,

$$\phi(h) \to 0 \text{ as } h \to 0.$$

Hence
$$\frac{\exp(x+h) - \exp x}{h} = \exp x \left(\frac{\exp h - 1}{h}\right),$$

and by taking the limit as $h \to 0$,

$$\frac{d}{dx}\exp x = \exp x \quad\dots\dots\dots(3).$$

Incidentally we have implicitly proved that $\exp x$ is a continuous function of x. From (2), if m and n are positive integers,

$$(\exp x)^n = \exp nx, \quad (\exp 1)^n = \exp n,$$
$$\{\exp(m/n)\}^n = \exp m = (\exp 1)^m,$$

and so $\exp(m/n)$ is the positive value of $(\exp 1)^{m/n}$.

Also
$$\exp x \,.\, \exp(-x) = 1,$$

and so for all ***rational*** values of x we have

$$\exp x = (\exp 1)^x = e^x,$$

where
$$e = \exp 1 = 1 + 1 + \frac{1}{2!} + \frac{1}{3!} + \dots.$$

We define e^x when x is irrational as being equal to $\exp x$. The logarithm is then defined as the function inverse to $\exp x$ or e^x.

The method adopted in this book is preferable, because it depends upon the concept of a definite integral, and this itself may be defined as soon as the concept of a bound is understood. The alternative method sketched here depends upon a knowledge of operations on infinite series, and infinite series themselves require a knowledge of limiting operations.

10·13. The identification of the exponential function e^x with $\lim_{n\to\infty}\left(1+\frac{x}{n}\right)^n$.

In § 2·8 we have already proved that if s be the sum of the convergent series

$$1+x+\frac{x^2}{2!}+\frac{x^3}{3!}+\dots,$$

then

$$s=\lim_{n\to\infty}\left(1+\frac{x}{n}\right)^n.$$

It remains to shew that if e^x be the exponential function as defined above*,

$$\lim_{n\to\infty}\left(1+\frac{x}{n}\right)^n=\lim_{n\to\infty}\left(1-\frac{x}{n}\right)^{-n}=e^x \quad \text{.............(1).}$$

Now by § 10·11 (1)

$$\lim_{h\to 0}\frac{\log(1+xh)}{h}=x.$$

If $h=1/t$,

$$\lim t\log\left(1+\frac{x}{t}\right)=x,$$

as $t\to\infty$ or as $t\to-\infty$. Since the exponential function is continuous, we get

$$\left(1+\frac{x}{t}\right)^t=e^{t\log\{1+(x/t)\}}\to e^x,$$

as $t\to\infty$ or as $t\to-\infty$.

Hence

$$\lim_{t\to\infty}\left(1+\frac{x}{t}\right)^t=\lim_{t\to-\infty}\left(1+\frac{x}{t}\right)^t=e^x,$$

and (1) follows at once by making t range through *integral* values only.

The function $\log x$ may also be expressed as a limit,

$$\log x=\lim_{n\to\infty}n(1-x^{-1/n})=\lim_{n\to\infty}n(x^{1/n}-1);$$

for

$$n(x^{1/n}-1)-n(1-x^{-1/n})=n(x^{1/n}-1)(1-x^{-1/n}),$$

the right-hand side of which may be shewn to tend to zero.

Also, if n is a positive integer and $x>1$,

$$\int_1^x\frac{dt}{t^{1+(1/n)}}<\int_1^x\frac{dt}{t}<\int_1^x\frac{dt}{t^{1-(1/n)}},$$

or

$$n(1-x^{-1/n})<\log x<n(x^{1/n}-1).$$

* The proof here given is on the lines of one given in Hardy's *Pure Mathematics*, p. 368.

10·2. Differentiation of implicit functions.

To pass from the consideration of explicit to that of implicit functions, let $u = u(x, y, \ldots)$ be a function of the independent variables $x, y, \ldots$. If the function u reduces to a constant, then Δu, and therefore also du, is zero. Conversely, if

$$du = \frac{\partial u}{\partial x} dx + \frac{\partial u}{\partial y} dy + \ldots = 0,$$

since $dx, dy, \ldots$ are arbitrary it follows that

$$\frac{\partial u}{\partial x} = \frac{\partial u}{\partial y} = \ldots = 0,$$

in other words u is a constant.

Hence the necessary and sufficient condition that the function u should reduce to a constant is that $du = 0$.

Now suppose that the set of variables $x, y, \ldots$, *independent or not*, satisfy a functional relation

$$f(x, y, \ldots) = 0.$$

This equation is said to be *differentiable*, if the function $f(x, y, \ldots)$ is differentiable. Hence, if the variables concerned, $x, y, \ldots$ are differentiable we have, since $df = 0$,

$$\frac{\partial f}{\partial x} dx + \frac{\partial f}{\partial y} dy + \ldots = 0.$$

This result is fundamental, and may be expressed as follows.

Given a differentiable equation between a certain number of variables, it is always permissible to differentiate the equation, whether the variables concerned are the independent variables or not.

The above principle enables us to deal with implicit functions by differentiating the equations which define them. It must be emphasised, however, that the results of differentiating implicit functions depend for their interpretation on a knowledge of which are the independent and which the dependent variables.

(1) To fix the ideas consider two variables only. Let x and y be connected by the differentiable equation

$$F(x, y) = 0 \quad \ldots\ldots\ldots\ldots\ldots\ldots\ldots\ldots(1).$$

On differentiating we get

$$\frac{\partial F}{\partial x} dx + \frac{\partial F}{\partial y} dy = 0 \quad \ldots\ldots\ldots\ldots\ldots\ldots(2).$$

If y is considered as a function of the independent variable x chosen so as to satisfy equation (1), then the derivative of this function, $\frac{dy}{dx}$, is deducible from equation (2),

$$\frac{dy}{dx} = -\frac{F_x}{F_y}.$$

The same result can be obtained by using the rule for the derivation of composite functions given in § 9·51. $F(x, y)$ is a composite function of x, and if we write

$$u = F(x, y) = 0,$$

since $du = 0$, we get
$$0 = \frac{\partial F}{\partial x} + \frac{\partial F}{\partial y}\frac{dy}{dx}.$$

(2) This argument can be applied to the case where u is an implicit function of several independent variables $x, y, \ldots$ defined by the differentiable equation

$$F(x, y, \ldots, u) = 0.$$

We get, by differentiating the equation,

$$\frac{\partial F}{\partial x}dx + \frac{\partial F}{\partial y}dy + \ldots + \frac{\partial F}{\partial u}du = 0,$$

and provided that $F_u \neq 0$,

$$du = -\frac{F_x dx + F_y dy + \ldots}{F_u},$$

and the partial derivatives of u with respect to $x, y, \ldots$ are the coefficients of $dx, dy, \ldots$

$$\frac{\partial u}{\partial x} = -\frac{F_x}{F_u}, \quad \frac{\partial u}{\partial y} = -\frac{F_y}{F_u}, \ldots.$$

(3) In the general case, suppose that m implicit functions $u, v, \ldots$ of n independent variables $x, y, \ldots$ are defined by the m differentiable equations

$$F_p(x, y, \ldots; u, v, \ldots) = 0, \quad (p = 1, 2, \ldots, m) \quad \ldots\ldots(3).$$

On differentiation we get m equations of the type

$$\frac{\partial F_p}{\partial x}dx + \frac{\partial F_p}{\partial y}dy + \ldots + \frac{\partial F_p}{\partial u}du + \frac{\partial F_p}{\partial v}dv + \ldots = 0 \quad \ldots\ldots(4),$$

where p takes the values $1, 2, \ldots, m$.

If the determinant

$$J = \begin{vmatrix} \frac{\partial F_1}{\partial u}, & \frac{\partial F_1}{\partial v}, & \dots \\ \frac{\partial F_2}{\partial u}, & \frac{\partial F_2}{\partial v}, & \dots \\ \dots\dots & \dots\dots & \dots \\ \frac{\partial F_m}{\partial u}, & \frac{\partial F_m}{\partial v}, & \dots \end{vmatrix}$$

does not vanish identically, the equations (4) can be solved for $du, dv, \dots$. Each of these differentials is obtained in the form of a fraction having J for denominator, and each numerator is linear in $dx, dy, \dots$. Thus the partial derivatives of any one of the functions $u, v, \dots$ with respect to $x, y, \dots$ are respectively the coefficients of $dx, dy, \dots$ in its differential.

10·21. The choice of independent variables.

In the preceding section it has been assumed that the independent variables have been definitely specified, so that it is known which of the variables concerned are dependent and which are independent. Suppose now that we are given a differentiable equation

$$F(x, y, z) = 0 \quad \dots\dots\dots\dots(1),$$

and nothing further is specified about the variables x, y, z.

By the fundamental principle of the last section, the equation (1) can be differentiated, and so

$$\frac{\partial F}{\partial x} dx + \frac{\partial F}{\partial y} dy + \frac{\partial F}{\partial z} dz = 0 \quad \dots\dots\dots\dots(2).$$

Before we can proceed further the independent variables must be chosen, and there are three cases to consider.

(i) Let z be chosen as the variable dependent on the two independent variables x and y in such a way that equation (1) is satisfied. Since z is a function of the two independent variables x and y, say

$$z = z(x, y),$$

we have

$$dz = \frac{\partial z}{\partial x} dx + \frac{\partial z}{\partial y} dy \quad \dots\dots\dots\dots(3).$$

Now equation (2) may be written

$$dz = -\frac{F_x}{F_z}dx - \frac{F_y}{F_z}dy, \quad (F_z \neq 0) \quad \text{.............(4)},$$

and by comparing (3) and (4)

$$\frac{\partial z}{\partial x} = -\frac{F_x}{F_z}, \quad \frac{\partial z}{\partial y} = -\frac{F_y}{F_z} \quad \text{..................(A)}.$$

(ii) If y had been chosen as a function of the two independent variables x and z so as to satisfy (1), instead of equation (3) we should have

$$dy = \frac{\partial y}{\partial x}dx + \frac{\partial y}{\partial z}dz \quad \text{....................(5)},$$

and on comparing with equation (2), which is now written in the form

$$dy = -\frac{F_x}{F_y}dx - \frac{F_z}{F_y}dz, \quad (F_y \neq 0),$$

we get

$$\frac{\partial y}{\partial x} = -\frac{F_x}{F_y}, \quad \frac{\partial y}{\partial z} = -\frac{F_z}{F_y} \quad \text{..................(B)}.$$

(iii) Similarly when x is the dependent and y and z the independent variables we get

$$\frac{\partial x}{\partial y} = -\frac{F_y}{F_x}, \quad \frac{\partial x}{\partial z} = -\frac{F_z}{F_x}, \quad (F_x \neq 0) \text{...........(C)}.$$

The reader will see that in dealing with implicit functions, the partial derivatives of one variable with respect to another can only be calculated when it is known which are the independent variables. Thus $\frac{\partial y}{\partial x}$ only has a meaning in case (ii) when y is the dependent and x is one of the independent variables: in cases (i) and (iii) $\frac{\partial y}{\partial x}$ is meaningless.

10·22. Illustration.

If $u = f(x, y)$ *and* $y = \phi(x, z)$, *explain the meanings of* $\frac{\partial u}{\partial x}$.

Here we are given two functional relations

$$u = f(x, y) \quad \text{........................(1)},$$

$$y = \phi(x, z) \quad \text{........................(2)},$$

connecting the four variables u, x, y, z. Any two of the four can be chosen as the independent variables, and, as we have already stated, the meaning of any partial derivative will depend upon our choice of independent variables. The choice of two out of four can of course be made in six ways, but if $\frac{\partial u}{\partial x}$ is to have any meaning, (i) x must be one of the independent variables, and (ii) u must be one of the dependent variables. Hence in this problem we have only to consider the cases when *the independent variables are either* (1) *x and y, or* (2) *x and z.*

Since we are assuming equations (1) and (2) to be differentiable we have the two permanent equations

$$du = \frac{\partial f}{\partial x}\,dx + \frac{\partial f}{\partial y}\,dy \quad \ldots\ldots\ldots\ldots\ldots\ldots(3),$$

$$dy = \frac{\partial \phi}{\partial x}\,dx + \frac{\partial \phi}{\partial z}\,dz \quad \ldots\ldots\ldots\ldots\ldots\ldots(4),$$

which hold no matter which two of the variables are chosen as the independent variables.

(1) *Let the independent variables be x and y.* In this case we must express du and dz as linear functions of dx and dy. The permanent equations (3) and (4) do so express du and dz, and so from (3)

$$\frac{\partial u}{\partial x} = \frac{\partial f}{\partial x}.$$

(2) *Independent variables x and z.*

To find $\frac{\partial u}{\partial x}$ we must express du as a linear function of dx and dz and then select the coefficient of dx. In this case also the permanent equations (3) and (4) are sufficient, for the fact that y is a function of x and z is expressed by the given equation (2). Thus, from (3) and (4), on eliminating dy,

$$du = \frac{\partial f}{\partial x}\,dx + \frac{\partial f}{\partial y}\left(\frac{\partial \phi}{\partial x}\,dx + \frac{\partial \phi}{\partial z}\,dz\right),$$

and so

$$\frac{\partial u}{\partial x} = \frac{\partial f}{\partial x} + \frac{\partial f}{\partial y}\frac{\partial \phi}{\partial x} \quad \ldots\ldots\ldots\ldots\ldots\ldots(5).$$

In order to indicate which case is under consideration the other independent variables are sometimes written as suffixes: if this is done the equation (5) would read

$$\left.\frac{\partial u}{\partial x}\right]_z = \left.\frac{\partial u}{\partial x}\right]_y + \left.\frac{\partial u}{\partial y}\right]_x \cdot \left.\frac{\partial y}{\partial x}\right]_z .$$

10·3. Differentials of higher orders.

If all the equations concerned are differentiable, the differentials of the second, third and higher orders are obtained in the same way by differentiating twice, three times, and so on. It is of course necessary to know which are the independent variables, and to remember that if the independent variables are $x, y, \ldots$ the differentials $dx, dy, \ldots$ are constants, and so $d^2x, d^2y, \ldots, d^3x, d^3y, \ldots$ all vanish.

Suppose that the m dependent variables $u, v, \ldots$ are connected with the n independent variables $x, y, \ldots$ by the system of equations

$$F_p(x, y, \ldots;\ u, v, \ldots) = 0, \quad (p = 1, 2, \ldots, m) \ \ldots\ldots(1).$$

On differentiating we get the systems of equations

$$\frac{\partial F_p}{\partial x} dx + \frac{\partial F_p}{\partial y} dy + \ldots + \frac{\partial F_p}{\partial u} du + \frac{\partial F_p}{\partial v} dv + \ldots = 0, \quad (p = 1, 2, \ldots, m) \ \ldots\ldots\ldots(2),$$

$$\left(\frac{\partial}{\partial x} dx + \frac{\partial}{\partial y} dy + \ldots + \frac{\partial}{\partial u} du + \ldots\right)^2 F_p + \frac{\partial F_p}{\partial u} d^2u + \frac{\partial F_p}{\partial v} d^2v + \ldots = 0, \quad (p = 1, 2, \ldots, m) \ldots\ldots\ldots(3).$$

From the system of m equations (2) we can express the m differentials $du, dv, \ldots$ as linear functions of the differentials $dx, dy, \ldots$, and after substituting for $du, dv, \ldots$ from these equations in the system of equations (3) we have m equations to determine the m second order differentials $d^2u, d^2v, \ldots$ as quadratic functions of the differentials $dx, dy, \ldots$, and so on.

It should be noticed that J, the determinant of the coefficients of the differentials of the dependent variables (which determinant is assumed not to be identically zero), is the same for each system of equations such as (2), (3), ...

10·31. Illustration.

Suppose that we have three variables x, y, z connected by the functional relation $F(x, y, z) = 0$. If z be chosen as the dependent variable so that $z = z(x, y)$, it is usual to denote the partial derivatives* $\frac{\partial z}{\partial x}$, $\frac{\partial z}{\partial y}$, $\frac{\partial^2 z}{\partial x^2}$, $\frac{\partial^2 z}{\partial x \partial y}$ and $\frac{\partial^2 z}{\partial y^2}$ by p, q, r, s and t. If we now suppose that x is the dependent variable, so that $x = x(y, z)$, we shall shew how to express the partial derivatives of the first and second orders of x with respect to y and z in terms of p, q, r, s and t.

The problem reduces to the calculation of dx and d^2x. The equation involving first differentials is permanent, no matter what the variables may be, and so, since $z = z(x, y)$,

$$dz = \frac{\partial z}{\partial x} dx + \frac{\partial z}{\partial y} dy \text{..........................(1)}.$$

We now differentiate equation (1), taking x as the dependent variable: thus dy and dz are constants, and so

$$0 = \frac{\partial^2 z}{\partial x^2} dx^2 + 2 \frac{\partial^2 z}{\partial x \partial y} dx\, dy + \frac{\partial^2 z}{\partial y^2} dy^2 + \frac{\partial z}{\partial x} d^2x \text{ ......(2)}.$$

Equations (1) and (2) may be written

$$dz = p\, dx + q\, dy \text{..........................(1')},$$

$$0 = r\, dx^2 + 2s\, dx\, dy + t\, dy^2 + p\, d^2x \text{ ....(2')}.$$

From (1′) we get

$$dx = \frac{1}{p}(dz - q\, dy) \text{ ....(3)},$$

and on substituting for dx in (2′) we have

$$d^2x = -\frac{1}{p^3}\{r\, dz^2 + 2(ps - qr)\, dy\, dz + (q^2r - 2pqs + p^2t)\, dy^2\} \text{ ....(4)}.$$

From (3) and (4) it follows, by selecting coefficients, that

$$\frac{\partial x}{\partial z} = \frac{1}{p}, \quad \frac{\partial x}{\partial y} = -\frac{q}{p},$$

$$\frac{\partial^2 x}{\partial z^2} = -\frac{r}{p^3}, \quad \frac{\partial^2 x}{\partial z \partial y} = \frac{qr - ps}{p^3} \quad \frac{\partial^2 x}{\partial y^2} = \frac{2pqs - p^2t - q^2r}{p^3}.$$

* We are assuming here that $\frac{\partial^2 z}{\partial x \partial y} = \frac{\partial^2 z}{\partial y \partial x}$.

10·4. Change of variables.

One example of changing the independent variables has already been given in § 9·61. On account of its importance in practical applications we now give two methods of changing the independent variables from Cartesian to polar coordinates in Laplace's operator.

Example 1. *Let V be a function of the two variables x and y: transform the expression*

$$\frac{\partial^2 V}{\partial x^2}+\frac{\partial^2 V}{\partial y^2}$$

by the formulae of plane polar transformation $x=u\cos v,\ y=u\sin v$.

On differentiation of these two equations we get

$$\left.\begin{aligned} dx&=\cos v\,du-u\sin v\,dv\\ dy&=\sin v\,du+u\cos v\,dv\end{aligned}\right\}\ \text{..........................(1)},$$

and hence

$$\left.\begin{aligned} du&=\ \ \cos v\,dx+\sin v\,dy\\ u\,dv&=-\sin v\,dx+\cos v\,dy\end{aligned}\right\}\ \text{..........................(2)}.$$

Method (i).

Now

$$\frac{\partial V}{\partial x}=\frac{\partial V}{\partial u}\frac{\partial u}{\partial x}+\frac{\partial V}{\partial v}\frac{\partial v}{\partial x}=\left(\cos v\frac{\partial}{\partial u}-\frac{\sin v}{u}\frac{\partial}{\partial v}\right)V\ \text{...............(3)}.$$

Similarly

$$\frac{\partial V}{\partial y}=\left(\sin v\frac{\partial}{\partial u}+\frac{\cos v}{u}\frac{\partial}{\partial v}\right)V\ \text{.........................(4)}.$$

Hence

$$\frac{\partial^2 V}{\partial x^2}=\left(\cos v\frac{\partial}{\partial u}-\frac{\sin v}{u}\frac{\partial}{\partial v}\right)\left(\cos v\frac{\partial V}{\partial u}-\frac{\sin v}{u}\frac{\partial V}{\partial v}\right)$$

$$=\cos^2 v\frac{\partial^2 V}{\partial u^2}-\frac{\cos v\sin v}{u}\frac{\partial^2 V}{\partial u\,\partial v}+\frac{1}{u^2}\cos v\sin v\frac{\partial V}{\partial v}$$

$$-\frac{\sin v\cos v}{u}\frac{\partial^2 V}{\partial v\,\partial u}+\frac{\sin^2 v}{u}\frac{\partial V}{\partial u}+\frac{\sin^2 v}{u^2}\frac{\partial^2 V}{\partial v^2}+\frac{\sin v\cos v}{u^2}\frac{\partial V}{\partial v}\ \text{......(5)}.$$

Similarly we have, by direct calculation, or by replacing v by $v-\frac{1}{2}\pi$ in (5),

$$\frac{\partial^2 V}{\partial y^2}=\sin^2 v\frac{\partial^2 V}{\partial u^2}+\frac{\sin v\cos v}{u}\frac{\partial^2 V}{\partial u\,\partial v}-\frac{\cos v\sin v}{u^2}\frac{\partial V}{\partial v}$$

$$+\frac{\cos v\sin v}{u}\frac{\partial^2 V}{\partial v\,\partial u}+\frac{\cos^2 v}{u}\frac{\partial V}{\partial u}+\frac{\cos^2 v}{u^2}\frac{\partial^2 V}{\partial v^2}-\frac{\sin v\cos v}{u^2}\frac{\partial V}{\partial v}\ \text{......(6)}.$$

On adding (5) and (6) we get

$$\frac{\partial^2 V}{\partial x^2}+\frac{\partial^2 V}{\partial y^2}=\frac{\partial^2 V}{\partial u^2}+\frac{1}{u^2}\frac{\partial^2 V}{\partial v^2}+\frac{1}{u}\frac{\partial V}{\partial u}.$$

Method (ii).

Since x and y are the independent variables

$$d^2V=\frac{\partial^2 V}{\partial u^2}du^2+2\frac{\partial^2 V}{\partial u\,\partial v}du\,dv+\frac{\partial^2 V}{\partial v^2}dv^2+\frac{\partial V}{\partial u}d^2u+\frac{\partial V}{\partial v}d^2v\ \text{......(7)}.$$

On differentiating equations (2), since dx and dy are constants, we get

$$d^2u = -\sin v\,dv\,dx + \cos v\,dv\,dy,$$

$$du\,dv + u\,d^2v = -\cos v\,dv\,dx - \sin v\,dv\,dy:$$

these become, on using equations (2) to eliminate du, dv,

$$u\,d^2u = \sin^2 v\,dx^2 - 2\sin v\cos v\,dx\,dy + \cos^2 v\,dy^2,$$

$$u^2\,d^2v = 2\sin v\cos v\,(dx^2 - dy^2) + 2(\sin^2 v - \cos^2 v)\,dx\,dy.$$

We can now express (7) as a quadratic function of the differentials dx, dy, and on writing down the coefficients of dx^2 and dy^2 we obtain the same expressions as (5) and (6).

Example 2. Transform the expression

$$\frac{\partial^2 V}{\partial x^2} + \frac{\partial^2 V}{\partial y^2} + \frac{\partial^2 V}{\partial z^2}$$

by the formulae of spherical polar transformation

$$x = r\sin\theta\cos\phi, \quad y = r\sin\theta\sin\phi, \quad z = r\cos\theta.$$

The problem is simplified by observing that the given transformation is a combination of the two plane polar transformations

$$\left.\begin{aligned} x &= u\cos\phi \\ y &= u\sin\phi \end{aligned}\right\}; \qquad \left.\begin{aligned} z &= r\cos\theta \\ u &= r\sin\theta \end{aligned}\right\}.$$

From (1), by writing ϕ for v, we deduce at once that

$$\frac{\partial^2 V}{\partial x^2} + \frac{\partial^2 V}{\partial y^2} = \frac{\partial^2 V}{\partial u^2} + \frac{1}{u^2}\frac{\partial^2 V}{\partial \phi^2} + \frac{1}{u}\frac{\partial V}{\partial u} \quad \text{..................(8).}$$

Similarly we see that

$$\frac{\partial^2 V}{\partial z^2} + \frac{\partial^2 V}{\partial u^2} = \frac{\partial^2 V}{\partial r^2} + \frac{1}{r^2}\frac{\partial^2 V}{\partial \theta^2} + \frac{1}{r}\frac{\partial V}{\partial r} \quad \text{..................(9).}$$

From equation (4), on writing u, r, θ for y, u, v respectively,

$$\frac{\partial V}{\partial u} = \sin\theta\frac{\partial V}{\partial r} + \frac{\cos\theta}{r}\frac{\partial V}{\partial \theta} \quad \text{..................(10).}$$

Add (8) and (9) and use (10) to replace $\dfrac{\partial V}{\partial u}$ which occurs on the right-hand side of (8), and we get

$$\frac{\partial^2 V}{\partial x^2} + \frac{\partial^2 V}{\partial y^2} + \frac{\partial^2 V}{\partial z^2} = \frac{\partial^2 V}{\partial r^2} + \frac{1}{r^2}\frac{\partial^2 V}{\partial \theta^2} + \frac{1}{r^2\sin^2\theta}\frac{\partial^2 V}{\partial \phi^2} + \frac{2}{r}\frac{\partial V}{\partial r} + \frac{\cot\theta}{r^2}\frac{\partial V}{\partial \theta}$$

$$= \frac{1}{r^2}\left\{\frac{\partial}{\partial r}\left(r^2\frac{\partial V}{\partial r}\right) + \frac{1}{\sin\theta}\frac{\partial}{\partial \theta}\left(\sin\theta\frac{\partial V}{\partial \theta}\right) + \frac{1}{\sin^2\theta}\frac{\partial^2 V}{\partial \phi^2}\right\}.$$

10·5. Extreme values.

In § 4·9 the theory of extreme values for functions of one variable was considered. We now investigate the theory for functions of more than one variable. There are two cases to consider, (1) the investigation of extreme values of an explicit function, and (2) the investigation of the extreme values of an implicit function of several variables when these variables are connected by a number of given "equations of condition."

Explicit functions.

Let $u=f(x, y)$ be the equation which defines u as a function of the *two* independent variables x and y. For *three* independent variables the *sufficient* conditions for maxima and minima are different from those in the case of two.

The function $f(x, y)$ has an extreme value at the point (a, b) when the increment

$$\Delta f=f(a+h, b+k)-f(a, b) \quad \text{..............(1)}$$

preserves the same sign for all values of h and k whose moduli do not exceed a sufficiently small positive number δ. If Δf be negative the extreme value is a maximum, and if Δf be positive it is a minimum.

A *necessary* condition that $f(a, b)$ should be an extreme value is that both f_x and f_y should be zero at the point (a, b), for $f(a, b)$ cannot be an extreme value of $f(x, y)$ unless it be an extreme value of the function $f(x, b)$ and also of the function $f(a, y)$. In other words we must have $f_x(x, b)=0$ when $x=a$, and $f_y(a, y)=0$ when $y=b$. Values of (x, y) at which $df=0$ are called *stationary values.*

To investigate *sufficient* conditions we must consider the sign of Δf. Let us suppose that the first partial derivatives of $f(x, y)$ are differentiable at (a, b) and hence existent in the neighbourhood of this point, then, if $|h|$ and $|k|$ are sufficiently small, and $0<\theta<1$,

$$\Delta f=hf_a(a+\theta h, b+\theta k)+kf_b(a+\theta h, b+\theta k) \quad \text{...(2)},$$

by Taylor's theorem with remainder when $n=1$.

Let $\rho=\sqrt{(h^2+k^2)}$, $h=\rho\sin\phi$, $k=\rho\cos\phi$, so that ρ tends to zero as h and k tend separately to zero, and ϕ is arbitrary. Let us write A, B, C respectively for $\dfrac{\partial^2 f}{\partial a^2}, \dfrac{\partial^2 f}{\partial a\partial b}, \dfrac{\partial^2 f}{\partial b^2}$.

Now since f_a and f_b are differentiable at (a, b), and also vanish at this point,

$$\left.\begin{aligned} f_a(a+\theta h, b+\theta k)&=\theta(Ah+Bk+\epsilon_1\rho)\\ f_b(a+\theta h, b+\theta k)&=\theta(Bh+Ck+\epsilon_2\rho)\end{aligned}\right\} \quad \text{.........(3)},$$

where ϵ_1 and ϵ_2 tend to zero as $\rho\to 0$.

Thus $\Delta f=\theta\rho^2[A\sin^2\phi+2B\sin\phi\cos\phi+C\cos^2\phi+\eta]$ ...(4),

where $\eta\to 0$ as $\rho\to 0$. η is of unknown sign.

Since $\theta\rho^2$ is positive, the sign of Δf depends upon the sign of the expression $E+\eta$, where

$$E \equiv A \sin^2\phi + 2B \sin\phi \cos\phi + C \cos^2\phi \quad \text{.........(5).}$$

There are several cases to consider:

(1) If E never vanishes it keeps a constant sign, and there must be a certain positive number m which it exceeds in absolute value. Thus, as soon as $|\eta| < m$, that is for a small enough value of ρ, the sign of Δf is the same as the sign of E.

Hence there will be a maximum or a minimum value according as E is negative or positive.

(2) If E can change sign, since Δf and E have the same sign when ρ is small enough, there will be no extreme value.

(3) If E, without ever changing sign, may vanish for certain values of ϕ the sign of Δf depends upon η, which is of unknown sign, and so no conclusion can be drawn. This is called *the doubtful case*.

E may be written in the form

$$E = \frac{(A \sin\phi + B \cos\phi)^2 + (AC - B^2)\cos^2\phi}{A}.$$

Suppose first that $A \neq 0$.

(1) If $AC - B^2 > 0$ the numerator of E is the sum of two squares and it never vanishes. Hence E never vanishes and it has the same sign as A.

Thus there is a maximum value if $A < 0$, a minimum value if $A > 0$.

(2) If $AC - B^2 < 0$ we can find two values of ϕ for which the numerator of E has different signs, namely (i) $\cos\phi = 0$ and (ii) $\tan\phi = -B/A$. Hence E can change sign and there is no extreme value.

(3) If $AC - B^2 = 0$ the numerator of E is a perfect square, and E, without changing sign, may be zero.

This is the doubtful case, in which the sign of Δf depends upon η. If $A = 0$, then

$$E = \cos\phi\,(2B \sin\phi + C \cos\phi),$$

and if $B \neq 0$, E changes sign with $\cos\phi$ and there is no extreme value.

If $A = B = 0$, then

$$E = C\cos^2\phi,$$

and E may vanish but cannot change sign. This is therefore the doubtful case.

In the above discussion we have assumed only the existence of A, B and C *at* the point (a, b). We have not assumed their continuity nor their existence near (a, b).

To summarise, it is convenient to state the criteria in the form: *The value $f(a, b)$ is an extreme value of $f(x, y)$ if*

$$f_a(a, b) = f_b(a, b) = 0,$$

and if

$$f_{aa} f_{bb} > f_{ab}^2,$$

and the value is a maximum or a minimum according as f_{aa} (or *f_{bb}*) *is negative or positive.*

10·51. Discussion of the doubtful case.

The general discussion of the doubtful case involves the consideration of terms of higher order than the second in the Taylor expansion of $f(a + h, b + k)$; in general this is not easy, and it will not be considered here.

When $f_{aa} f_{bb} = f_{ab}^2$ it is sometimes possible to decide whether $f(x, y)$ has maximum or a minimum at (a, b) by geometrical considerations, without considering the terms of higher order in the Taylor expansion. The two examples which follow shew that, in the doubtful case, $f(x, y)$ may or may not have an extreme value at (a, b).

Example 1. Let

$$f(x, y) = x^2 - 3xy^2 + 2y^4 = (x - y^2)(x - 2y^2),$$

and let $a = 0$, $b = 0$.

It is easy to verify that $f_x = 0$, $f_y = 0$ at $(0, 0)$, and that $f_{xx} = 2$, $f_{xy} = 0$, $f_{yy} = 0$ at this point. Consequently

$$f_{xx} f_{yy} = f_{xy}^2.$$

If L be the parabola $x = y^2$ and M the parabola $x = 2y^2$, we easily see that $f(x, y) < 0$ between the two curves L and M, $f(x, y) = 0$ on the curves, and

$f(x, y) > 0$ everywhere else in the neighbourhood of the origin. Thus the origin is not a point at which $f(x, y)$ has an extreme value, so that there can be neither maximum nor minimum there. See Fig. 21.

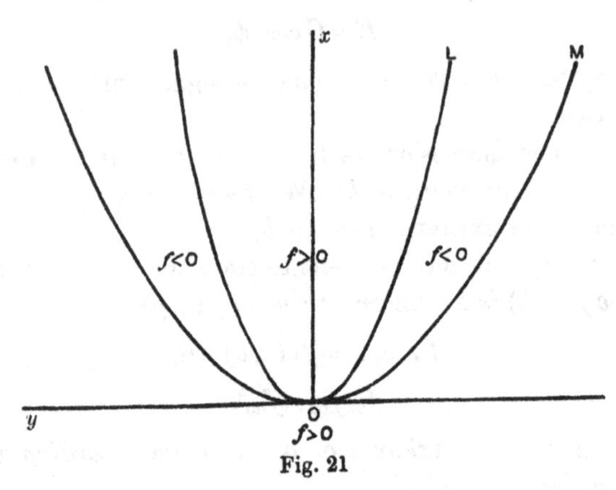

Fig. 21

Example 2. Let

$$f(x, y) = y^2 + x^2y + x^4.$$

In this case it is easy to verify that, at the origin,

$$f_x = 0, \quad f_y = 0; \quad f_{xx} = 0, \quad f_{xy} = 0, \quad f_{yy} = 2.$$

Hence again, at the point (0, 0) we have $f_{xx}f_{yy} = f_{xy}^2$. However, on writing

$$y^2 + x^2y + x^4 = (y + \tfrac{1}{2}x^2)^2 + \frac{3x^4}{4},$$

it is clear that $f(x, y)$ has a minimum value at the origin, since

$$\Delta f = f(h, k) - f(0, 0) = \left(k + \frac{h^2}{2}\right)^2 + \frac{3h^4}{4}$$

is greater than zero for all values of h and k.

10·52. Stationary values of implicit functions.

Let us consider the general problem of finding the stationary values of a given differentiable function $f(x, y, \ldots; u, v, \ldots)$ of $m + n$ variables which are connected by n independent differentiable relations

$$F_r(x, y, \ldots; u, v, \ldots) = 0, \quad (r = 1, 2, \ldots, n) \ldots\ldots\ldots(1),$$

so that f really depends on m independent variables.

At a stationary value of f, $df = 0$, so every system of values of the m independent variables must satisfy $df = 0$.

Also, since the n equations (1) are differentiable, we have the following $n+1$ equations:

$$\left.\begin{aligned}\frac{\partial f}{\partial x}dx+\frac{\partial f}{\partial y}dy+\ldots+\frac{\partial f}{\partial u}du+\ldots&=0\\ \frac{\partial F_1}{\partial x}dx+\frac{\partial F_1}{\partial y}dy+\ldots+\frac{\partial F_1}{\partial u}du+\ldots&=0\\ \ldots\ldots\ldots\ldots\ldots\ldots\ldots\ldots\ldots\ldots\ldots&\\ \frac{\partial F_n}{\partial x}dx+\frac{\partial F_n}{\partial y}dy+\ldots+\frac{\partial F_n}{\partial u}du+\ldots&=0\end{aligned}\right\}\ldots\ldots\ldots(2).$$

From these the differentials $du, dv, \ldots$ of the n dependent variables can be eliminated, yielding a result of the form

$$Mdx+Ndy+\ldots=0,$$

and since the differentials $dx, dy, \ldots$ of the independent variables are arbitrary, we get m equations $M=0, N=0, \ldots$, which, together with the n equations (1), form a system of $m+n$ simultaneous equations between the $m+n$ unknowns $x, y, \ldots, u, v, \ldots$. On solving these we get the system of values for which the function f has a stationary value.

By this method we cannot discriminate between maxima and minima.

10·53. Lagrange's multipliers.

The elimination of the n differentials $du, dv, \ldots$ from the equations (2) above is most conveniently effected in practice by the use of multipliers, a method due to Lagrange. The process is as follows. Multiply each of the equations (2), except the first, by constant factors $\lambda_1, \lambda_2, \ldots, \lambda_n$ which are to be determined. On addition we obtain a result of the form

$$Adx+Bdy+\ldots+Udu+Vdv+\ldots=0\ \ldots\ldots\ldots(3).$$

Now assume that the n multipliers have been so chosen that the n coefficients $U, V, \ldots$ of the differentials $du, dv, \ldots$ all vanish. Then in equation (3) there remain only the differentials $dx, dy, \ldots$ of the independent variables, and, since these are arbitrary, it follows that

$$A=0, \quad B=0, \quad \ldots.$$

Hence, all the coefficients in equation (3) vanish and we have the following $m+n$ equations:

$$\left.\begin{array}{l}\dfrac{\partial f}{\partial x}+\lambda_1\dfrac{\partial F_1}{\partial x}+\ldots+\lambda_n\dfrac{\partial F_n}{\partial x}=0\\ \ldots\ldots\ldots\ldots\ldots\ldots\ldots\ldots\ldots\ldots\\ \dfrac{\partial f}{\partial u}+\lambda_1\dfrac{\partial F_1}{\partial u}+\ldots+\lambda_n\dfrac{\partial F_n}{\partial u}=0\\ \ldots\ldots\ldots\ldots\ldots\ldots\ldots\ldots\ldots\ldots\end{array}\right\} \qquad \ldots\ldots\ldots\ldots(4).$$

The systems (1) and (4) contain in all $2n+m$ equations, and these suffice to determine the values of the n multipliers $\lambda_1, \lambda_2, \ldots, \lambda_n$ and the values of the $m+n$ variables $x, y, \ldots; u, v, \ldots$ for which the function f has a stationary value.

Example. Find the lengths of the axes of the section of the ellipsoid $\dfrac{x^2}{a^2}+\dfrac{y^2}{b^2}+\dfrac{z^2}{c^2}=1$ *by the plane* $lx+my+nz=0$.

The problem can be solved by finding the stationary values of the function r^2 where

$$r^2=x^2+y^2+z^2,$$

subject to the two equations of condition

$$\frac{x^2}{a^2}+\frac{y^2}{b^2}+\frac{z^2}{c^2}-1=0,$$

$$lx+my+nz=0.$$

The method of multipliers gives the set of equations

$$x+\lambda_1\frac{x}{a^2}+\lambda_2 l\ =0,$$

$$y+\lambda_1\frac{y}{b^2}+\lambda_2 m=0,$$

$$z+\lambda_1\frac{z}{c^2}+\lambda_2 n\ =0.$$

Now λ_1 is easily found by multiplying the above three equations by x, y, z respectively and adding, for on using the given equations we get

$$\lambda_1=-r^2,$$

hence $$x=\frac{\lambda_2 l}{\dfrac{r^2}{a^2}-1},\quad y=\frac{\lambda_2 m}{\dfrac{r^2}{b^2}-1},\quad z=\frac{\lambda_2 n}{\dfrac{r^2}{c^2}-1}:$$

but $$0=lx+my+nz=\lambda_2\left\{\frac{l^2a^2}{r^2-a^2}+\frac{m^2b^2}{r^2-b^2}+\frac{n^2c^2}{r^2-c^2}\right\},$$

and, since $\lambda_3 \neq 0$, we get the quadratic in r^2 giving the stationary values:

$$\frac{l^2a^2}{r^2-a^2}+\frac{m^2b^2}{r^2-b^2}+\frac{n^2c^2}{r^2-c^2}=0.$$

By geometrical considerations it is clear that one value of r^2 is a maximum and the other a minimum.

10·6. Jacobians.

If $F_1, F_2, \ldots, F_n$ denote n differentiable functions of the $n+p$ variables $u_1, u_2, \ldots, u_n; x_1, x_2, \ldots, x_p$, the determinant

$$J \equiv \begin{vmatrix} \frac{\partial F_1}{\partial u_1}, & \frac{\partial F_1}{\partial u_2}, & \ldots, & \frac{\partial F_1}{\partial u_n} \\ \frac{\partial F_2}{\partial u_1}, & \frac{\partial F_2}{\partial u_2}, & \ldots, & \frac{\partial F_2}{\partial u_n} \\ \cdots & \cdots & \cdots & \cdots \\ \frac{\partial F_n}{\partial u_1}, & \frac{\partial F_n}{\partial u_2}, & \ldots, & \frac{\partial F_n}{\partial u_n} \end{vmatrix}$$

is called the Jacobian* of the n functions with respect to the n variables $u_1, u_2, \ldots, u_n$. In the following sections we shall consider briefly some of the applications of Jacobians to important results in the theory of implicit functions, and to the determination of the question whether a given set of functions are or are not independent. A few of the more important properties of Jacobians are also discussed.

The notation for Jacobians varies with different writers, but we shall adopt the notation† that

$$J \equiv \frac{\partial(F_1, F_2, \ldots, F_n)}{\partial(u_1, u_2, \ldots, u_n)}.$$

10·61. An existence theorem.

We now prove a general theorem analogous to the existence theorem for implicit functions which was proved in § 10·1. To fix the ideas we consider a system of two equations involving three independent variables x, y, z and two unknowns u, v.

THEOREM. *Let the equations*

$$F_1(x, y, z, u, v)=0, \quad F_2(x, y, z, u, v)=0 \text{ ....(A)},$$

* Jacobians are also called "Functional determinants."

† Sometimes the ∂ is replaced by D or d, and the notation $J\begin{pmatrix} F_1, F_2, \ldots, F_n \\ u_1, u_2, \ldots, u_n \end{pmatrix}$ is also used.

be satisfied for $x=x_0$, $y=y_0$, $z=z_0$, $u=u_0$, $v=v_0$, and suppose that the Jacobian $\dfrac{\partial(F_1, F_2)}{\partial(u,v)}$ does not vanish for this set of values.

If the functions F_1, F_2 are continuous, and possess continuous first partial derivatives in the neighbourhood of these values, then there exists one and only one pair of continuous functions

$$u=\phi_1(x,y,z), \quad v=\phi_2(x,y,z),$$

which satisfy the equations (A) and reduce to u_0 and v_0 when

$$x=x_0, \; y=y_0, \; z=z_0$$

By hypothesis the equations

$$F_1(x,y,z,u,v)=0 \quad \text{.....................(1)},$$

$$F_2(x,y,z,u,v)=0 \quad \text{.....................(2)},$$

are satisfied for $x=x_0$, $y=y_0$, $z=z_0$, $u=u_0$, $v=v_0$; and since $\dfrac{\partial(F_1, F_2)}{\partial(u,v)} \neq 0$ for these values, one at least of the partial derivatives $\dfrac{\partial F_1}{\partial v}$, $\dfrac{\partial F_2}{\partial v}$ does not vanish for these values. Suppose that $\dfrac{\partial F_1}{\partial v} \neq 0$. Then by Theorem 1, § 10·1, the equation (1) defines a function

$$v=f(x,y,z,u) \quad \text{........................(3)},$$

which reduces to v_0 when $x=x_0$, $y=y_0$, $z=z_0$, $u=u_0$.

Replace v in the equation (2) by this value, and we obtain an equation

$$G(x,y,z,u)=F_2(x,y,z,u,f)=0 \quad \text{...........(4)},$$

which is satisfied by the values $x=x_0$, $y=y_0$, $z=z_0$, $u=u_0$.

Now $$\frac{\partial G}{\partial u}=\frac{\partial F_2}{\partial u}+\frac{\partial F_2}{\partial v}\frac{\partial f}{\partial u} \quad \text{.....................(5)},$$

and from (1) $$\frac{\partial F_1}{\partial u}+\frac{\partial F_1}{\partial v}\frac{\partial f}{\partial u}=0 \quad \text{.....................(6)};$$

hence on substituting for $\dfrac{\partial f}{\partial u}$ in (5) the value obtained from (6) we get

$$\frac{\partial G}{\partial u}=-\frac{\dfrac{\partial(F_1, F_2)}{\partial(u,v)}}{\dfrac{\partial F_1}{\partial v}} \quad \text{.....................(7)},$$

and this derivative clearly does not vanish for the values x_0, y_0, z_0, u_0. Now the equation (4) is satisfied when u is replaced by a certain continuous function $u=\phi_1(x, y, z)$ which reduces to u_0 when $x=x_0$, $y=y_0$, $z=z_0$. Thus, if we replace u by $\phi_1(x,y,z)$ in $f(x, y, z, u)$ we get from the equation (3) a certain continuous function

$$v=\phi_2(x, y, z).$$

This proves the theorem, since $v=\phi_2(x,y,z)$ reduces to v_0 when $x=x_0$, $y=y_0$, $z=z_0$.

COROLLARY. *It can be shewn that the functions*

$$u=\phi_1(x,y,z), \quad v=\phi_2(x,y,z),$$

possess first order partial derivatives with respect to x, y and z.

To prove this, keep y and z fixed and let Δu and Δv be the increments of u and v corresponding to an increment Δx of x. Then, from (1) and (2),

$$\left.\begin{aligned} \Delta x\left(\frac{\partial F_1}{\partial x}+\epsilon\right)+\Delta u\left(\frac{\partial F_1}{\partial u}+\epsilon_1\right)+\Delta v\left(\frac{\partial F_1}{\partial v}+\epsilon_2\right)=0 \\ \Delta x\left(\frac{\partial F_2}{\partial x}+\eta\right)+\Delta u\left(\frac{\partial F_2}{\partial u}+\eta_1\right)+\Delta v\left(\frac{\partial F_2}{\partial v}+\eta_2\right)=0 \end{aligned}\right\} \quad \ldots(8),$$

where the ϵ's and the η's tend to zero with Δx, Δu and Δv. On solving the equations (8) for the ratios $\Delta u:\Delta x$ and $\Delta v:\Delta x$ and taking the limit as $\Delta x\to 0$ we get

$$\frac{\partial u}{\partial x}=-\frac{\dfrac{\partial(F_1, F_2)}{\partial(x, v)}}{\dfrac{\partial(F_1, F_2)}{\partial(u, v)}} \quad \ldots\ldots\ldots\ldots\ldots\ldots\ldots\ldots(9),$$

and a similar formula for $\dfrac{\partial v}{\partial x}$.

Similarly we can find the partial derivatives of u and v with respect to y and z.

10·62. Dependence of functions.

We now prove the fundamental theorems on the dependence of functions. These important theorems illustrate the way in which Jacobians enter into the theory of dependence and independence of functions.

THEOREM 1. *If $u_1, u_2, \ldots, u_n$ are n differentiable functions of the n independent variables $x_1, x_2, \ldots, x_n$, and there exists an identical differentiable functional relation $\phi(u_1, u_2, \ldots, u_n)=0$ which does not involve the x's explicitly, then the Jacobian*

$$\frac{\partial(u_1, u_2, \ldots, u_n)}{\partial(x_1, x_2, \ldots, x_n)}$$

vanishes identically provided that ϕ, as a function of the u's, has no stationary values in the domain considered.

Since
$$\phi(u_1, u_2, \ldots, u_n)=0,$$
we have
$$\frac{\partial\phi}{\partial u_1}du_1+\frac{\partial\phi}{\partial u_2}du_2+\ldots+\frac{\partial\phi}{\partial u_n}du_n=0 \quad\ldots\ldots\ldots\ldots(1);$$
but
$$\left.\begin{array}{l} du_1=\dfrac{\partial u_1}{\partial x_1}dx_1+\dfrac{\partial u_1}{\partial x_2}dx_2+\ldots+\dfrac{\partial u_1}{\partial x_n}dx_n \\ \cdots\cdots\cdots\cdots\cdots\cdots\cdots\cdots\cdots\cdots\cdots \\ du_n=\dfrac{\partial u_n}{\partial x_1}dx_1+\dfrac{\partial u_n}{\partial x_2}dx_2+\ldots+\dfrac{\partial u_n}{\partial x_n}dx_n \end{array}\right\} \quad\ldots\ldots\ldots(2),$$
and on substituting these values in (1) we get an equation of the form
$$A_1dx_1+A_2dx_2+\ldots+A_ndx_n=0 \quad\ldots\ldots\ldots\ldots(3),$$
and since $dx_1, dx_2, \ldots, dx_n$ are the arbitrary differentials of the *independent* variables, it follows that
$$A_1=0,\ A_2=0,\ \ldots,\ A_n=0;$$
in other words,
$$\left.\begin{array}{l} \dfrac{\partial\phi}{\partial u_1}\dfrac{\partial u_1}{\partial x_1}+\dfrac{\partial\phi}{\partial u_2}\dfrac{\partial u_2}{\partial x_1}+\ldots+\dfrac{\partial\phi}{\partial u_n}\dfrac{\partial u_n}{\partial x_1}=0 \\ \dfrac{\partial\phi}{\partial u_1}\dfrac{\partial u_1}{\partial x_2}+\dfrac{\partial\phi}{\partial u_2}\dfrac{\partial u_2}{\partial x_2}+\ldots+\dfrac{\partial\phi}{\partial u_n}\dfrac{\partial u_n}{\partial x_2}=0 \\ \cdots\cdots\cdots\cdots\cdots\cdots\cdots\cdots\cdots\cdots\cdots \\ \dfrac{\partial\phi}{\partial u_1}\dfrac{\partial u_1}{\partial x_n}+\dfrac{\partial\phi}{\partial u_2}\dfrac{\partial u_2}{\partial x_n}+\ldots+\dfrac{\partial\phi}{\partial u_n}\dfrac{\partial u_n}{\partial x_n}=0 \end{array}\right\} \quad\ldots\ldots\ldots\ldots(4);$$
and since, by hypothesis, we cannot have
$$\frac{\partial\phi}{\partial u_1}=\frac{\partial\phi}{\partial u_2}=\ldots=\frac{\partial\phi}{\partial u_n}=0,$$
on eliminating the partial derivatives of ϕ from the set of equations (4) we get
$$\frac{\partial(u_1, u_2, \ldots, u_n)}{\partial(x_1, x_2, \ldots, x_n)}=0.$$

The theorem is therefore proved.

The vanishing of the Jacobian is also a sufficient condition for the existence of a functional relation between the u's. The proof of this result in the general case is not easy. We first give a proof by induction, and then indicate a method of proof which is easily applied in the case of three variables, but suffers from the defect that it cannot easily be extended to more than three variables: the latter proof is a good illustration of the use of differentials.

THEOREM 2. *If* $u_1, u_2, \ldots, u_n$ *are* n *functions of the* n *variables* $x_1, x_2, \ldots, x_n$, *say* $u_m = f_m(x_1, x_2, \ldots, x_n)$, $(m = 1, 2, \ldots, n)$,

and if
$$\frac{\partial(u_1, u_2, \ldots, u_n)}{\partial(x_1, x_2, \ldots, x_n)} \equiv 0,$$
then, if all the differential coefficients concerned are continuous, there exists a functional relation connecting some or all of the variables $u_1, u_2, \ldots, u_n$ *which is independent of* $x_1, x_2, \ldots, x_n$.

To prove the theorem when $n = 2$.

We have $u = f(x, y)$, $v = g(x, y)$ and
$$\begin{vmatrix} \frac{\partial u}{\partial x}, & \frac{\partial u}{\partial y} \\ \frac{\partial v}{\partial x}, & \frac{\partial v}{\partial y} \end{vmatrix} = 0.$$

If v does not depend on y, then $\frac{\partial v}{\partial y} = 0$, and so either $\frac{\partial u}{\partial y} = 0$ or else $\frac{\partial v}{\partial x} = 0$. In the former case u and v are functions of x only, and the functional relation sought is obtained from
$$u = f(x), \quad v = g(x),$$
by regarding x as a function of v and substituting in $u = f(x)$. In the latter case v is a constant, and the functional relation is
$$v = a.$$

If v does depend on y, since $\frac{\partial v}{\partial y} \neq 0$ the equation $v = g(x, y)$ defines y as a function of x and v, say
$$y = \psi(x, v),$$
and on substituting in the other equation we get an equation of the form*
$$u = F(x, v).$$

* The function $F\{x, g(x, y)\}$ is the same function of x and y as $f(x, y)$.

Then

$$0 = \begin{vmatrix} \dfrac{\partial u}{\partial x}, & \dfrac{\partial u}{\partial y} \\ \dfrac{\partial v}{\partial x}, & \dfrac{\partial v}{\partial y} \end{vmatrix} = \begin{vmatrix} \dfrac{\partial F}{\partial x} + \dfrac{\partial F}{\partial v}\dfrac{\partial v}{\partial x}, & \dfrac{\partial F}{\partial v}\dfrac{\partial v}{\partial y} \\ \dfrac{\partial v}{\partial x}, & \dfrac{\partial v}{\partial y} \end{vmatrix} = \begin{vmatrix} \dfrac{\partial F}{\partial x}, & 0 \\ \dfrac{\partial v}{\partial x}, & \dfrac{\partial v}{\partial y} \end{vmatrix}$$

and so, either $\dfrac{\partial v}{\partial y} = 0$, which is contrary to hypothesis, or else $\dfrac{\partial F}{\partial x} = 0$, so that F is a function of v only; hence the functional relation is

$$u = F(v).$$

Now assume that the theorem holds for $n-1$.

Now u_n must involve one of the variables at least, for if not there is a functional relation $u_n = a$. Let one such variable be called x_n*. Since $\dfrac{\partial u_n}{\partial x_n} \neq 0$ we can solve the equation

$$u_n = f_n(x_1, x_2, \ldots, x_n) \quad \ldots\ldots\ldots\ldots\ldots\ldots(1)$$

for x_n in terms of $x_1, x_2, \ldots, x_{n-1}$ and u_n, and on substituting this value in each of the other equations we get $n-1$ equations of the form

$$u_r = g_r(x_1, x_2, \ldots, x_{n-1}, u_n), \quad (r = 1, 2, \ldots, n-1) \ldots(2).$$

If now we substitute $f_n(x_1, x_2, \ldots, x_n)$ for u_n the functions $g_r(x_1, x_2, \ldots, x_{n-1}, u_n)$ become

$$f_r(x_1, x_2, \ldots, x_{n-1}, x_n), \quad (r = 1, 2, \ldots, n-1).$$

Then

$$0 = \begin{vmatrix} \dfrac{\partial f_1}{\partial x_1}, & \dfrac{\partial f_1}{\partial x_2}, & \ldots, & \dfrac{\partial f_1}{\partial x_n} \\ \dfrac{\partial f_2}{\partial x_1}, & \dfrac{\partial f_2}{\partial x_2}, & \ldots, & \dfrac{\partial f_2}{\partial x_n} \\ \cdots & \cdots & \cdots & \cdots \\ \dfrac{\partial f_n}{\partial x_1}, & \dfrac{\partial f_n}{\partial x_2}, & \ldots, & \dfrac{\partial f_n}{\partial x_n} \end{vmatrix}$$

$$= \begin{vmatrix} \dfrac{\partial g_1}{\partial x_1} + \dfrac{\partial g_1}{\partial u_n}\dfrac{\partial u_n}{\partial x_1}, & \ldots, & \dfrac{\partial g_1}{\partial x_{n-1}} + \dfrac{\partial g_1}{\partial u_n}\dfrac{\partial u_n}{\partial x_{n-1}}, & \dfrac{\partial g_1}{\partial u_n}\dfrac{\partial u_n}{\partial x_n} \\ \dfrac{\partial g_2}{\partial x_1} + \dfrac{\partial g_2}{\partial u_n}\dfrac{\partial u_n}{\partial x_1}, & \ldots, & \dfrac{\partial g_2}{\partial x_{n-1}} + \dfrac{\partial g_2}{\partial u_n}\dfrac{\partial u_n}{\partial x_{n-1}}, & \dfrac{\partial g_2}{\partial u_n}\dfrac{\partial u_n}{\partial x_n} \\ \cdots & \cdots & \cdots & \cdots \\ \dfrac{\partial u_n}{\partial x_1} & , \ldots, & \dfrac{\partial u_n}{\partial x_{n-1}} & , \dfrac{\partial u_n}{\partial x_n} \end{vmatrix}$$

* The interchanging of the names of two variables merely changes the sign of the Jacobian, so that its vanishing is unaffected.

$$= \begin{vmatrix} \frac{\partial g_1}{\partial x_1}, \ldots, & \frac{\partial g_1}{\partial x_{n-1}}, & 0 \\ \frac{\partial g_2}{\partial x_1}, \ldots, & \frac{\partial g_2}{\partial x_{n-1}}, & 0 \\ \cdots & \cdots & \cdots \\ \frac{\partial u_n}{\partial x_1}, \ldots, & \frac{\partial u_n}{\partial x_{n-1}}, & \frac{\partial u_n}{\partial x_n} \end{vmatrix}$$

by subtracting the elements of the last row multiplied by

$$\frac{\partial g_1}{\partial u_n}, \frac{\partial g_2}{\partial u_n}, \ldots, \frac{\partial g_{n-1}}{\partial u_n}$$

from each of the others. Hence

$$\frac{\partial u_n}{\partial x_n} \cdot \frac{\partial (g_1, g_2, \ldots, g_{n-1})}{\partial (u_1, u_2, \ldots, u_{n-1})} = 0.$$

Since $\frac{\partial u_n}{\partial x_n} \neq 0$ we must have $\frac{\partial (g_1, g_2, \ldots, g_{n-1})}{\partial (u_1, u_2, \ldots, u_{n-1})} = 0$, and so by hypothesis there is a functional relation between $g_1, g_2, \ldots, g_{n-1}$, that is between $u_1, u_2, \ldots, u_{n-1}$, into which u_n may enter, because u_n may occur in the set of equations (2) as an auxiliary variable. We have therefore proved by induction that there is a relation between $u_1, u_2, \ldots, u_n$.

10·63. Alternative proof for three functions of three variables.

Let

$$u_1 = f_1(x_1, x_2, x_3) \quad \ldots\ldots (1),$$

$$u_2 = f_2(x_1, x_2, x_3) \quad \ldots\ldots (2),$$

$$u_3 = f_3(x_1, x_2, x_3) \quad \ldots\ldots (3),$$

and suppose that

$$J \equiv \frac{\partial (u_1, u_2, u_3)}{\partial (x_1, x_2, x_3)} = 0.$$

(i) Suppose that one at least of the first minors of J, say $\frac{\partial (u_1, u_2)}{\partial (x_1, x_2)} \neq 0$, then equations (1) and (2) can be solved for x_1 and x_2 so that

$$x_1 = g_1(u_1, u_2, x_3) \quad \ldots\ldots (4),$$

$$x_2 = g_2(u_1, u_2, x_3) \quad \ldots\ldots (5),$$

and hence

$$u_3 = f_3(x_1, x_2, x_3) = G(u_1, u_2, x_3) \quad \ldots\ldots (6).$$

If we now prove that $\frac{\partial G}{\partial x_3} = 0$, then u_3 does not depend on x_3, and $u_3 = G(u_1, u_2)$ is the functional relation sought.

Consider the determinant

$$\Delta=\begin{vmatrix} \dfrac{\partial f_1}{\partial x_1}, & \dfrac{\partial f_1}{\partial x_2}, & du_1 \\ \dfrac{\partial f_2}{\partial x_1}, & \dfrac{\partial f_2}{\partial x_2}, & du_2 \\ \dfrac{\partial f_3}{\partial x_1}, & \dfrac{\partial f_3}{\partial x_2}, & du_3 \end{vmatrix}$$

and replace in it du_1, du_2 and du_3 by their values

$$du_r=\frac{\partial f_r}{\partial x_1}dx_1+\frac{\partial f_r}{\partial x_2}dx_2+\frac{\partial f_r}{\partial x_3}dx_3 \qquad (r=1, 2, 3)$$

obtained by differentiating the equations (1), (2) and (3).

It follows that $\Delta=0$, since the determinants which are the coefficients of dx_1 and dx_2 each have two columns identical and so vanish, and the coefficient of dx_3 is the determinant J, which vanishes by hypothesis.

If, however, the determinant Δ is expanded from the last column the coefficient of du_3 is $\dfrac{\partial(f_1, f_2)}{\partial(x_1, x_2)}$, which is not zero, and we get an equation of the form

$$du_3=A_1du_1+A_2du_2 \text{ ....(7).}$$

From equation (6)

$$du_3=\frac{\partial G}{\partial u_1}du_1+\frac{\partial G}{\partial u_2}du_2+\frac{\partial G}{\partial x_3}dx_3 \text{ ....(8),}$$

and on comparing (7) and (8) we see that $\dfrac{\partial G}{\partial x_3}=0$, and so

$$u_3=G(u_1, u_2) \text{ ....(9).}$$

Further, no other relation can exist which is distinct from the one just found, for if another relation $H(u_1, u_2, u_3)=0$ existed, on substituting for u_3 from (9) we should have

$$H\{u_1, u_2, G(u_1, u_2)\}=0,$$

which contradicts the hypothesis that $\dfrac{\partial(f_1, f_2)}{\partial(x_1, x_2)}\neq 0$.

(ii) If we suppose that all the first minors of J are zero, but say $\dfrac{\partial f_1}{\partial x_1}\neq 0$, then on solving (1) for x_1 we get

$$x_1=g_1(u_1, x_2, x_3),$$

$$u_2=f_2(g_1, x_2, x_3)=G_2(u_1, x_2, x_3),$$

$$u_3=f_3(g_1, x_2, x_3)=G_3(u_1, x_2, x_3);$$

as before we shew that the determinant

$$\Delta'=\begin{vmatrix} \dfrac{\partial f_1}{\partial x_1}, & du_1 \\ \dfrac{\partial f_2}{\partial x_1}, & du_2 \end{vmatrix}$$

vanishes, and deduce that $\frac{\partial G_2}{\partial x_2}=0,\quad \frac{\partial G_2}{\partial x_3}=0.$

Similarly we shew that $\frac{\partial G_3}{\partial x_2}=0,\quad \frac{\partial G_3}{\partial x_3}=0,$

and so

$$u_2=G_2(u_1),$$
$$u_3=G_3(u_1).$$

The above proof is elegant and straightforward, but does not easily extend to the general case.

10·64. Properties of Jacobians.

Jacobians have the remarkable property of behaving like the derivatives of functions of one variable. A few of the important relations are given here, and the proofs all depend upon the algebra of determinants. For simplicity the properties are stated in terms of two or three variables only, but they are evidently true in general, however many variables may be involved.

I. *Let x and y be two given variables, and u and v two functions*

$$u=u(x, y),\quad v=v(x, y),$$

and suppose that x and y are themselves functions of the two independent variables ξ and η, then, if all the differential coefficients concerned exist,

$$\frac{\partial(u, v)}{\partial(\xi, \eta)}=\frac{\partial(u, v)}{\partial(x, y)}\frac{\partial(x, y)}{\partial(\xi, \eta)} \quad\text{..................(I).}$$

This is immediate, for

$$\begin{vmatrix} \frac{\partial u}{\partial \xi}, & \frac{\partial u}{\partial \eta} \\ \frac{\partial v}{\partial \xi}, & \frac{\partial v}{\partial \eta} \end{vmatrix} = \begin{vmatrix} \frac{\partial u}{\partial x}\frac{\partial x}{\partial \xi}+\frac{\partial u}{\partial y}\frac{\partial y}{\partial \xi}, & \frac{\partial u}{\partial x}\frac{\partial x}{\partial \eta}+\frac{\partial u}{\partial y}\frac{\partial y}{\partial \eta} \\ \frac{\partial v}{\partial x}\frac{\partial x}{\partial \xi}+\frac{\partial v}{\partial y}\frac{\partial y}{\partial \xi}, & \frac{\partial v}{\partial x}\frac{\partial x}{\partial \eta}+\frac{\partial v}{\partial y}\frac{\partial y}{\partial \eta} \end{vmatrix},$$

and the second determinant is the product of the two determinants on the right-hand side of (I), by the rule for the multiplication of two determinants.

II. *If $\xi=u$ and $\eta=v$, then from* (I), *on assuming the existence of the two inverse functions x and y,*

$$x=x(u, v),\quad y=y(u, v),$$

we get

$$\frac{\partial(u, v)}{\partial(x, y)}=1\Big/\frac{\partial(x, y)}{\partial(u, v)} \quad\text{..................(II).}$$

By the theorem in § 10·61 we know that if the functions u and v take the values u_0 and v_0 at the point (x_0, y_0) and if their Jacobian does not vanish at this point, then x and y are functions of u and v which are continuous in the neighbourhood of the point (u_0, v_0); and further, these functions are single-valued and possess first order derivatives.

III. *Let $u = u(r, s, t)$, $v = v(r, s, t)$, where the variables r, s, t are themselves functions of the independent variables x and y, then*

$$\frac{\partial(u, v)}{\partial(x, y)} = \frac{\partial(u, v)}{\partial(r, s)}\frac{\partial(r, s)}{\partial(x, y)} + \frac{\partial(u, v)}{\partial(s, t)}\frac{\partial(s, t)}{\partial(x, y)} + \frac{\partial(u, v)}{\partial(t, r)}\frac{\partial(t, r)}{\partial(x, y)} \quad \text{......(1).}$$

We have

$$\frac{\partial u}{\partial x} = \frac{\partial u}{\partial r}\frac{\partial r}{\partial x} + \frac{\partial u}{\partial s}\frac{\partial s}{\partial x} + \frac{\partial u}{\partial t}\frac{\partial t}{\partial x} \quad \text{...............(2),}$$

$$\frac{\partial u}{\partial y} = \frac{\partial u}{\partial r}\frac{\partial r}{\partial y} + \frac{\partial u}{\partial s}\frac{\partial s}{\partial y} + \frac{\partial u}{\partial t}\frac{\partial t}{\partial y} \quad \text{...............(3),}$$

and if we substitute these values in the Jacobian $\frac{\partial(u, v)}{\partial(x, y)}$ we get

$$\frac{\partial(u, v)}{\partial(x, y)} = \frac{\partial u}{\partial r}\frac{\partial(r, v)}{\partial(x, y)} + \frac{\partial u}{\partial s}\frac{\partial(s, v)}{\partial(x, y)} + \frac{\partial u}{\partial t}\frac{\partial(t, v)}{\partial(x, y)} \quad \text{......(4),}$$

which is a linear expression of the Jacobians of (r, v), (s, v) and (t, v) with respect to x and y.

Now in each Jacobian on the right-hand side of (4) substitute the expressions for $\frac{\partial v}{\partial x}$ and $\frac{\partial v}{\partial y}$ which are analogous to (2) and (3): each of these Jacobians will be given as a linear expression of the Jacobians of (r, s), (s, t) and (t, r), since those of (r, r), (s, s) and (t, t) of course vanish. Thus we see that the terms which involve the Jacobian of (r, s) are

$$\frac{\partial u}{\partial r}\frac{\partial v}{\partial s}\frac{\partial(r, s)}{\partial(x, y)} + \frac{\partial u}{\partial s}\frac{\partial v}{\partial r}\frac{\partial(s, r)}{\partial(x, y)},$$

and this is equal to $\frac{\partial(u, v)}{\partial(r, s)}\frac{\partial(r, s)}{\partial(x, y)}$, the first term on the right of (1). Similarly we obtain the remaining two terms, and the formula is established.

Note. Formula (I) is analogous to the formula for the derivation of a function of a function; if $y = f(x)$ and $x = g(t)$, then

$$\frac{dy}{dt} = \frac{dy}{dx}\frac{dx}{dt};$$

formula (II) corresponds to the relation

$$\frac{dy}{dx}=1\Big/\frac{dx}{dy};$$

and formula (1) in (III) is the analogue in Jacobians of the formula (2) which involves ordinary partial derivatives.

EXAMPLES X.

1. If $x=u+vu^v$, $y=v-uv^u$, find the values of $\frac{\partial u}{\partial x}$ in terms of u and v.

2. The equation $3y=z^3+3xz$ defines z implicitly as a function of x and y; prove that

$$x\frac{\partial^2 z}{\partial y^2}+\frac{\partial^2 z}{\partial x^2}=0,$$

3. Shew that all the functions $z=\phi(x, y)$ which satisfy the differential equation

$$q^2r+p^2t=2pqs$$

may be expressed by the solution of the equation

$$x=yf(z)+g(z),$$

where f and g are arbitrary functions. Give the geometrical significance of this result.

[See § 10·31.]

4. The three variables x, y, z are connected by a functional relation: prove that

$$\left(\frac{\partial x}{\partial z}\right)^{-3}\frac{\partial^2 x}{\partial y\partial z}=\frac{\partial z}{\partial y}\frac{\partial^2 z}{\partial x^2}-\frac{\partial z}{\partial x}\frac{\partial^2 z}{\partial x\partial y}$$

where on the right-hand side the independent variables are x and y and on the left-hand side they are y and z.

5. (i) If $u=\sin x\cosh y$, $\log(x+y)+2y^2-3\tan z=4$, explain all the meanings of $\frac{\partial u}{\partial x}$ and calculate it in each case.

(ii) If $z=u^3v^2$, $u^2-v+x=0$, $u+v^2-y=0$, find $\frac{\partial x}{\partial z}$ and $\frac{\partial y}{\partial z}$ whenever they have a meaning.

6. The variables z and u are each functions of x and y defined by the equations

$$\{z-f(u)\}^2=x^2(y^2-u^2),\quad \{z-f(u)\}f'(u)=ux^2;$$

prove that

$$\frac{\partial z}{\partial x}\frac{\partial z}{\partial y}=xy.$$

7. If p, v and t are three variables connected by a functional relation $\phi(p, v, t)=0$, prove that

$$\frac{\partial p}{\partial t}\frac{\partial t}{\partial v}\frac{\partial v}{\partial p}=-1,$$

and explain the meaning of each of the partial derivatives.

8. If z is a function of x and y defined by the equations

$$z=ax+yf(a)+\phi(a),$$

$$0=x+yf'(a)+\phi'(a),$$

and a is an auxiliary variable, prove that, whatever the functions f and ϕ may be, the relation

$$rt-s^2=0$$

is satisfied. Interpret the result geometrically.

9. If $y=\dfrac{ax+b}{cx+d}$ and x is a function of t while a, b, c and d are constants, prove that

$$\frac{x'''}{x'}-\frac{3}{2}\left(\frac{x''}{x'}\right)^2=\frac{y'''}{y'}-\frac{3}{2}\left(\frac{y''}{y'}\right)^2,$$

where dashes denote derivation with respect to t.

10. Prove that $\log x$ is not an algebraic function of x.

[If $y=\log x=\int\dfrac{dx}{x}$ were an algebraic function, y would have to satisfy an irreducible algebraic equation of the form

$$f(x, y)=0 \quad\text{.................................(1).}$$

Now

$$dy=dx/x \quad\text{.................................(2),}$$

and on differentiating (1) we get

$$f_x dx+f_y dy=0,$$

and so from (2)

$$xf_x+f_y=0 \quad\text{.................................(3).}$$

Now (1) and (3) must hold for all values of x, but since (1) is irreducible, if one of its roots satisfy (3), the others must all do so. Let $y_1, y_2, \ldots, y_n$ denote the n roots of (1), then

$$\frac{dx}{x}=dy_1,\ \frac{dx}{x}=dy_2,\ \ldots,\ \frac{dx}{x}=dy_n,$$

or

$$n\frac{dx}{x}=dy_1+dy_2+\ldots+dy_n,$$

and so

$$\int\frac{dx}{x}=\frac{y_1+y_2+\ldots+y_n}{n}=-\frac{R_1(x)}{n},$$

where $R_1(x)$ is a rational function of x. See now Theorem 3, § 3·1.]

11. (i) Prove that, if $a>0$, $\frac{\log x}{x^a}\to 0$ as $x\to\infty$.

[This is most easily proved by the theory of indeterminate forms, § 4·51. An alternative proof is as follows. Let $p>0$, then $1/t<1/t^{1-p}$ when $t>1$, and so

$$\log x=\int_1^x \frac{dt}{t}<\int_1^x \frac{dt}{t^{1-p}},$$

or

$$\log x<(x^p-1)/p<x^p/p,$$

when $x>1$. Since $a>0$, p can be chosen to be less than a, and

$$0<(\log x)/x^a<x^{p-a}/p \qquad (x>1):$$

but since $a>p$, the last term $\to 0$ as $x\to\infty$. Hence the theorem.]

(ii) Shew similarly that, if $a>0$, $\frac{x^a}{e^x}\to 0$ as $x\to\infty$.

[The above results are frequently stated loosely as follows: $\log x$ tends to infinity slower, and e^x tends to infinity faster than any positive power of x.]

12. (i) Transform the expression

$$\left\{1+\left(\frac{\partial z}{\partial x}\right)^2\right\}\frac{\partial^2 z}{\partial y^2}-2\frac{\partial z}{\partial x}\frac{\partial z}{\partial y}\frac{\partial^2 z}{\partial x\,\partial y}+\left\{1+\left(\frac{\partial z}{\partial y}\right)^2\right\}\frac{\partial^2 z}{\partial x^2}$$

by the substitution $\qquad x=lu+mv,\quad y=-mu+lv,$

where l and m are constants and $l^2+m^2=1$.

(ii) Transform the expression

$$\left(x\frac{\partial z}{\partial x}+y\frac{\partial z}{\partial y}\right)^2+(a^2-x^2-y^2)\left\{\left(\frac{\partial z}{\partial x}\right)^2+\left(\frac{\partial z}{\partial y}\right)^2\right\}$$

by the substitution $\qquad x=u\cos v,\quad y=u\sin v.$

13. By putting $G=x^nH$, and changing the independent variables x, y to u, v, where $u=y/x$ and $v=xy$, transform the equations

$$x\frac{\partial G}{\partial x}+y\frac{\partial G}{\partial y}=nG,\quad x\frac{\partial U}{\partial x}-y\frac{\partial U}{\partial y}=0.$$

Hence shew that $G=x^n\phi(y/x)$ and $U=\psi(xy)$, where ϕ and ψ denote arbitrary functions.

14. A set of three variables x, y, z is connected with another set u, v, w by the equations

$$x+y+z=u,\quad yz+zx+xy=v,\quad xyz=w;$$

prove that

$$\frac{\partial^2 x}{\partial w^2}=-\frac{2(2x-y-z)}{\{(x-y)(x-z)\}^3}.$$

15. If $x=cuv$, $y=c\{(1+u^2)(1-v^2)\}^{\frac{1}{2}}$, where c is a constant, shew that

$$\frac{1}{y}\left\{y\frac{\partial V}{\partial x}-x\frac{\partial V}{\partial y}\right\}=\left(v\frac{\partial V}{\partial u}+u\frac{\partial V}{\partial v}\right)\Big/c(u^2+v^2),$$

where V is any differentiable function of x and y.

16. Prove that the volume of the greatest rectangular parallelepiped that can be inscribed in the ellipsoid

$$\frac{x^2}{a^2}+\frac{y^2}{b^2}+\frac{z^2}{c^2}=1$$

is $\frac{8abc}{3\sqrt{3}}$.

17. Investigate the maxima and minima of the functions

$$\text{(i)}\quad u=21x-12x^2-2y^2+x^3+xy^2,$$
$$\text{(ii)}\quad u=2(x-y)^2-x^4-y^4,$$
$$\text{(iii)}\quad u=x^2y^2-5x^2-8xy-5y^2.$$

18. Find the stationary values of the function

$$u=e^{-\phi}(x-y+2z),$$

where
$$\phi=x^2+y^2+2z^2-2yz+2zx-xy.$$

19. The variables x, y, z satisfy the equation

$$\phi(x).\phi(y).\phi(z)=d^3;$$

shew that, if $\phi(a)=d\neq 0$ and $\phi'(a)\neq 0$, the expression

$$f(x)+f(y)+f(z)$$

is a maximum when $x=y=z=a$ provided that

$$f'(a)\left\{\frac{\phi''(a)}{\phi'(a)}-\frac{\phi'(a)}{\phi(a)}\right\}>f''(a).$$

20. Find the shortest distance between the points $P_1(x_1, y_1, z_1)$ and $P_2(x_2, y_2, z_2)$, if P_1 lies in the plane $x+y+z=2a$ and P_2 lies on the ellipsoid $x^2/a^2+y^2/b^2+z^2/c^2=1$.

21. Find the points of the circle $x^2+y^2+z^2=k^2$, $lx+my+nz=0$ at which the function

$$u=ax^2+by^2+cz^2+2fyz+2gzx+2hxy$$

attains its greatest and its least values.

22. Find the greatest and least distances from the origin to the curve of intersection of the surface

$$(x^2+y^2+z^2)^2=a^2x^2+b^2y^2+c^2z^2$$

and the plane
$$lx+my+nz=0.$$

23. If a, b, c are positive and

$$u=(a^2x^2+b^2y^2+c^2z^2)/x^2y^2z^2,\quad ax^2+by^2+cz^2=1,$$

shew that a stationary value of u is given by

$$x^2=\frac{\mu}{2a(\mu+a)},\quad y^2=\frac{\mu}{2b(\mu+b)},\quad z^2=\frac{\mu}{2c(\mu+c)},$$

where μ is the positive root of the cubic

$$\mu^3-(bc+ca+ab)\mu-2abc=0.$$

24. Find the condition that the expressions $px+qy+rz$, $p'x+q'y+r'z$ are connected with the expression

$$ax^2+by^2+cz^2+2fyz+2gzx+2hxy$$

by a functional relation.

What is the geometrical significance of the result?

25. If $u_m = \dfrac{x_m}{(1-x_1^2-\ldots-x_n^2)^{\frac{1}{2}}}$, $(m=1, 2, \ldots, n)$, prove that

$$\frac{\partial(u_1, u_2, \ldots, u_n)}{\partial(x_1, x_2, \ldots, x_n)} = (1-x_1^2-x_2^2-\ldots-x_n^2)^{-\frac{1}{2}n-1}.$$

26. If $x_1=\cos u_1$, $x_2=\cos u_2 \sin u_1$, $x_3=\cos u_3 \sin u_2 \sin u_1$, shew that

$$\frac{\partial(x_1, x_2, x_3)}{\partial(u_1, u_2, u_3)} = -\sin^3 u_1 \sin^2 u_2 \sin u_3.$$

Extend this result to the n-dimensional case.

27. Prove that if $f(0)=0$, $f'(x)=1/(1+x^2)$, then

$$f(x)+f(y)=f\left(\frac{x+y}{1-xy}\right).$$

[Use Theorem 2, § 10·62.]

28. Prove that the three functions u, v, w are connected by an identical functional relation if

(i) $u=x+y-z, \quad v=x-y+z, \quad w=x^2+y^2+z^2-2yz;$

(ii) $u=\dfrac{x}{y-z}, \quad v=\dfrac{y}{z-x}, \quad w=\dfrac{z}{x-y};$

(iii) $u=x^3+x^2y+x^2z-z^2(x+y+z), \quad v=x+z, \quad w=x^2-z^2+xy-zy.$

Find the functional relation in case (i).

29. If u, v, w are functions of the four independent variables x_1, x_2, x_3, x_4, and if

$$P \equiv \frac{\partial(u, v, w)}{\partial(x_2, x_3, x_4)}, \quad Q \equiv \frac{\partial(u, v, w)}{\partial(x_3, x_4, x_1)}, \quad R \equiv \frac{\partial(u, v, w)}{\partial(x_4, x_1, x_2)}, \quad S \equiv \frac{\partial(u, v, w)}{\partial(x_1, x_2, x_3)},$$

prove that

$$\frac{\partial P}{\partial x_1} - \frac{\partial Q}{\partial x_2} = \frac{\partial S}{\partial x_4} - \frac{\partial R}{\partial x_3}.$$

CHAPTER XI

DOUBLE INTEGRALS

11·1. Area and volume.

We have already considered the concept of the length of a plane curve, and the length of a curve was defined in § 8·6, but we have not yet given precise mathematical definitions of area and volume.

The definition of the integral of a function of one variable $f(x)$ with respect to x depends upon the sub-division of the range of integration (a, b) into a finite number of sub-intervals, and the integral may be described as the analytical substitute for a certain area, namely that area bounded by the curve $y=f(x)$, the axis of x and the two ordinates at $x=a$ and $x=b$.

The obvious extension of this definition to a function of two variables $f(x, y)$ follows by subdividing a rectangle R into sub-rectangles, and the double integral of $f(x, y)$ with respect to x and y can be defined in a similar way to that used for defining $\int_a^b f(x)\,dx$. Double integrals bear the same relation to volumes as single integrals bear to areas.

All our previous references to "area" have depended upon geometrical notions, and such areas as were required have been tacitly assumed to exist. We now give a precise mathematical definition of "area."

11·11. Definition of area.

Let S be any set of points in the xy-plane supposed bounded by a rectangle R. Subdivide R into sub-rectangles ρ_{rs} by lines respectively parallel to Ox and Oy*: from these sub-rectangles form two classes by selecting the ones with the following properties: (1) class U contains all the sub-rectangles which contain any points whatsoever of S, and (2) class L contains all the sub-rectangles which consist entirely of points of S. Class U evidently contains

* See Fig. 22, p. 280.

class L. Let the total area of all the sub-rectangles in U be denoted by A, and the total area of all those in L by α*.

The lower bound of the numbers A corresponding to all possible modes of subdivision of R is called the *outer content* of S, $\bar{C}S$, and the upper bound of the numbers α is called the *inner content*, $\underline{C}S$. Evidently it is true that $\bar{C}S \geqslant \underline{C}S$.

If the outer and inner contents are equal, the set S is said to have an *area*, which is measured by the common value of the two contents. This area, when it exists, can be shewn to have the fundamental properties, and to agree with our intuitive notions of area.

Curves which possess a length have been called rectifiable curves. Such curves have no spatial measure. This result is very important, and we therefore prove the

LEMMA. *A rectifiable curve possesses zero area.*

Let L be the length of the curve, and suppose that the curve is divided into n pieces of equal length. With each dividing point as centre describe a square whose sides are of length $2L/n$: there are $n+1$ such squares and they certainly enclose the curve. The total area of these squares is

$$(n+1)\left(\frac{2L}{n}\right)^2 = 4L^2\frac{n+1}{n^2},$$

and since this can be made as small as we please by increasing n, the area in question must be zero.

Thus the area of a set of points bounded by a closed rectifiable curve is unaffected by the inclusion or exclusion of the boundary.

The above definition of area of course applies only to a *plane* set of points: the concept of the area of a curved surface is a more difficult one to which reference will be made in the next chapter.

11·12. Definition of volume.

To avoid repetition it is merely necessary to state that volume can be defined in just the same way as area, the rectangles being replaced by rectangular parallelepipeds.

* The area of a rectangle is of course taken as fundamental, and is given by the product of the lengths of its sides. If the rectangle have sides of lengths a and b, the area is ab.

In the same way as rectifiable curves possess no area, a regular surface* possesses no volume.

Length, area and volume are special cases of the general concept of *measure*, which plays a very important part in the theory of Lebesgue integrals. It is one of the fundamental properties of measure, however defined, that if $m > n$, a set of points in n-dimensional space has no m-dimensional measure.

11·2. Double integrals.

Since the fundamental two-dimensional interval is a rectangle, we first define a double integral over a rectangle. It will be seen later that the extension of the concept of a double integral over domains other than rectangles is easily made, and that in practice the evaluation of such double integrals is effected by a simple artifice.

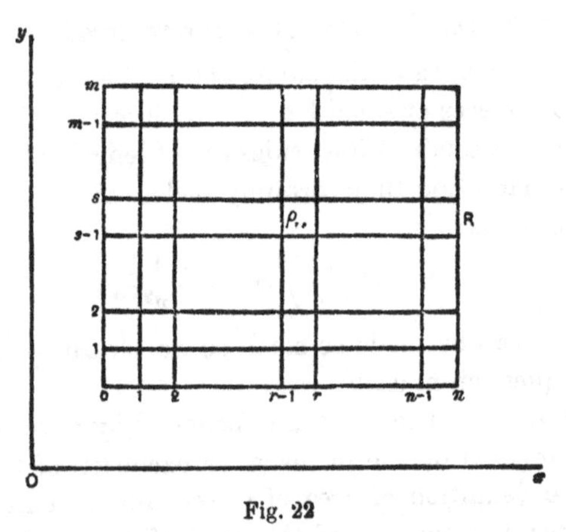

Fig. 22

In order to *define* a double integral the only notions required are those of a bound and of the area of a rectangle.

Definition. Let $f(x, y)$ be any bounded integrand, m and M its lower and upper bounds in the range of integration, which is a given rectangle R. Divide R, by lines parallel to the axes, into sub-rectangles ρ_{rs}, the areas of the latter being, for simplicity, denoted by the same symbol as the sub-rectangles themselves.

* A regular surface is defined in § 12·21.

Let M_{rs}, m_{rs} be the bounds of $f(x, y)$ in ρ_{rs}: form the sums

$$S = \sum_{r=1}^{n} \sum_{s=1}^{m} M_{rs}\,\rho_{rs},$$

$$s = \sum_{r=1}^{n} \sum_{s=1}^{m} m_{rs}\,\rho_{rs};$$

these are *the upper and lower approximative sums.*

Evidently $S \geqslant s$, and both S and s lie between $m\Delta$ and $M\Delta$, where Δ denotes the area of R: since they are roughly bounded the approximative sums S and s possess exact bounds. The lower bound of the upper sums is denoted by J, the upper bound of the lower sums by I: if $J = I$, then $f(x, y)$ is Riemann-integrable in R, and the common value of I and J is the double integral of $f(x, y)$ over R and is denoted by

$$\iint_R f(x, y)\, dx\, dy.$$

Exactly the same process may be applied to define triple integrals through a given rectangular parallelepiped T, so that we can similarly define

$$\iiint_T f(x, y, z)\, dx\, dy\, dz.$$

11·21. Properties of the double integral.

Since the definition of a double integral is analogous to the definition of a single integral, it is not necessary to repeat the proofs of theorems given in Chapter VII. For example, in order to shew that the definition given above is valid, it is necessary to prove that $J \geqslant I$. For convenience we state here some of the main properties of the double integral: let us denote the value of the double integral of $f(x, y)$ over R by D, then

(1) $$s \leqslant D \leqslant S,$$

(2) $$m\Delta \leqslant D \leqslant M\Delta,$$

where Δ denotes the area of R. The result (2) is *the first mean-value theorem for double integrals.*

If $f(x, y)$ is *continuous* in R, we have the theorem that

$$D = f(\xi, \eta)\,.\,\Delta,$$

where (ξ, η) is some point in R.

(3) *The condition of integrability.*

Since $J \geqslant I$ it is sufficient for integrability that $J - I < \epsilon$. As in Chapter VII it can be proved that it is *necessary and sufficient for integrability that there should exist a pair of approximative sums* S *and* s *such that*

$$S - s < \epsilon.$$

(4) *Important cases of integrability.*

(i) *Every continuous function is integrable.*

The proof of this theorem follows the same lines as the proof given in § 7·31 for single integrals. The reader is advised to write out the formal proof as an exercise*.

(ii) *A function is integrable in a given rectangle* R *if its discontinuities form a set of zero area.*

Since the set of discontinuities has area zero, it can be enclosed in a finite number of rectangles of total area less than ϵ: denote these rectangles by "d." In $R - d$ the function $f(x, y)$ is continuous, and so we can divide $R - d$ into a finite number of rectangles in each of which the oscillation of $f(x, y)$ is less than ϵ. When R is divided into sub-rectangles ρ_{rs} (some of which are the rectangles d mentioned above), we have

$$\begin{aligned} S - s &= \Sigma\Sigma\,(M_{rs} - m_{rs})\,\rho_{rs} \\ &= \sum_{(d)}\sum\,(M_{rs} - m_{rs})\,\rho_{rs} + \sum_{(R-d)}\sum\,(M_{rs} - m_{rs})\,\rho_{rs} \\ &\leqslant (M - m)\,A_d + \epsilon A_{R-d}, \end{aligned}$$

where A_d, A_{R-d} denote the areas of the rectangles d, and of the rectangles $R - d$; hence

$$S - s < (M - m)\,\epsilon + \epsilon\Delta$$

and this is small with ϵ; and so the condition of integrability is satisfied.

(iii) *If the discontinuities of* $f(x, y)$ *lie on a finite number of rectifiable curves,* $f(x, y)$ *is integrable.*

This is immediate in virtue of the Lemma in § 11·11.

* The theorems on continuous functions were proved in Chapter III for functions of one variable only, but they are all applicable to functions of more than one variable, generally with only slight changes of wording.

11·22. The calculation of a double integral.

We have defined a double integral in such a way that its properties are readily obtained, but we have not shewn how to *evaluate* a double integral. We now shew that the evaluation of a double integral can be reduced to the successive evaluations of two single integrals. The fundamental theorem is as follows.

THEOREM. *If* $\iint_R f(x, y)\,dx\,dy$ *exists, where* R *is the rectangle* $a \leqslant x \leqslant b$, $\alpha \leqslant y \leqslant \beta$, *and if also* $\int_a^b f(x, y)\,dx$ *exists for each value of* y *in* (α, β), *then the repeated integral*

$$\int_\alpha^\beta dy \int_a^b f(x, y)\,dx$$

exists and is identical with the double integral.

Subdivide R as usual into sub-rectangles ρ_{rs}. In ρ_{rs} we have

$$m_{rs} \leqslant f(x, y) \leqslant M_{rs} \;\ldots\ldots\ldots\ldots\ldots\ldots\ldots(1);$$

if y be fixed, we are concerned with a function of one variable only, and by the mean-value theorem for integrals

$$m_{rs}(x_r - x_{r-1}) \leqslant \int_{x_{r-1}}^{x_r} f(x, y)\,dx \leqslant M_{rs}(x_r - x_{r-1}) \;\ldots\ldots(2),$$

and this inequality holds for each value of y in (y_{s-1}, y_s). If the integral in (2) be denoted by $\phi(y)$, since the function $\phi(y)$ is bounded in (y_{s-1}, y_s), then

$$m_{rs}(x_r - x_{r-1})(y_s - y_{s-1}) \leqslant \begin{matrix} I' \\ J' \end{matrix} \leqslant M_{rs}(x_r - x_{r-1})(y_s - y_{s-1}) \ldots(3),$$

where I' and J' are respectively the lower and upper integrals of $\phi(y)$ in (y_{s-1}, y_s), namely

$$\underline{\int}_{y_{s-1}}^{y_s} \phi(y)\,dy \quad \text{and} \quad \overline{\int}_{y_{s-1}}^{y_s} \phi(y)\,dy.$$

Our object is to prove (i) that $I' = J'$ and (ii) that the repeated integral is identical with the double integral.

On summing the inequalities (3) we get*

$$s \leqslant \overline{\underline{\int}}_\alpha^\beta dy \int_a^b f(x, y)\,dx \leqslant S \;\ldots\ldots\ldots\ldots\ldots\ldots(4),$$

* The inequalities (4) are to be interpreted as representing concisely two distinct statements, namely that both $\underline{\int}_\alpha^\beta \phi(y)\,dy$ and $\overline{\int}_\alpha^\beta \phi(y)\,dy$ lie between the two sums s and S.

where s and S are the lower and upper approximative sums for the double integral over R. It follows that

$$I \leqslant \overline{\int_{\alpha}^{\beta}} dy \int_{a}^{b} f(x, y)\, dx \leqslant J,$$

but since, by hypothesis, the double integral exists, $I = J$ and so the unique integral

$$\int_{\alpha}^{\beta} dy \int_{a}^{b} f(x, y)\, dx$$

exists, and it must coincide with the double integral.

COROLLARY. *If the double integral exists, the two repeated integrals cannot exist without being equal.*

It is therefore *sufficient* for the changing of the order of the integrations in a repeated integral,

$$\int_{\alpha}^{\beta} dy \int_{a}^{b} f(x, y)\, dx = \int_{a}^{b} dx \int_{\alpha}^{\beta} f(x, y)\, dy,$$

that the integrand should be integrable with respect to both variables, in other words that

$$\iint_{R} f(x, y)\, dx\, dy$$

should exist.

11·221. Extension to triple integrals.

Suppose that $\iiint_{T} f(x, y, z)\, dx\, dy\, dz$ exists, where T is the rectangular parallelepiped $a \leqslant x \leqslant b$, $\alpha \leqslant y \leqslant \beta$, $A \leqslant z \leqslant B$. In a similar way it may be shewn that

(i) if $\int_{a}^{b} f(x, y, z)\, dx$ exists, then

$$\int_{A}^{B} \int_{\alpha}^{\beta} dz\, dy \int_{a}^{b} f(x, y, z)\, dx$$

also exists and is equal to the triple integral; or

(ii) if $\int_{\alpha}^{\beta} \int_{a}^{b} f(x, y, z)\, dx\, dy$ exists, then

$$\int_{A}^{B} dz \int_{\alpha}^{\beta} \int_{a}^{b} f(x, y, z)\, dx\, dy$$

also exists and is equal to the triple integral.

Thus if we can begin the calculation by a simple (double) integration we can complete it by a double (simple) integration. In any case if

$$\int_{A}^{B} dz \int_{\alpha}^{\beta} dy \int_{a}^{b} f(x, y, z)\, dx$$

exists, it gives the value of the triple integral.

11·23. Repeated integrals.

Although repeated integrals are used to evaluate double integrals, the two concepts are quite distinct. The double integral is defined to be the common value of the bounds I and J of two sets of approximative sums. The repeated integral

$$\int_{\alpha}^{\beta} dy \int_{a}^{b} f(x, y)\, dx$$

is merely a combination of two single integrals: if y be kept temporarily constant $\int_{a}^{b} f(x, y)\, dx$ is a function of y, say $\phi(y)$, and the repeated integral is then

$$\int_{\alpha}^{\beta} \phi(y)\, dy.$$

There is no *a priori* reason why the order of the integrations should be reversible, and in fact it is not always the case. In the Corollary above we have given a sufficient condition for reversing the order of the integrations, namely that the double integral of $f(x, y)$ over R should exist. The following example illustrates the fact that it is not always permissible to change the order of the integrations in a repeated integral. It is not possible to give many simple examples of this*.

Example. Prove that

$$\int_0^1 dx \int_0^1 \frac{x-y}{(x+y)^3}\, dy = \tfrac{1}{2}, \quad \int_0^1 dy \int_0^1 \frac{x-y}{(x+y)^3}\, dx = -\tfrac{1}{2}.$$

Consider the first of these. We have to evaluate

$$\int_0^1 \frac{x-y}{(x+y)^3}\, dy$$

when x is regarded as constant. Now

$$\int_0^1 \frac{x-y}{(x+y)^3}\, dy = \int_0^1 (y-x)\, d\{\tfrac{1}{2}(x+y)^{-2}\}$$

$$= \Big[\tfrac{1}{2}(x+y)^{-2}(y-x)\Big]_0^1 - \tfrac{1}{2}\int_0^1 (x+y)^{-2}\, dy,$$

on integrating by parts. Hence

$$\int_0^1 \frac{x-y}{(x+y)^3}\, dy = \tfrac{1}{2}\frac{1-x}{(1+x)^2} + \frac{1}{2x} + \frac{1}{2(x+1)} - \frac{1}{2x} = \frac{1}{(1+x)^2},$$

* Several examples are given by Hardy in a paper in *Proc. London Math. Soc.* XXIV (1926), *Records of Proceedings*, p. 50.

and so $$\int_0^1 \frac{dx}{(1+x)^2} = \left[-\frac{1}{1+x}\right]_0^1 = \tfrac{1}{2}.$$

Similarly we shew that $\int_0^1 \frac{x-y}{(x+y)^3}dx$ is $-\frac{1}{(1+y)^2}$, and hence the value of the second integral is found,

$$\int_0^1 dy \int_0^1 \frac{x-y}{(x+y)^3} dx = -\tfrac{1}{2}.$$

The reader may observe that in this case, if R is the square $0 \leqslant x \leqslant 1$, $0 \leqslant y \leqslant 1$,

$$\iint_R \frac{x-y}{(x+y)^3} dx\,dy$$

does not exist, since the integrand is undefined at the origin.

11·24. Double integral defined as a limit.

Just as in the case of a single integral, it can be shewn that a double integral, when it exists, is the limit of a certain summation.

If the rectangle R is divided into sub-rectangles ρ_{rs} and (ξ_{rs}, η_{rs}) are the coordinates of some point in ρ_{rs}, then clearly

$$m_{rs} \leqslant f(\xi_{rs}, \eta_{rs}) \leqslant M_{rs},$$

and so the summation $$\Sigma\Sigma f(\xi_{rs}, \eta_{rs})\,\rho_{rs} \quad \text{.................................(1)}$$

must lie between the approximative sums S and s for the mode of subdivision considered.

By a method of proof similar to that used for Darboux's theorem, § 7·22, it can be shewn that the sums S and s tend respectively to J and I as the number of sub-rectangles increases indefinitely in such a way that the dimensions of the greatest sub-rectangle tend to zero. The limit in question is of course a double limit.

Thus if $f(x, y)$ is integrable so that $J=I$, both S and s, and therefore also the summation (1) which lies between them, tend to the common value of J and I as the dimensions of the greatest sub-rectangle tend to zero.

Although it is not easy to justify, we therefore see that there is an alternative definition of a double integral by means of a limit as described above.

11·3. Dissection of a domain.

A two-dimensional domain is said to be *dissected relative to the axis of y* when it is divided by lines parallel to that axis into a finite number of pieces, each piece being bounded above by a curve $y=\phi_2(x)$, below by a curve $y=\phi_1(x)$ and at the sides by lines parallel to the axis of y. The functions $\phi_1(x)$ and $\phi_2(x)$ are assumed to be single-valued continuous functions of x.

Similarly dissection relative to the axis of x may be defined.

If a two-dimensional domain does not itself satisfy the condition that its boundary is met by a line parallel to either axis in not more than two points, by suitable dissection parallel to either or to both the axes, the domain may be divided into a finite number of pieces, each of which does satisfy the required condition.

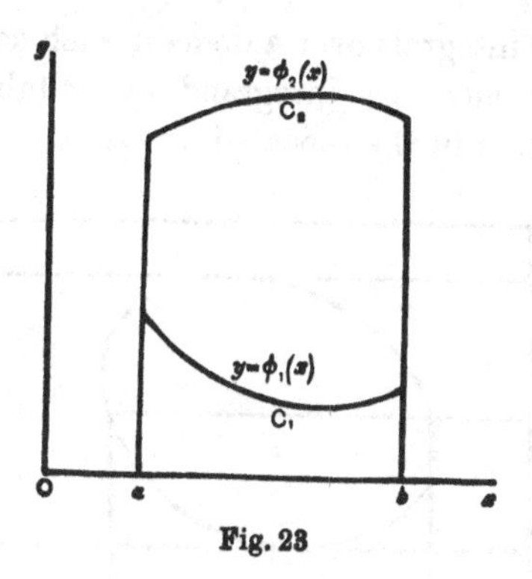

Fig. 23

11·4. THEOREM. *If $f(x, y)$ is a continuous function throughout a given domain D, which is bounded by a finite number of rectifiable curves, then* $\iint_D f(x, y)\, dx\, dy$ *exists.*

Let us define a function $f_D(x, y)$ as follows:

$$f_D(x, y) = f(x, y) \text{ at all points inside } D,$$
$$= 0 \text{ elsewhere.}$$

Now $f_D(x, y)$ is certainly integrable in any rectangle R which completely encloses the domain D, for the only discontinuities of $f_D(x, y)$ lie on the boundary of D which consists of rectifiable curves, and so, by case (iii) on page 282, $f_D(x, y)$ is integrable in R, and

$$\iint_D f(x, y)\, dx\, dy = \iint_R f_D(x, y)\, dx\, dy.$$

11·41. Double integrals over domains other than rectangles.

The artifice used in the preceding theorem is the one which enables us to define a double integral over a domain which is not a rectangle. In order to integrate $f(x, y)$ over any plane area A whose boundary is a simple closed rectifiable curve, we need only define an auxiliary integrand $f_A(x, y)$ which coincides with $f(x, y)$

at all points inside A, and which vanishes everywhere else. Then, if $f_A(x, y)$ is integrable in a rectangle R which completely encloses the area A, $f(x, y)$ is integrable in A and

$$\iint_A f(x, y)\,dx\,dy = \iint_R f_A(x, y)\,dx\,dy.$$

For *evaluating* integrals over a domain such as A, we avoid the formation of the auxiliary integrand by suitably adjusting the limits of integration in the repeated integrals.

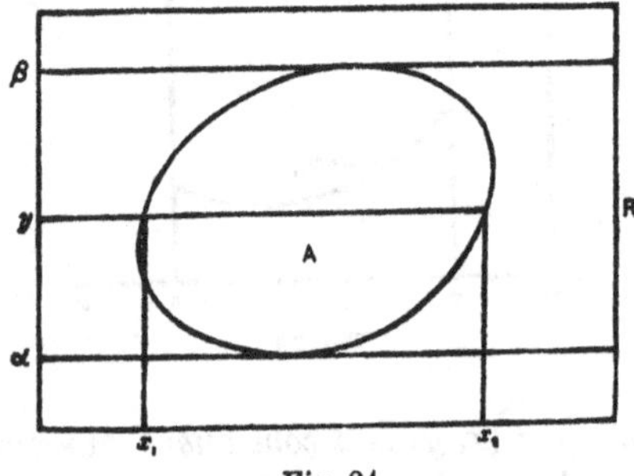

Fig. 24

Let α and β be the extreme values for y, and suppose that the area A is such that its boundary is cut by any line parallel to the x-axis in not more than two points. Let the points at which the line at height y, $(\alpha < y < \beta)$, cuts the boundary of the domain A have abscissae x_1 and x_2 which are, in general, functions of y, then

$$\iint_A f(x, y)\,dx\,dy = \int_\alpha^\beta dy \int_{x_1}^{x_2} f(x, y)\,dx.$$

Similarly by taking a line parallel to the axis of y, if y_1 and y_2 are the ordinates of the two points at which this line cuts the boundary of A, and a and b are the extreme values for x,

$$\iint_A f(x, y)\,dx\,dy = \int_a^b dx \int_{y_1}^{y_2} f(x, y)\,dy.$$

When the domain A does not itself satisfy the condition that its boundary must be cut in not more than two points by a line parallel to either axis, but can be dissected in such a way that each of the pieces so obtained shall satisfy the condition, the result for the entire domain follows by addition.

Example. Find the value of $\iint_A xy\,dx\,dy$, *where A is the domain bounded by the axis of x, the ordinate at* $x=2a$, *and the arc of the parabola* $x^2=4ay$.

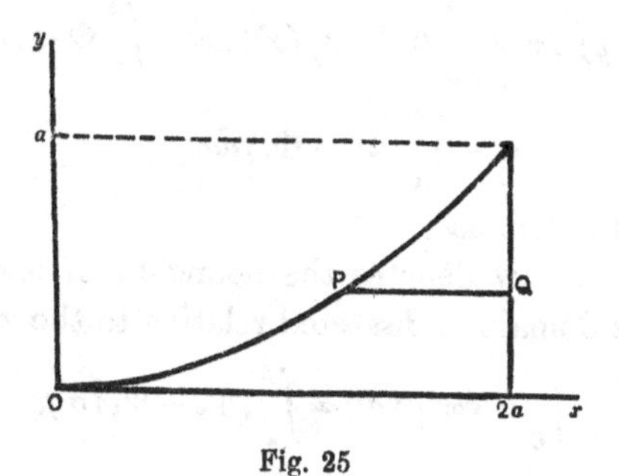

Fig. 25

Let PQ be the line at height y, then the abscissa of P is $\surd(4ay)$ and of Q is $2a$, hence

$$\iint_A xy\,dx\,dy=\int_0^a dy\int_{x_1}^{x_2} xy\,dx=\int_0^a dy\int_{\surd(4ay)}^{2a} xy\,dx$$

$$=\int_0^a y\left[\frac{x^2}{2}\right]_{\surd(4ay)}^{2a} dy=\int_0^a y\,(2a^2-2ay)\,dy=\tfrac{1}{3}a^4.$$

The reader should verify that the same result is obtained if PQ be taken parallel to the axis of y.

11·5. The connection between double and curvilinear integrals.

The fundamental theorem is the two-dimensional form of Green's theorem, and, as we shall see, this theorem plays a fundamental part in the subsequent development of the theory of double integrals. We first prove the

LEMMA. *Let C be the boundary of one of the pieces into which a given domain has been dissected relative to the axis of y, then if* $\Phi(x,y)$ *is continuous throughout the domain,*

$$\int_C \Phi(x,y)\,dx=-\int_a^b (\Phi_2-\Phi_1)\,dx,$$

where a and b are the abscissae of the straight boundary lines parallel to the y-axis, $\Phi_2\equiv\Phi\{x,\phi_2(x)\}$ *and* $\Phi_1\equiv\Phi\{x,\phi_1(x)\}$.

Now $\int_C \Phi(x,y)\,dx=\int_{C_1}\Phi(x,y)\,dx+\int_{C_2}\Phi(x,y)\,dx$

(see Fig. 23, p. 287), since curvilinear integrals with respect to x vanish along the straight boundary lines parallel to the y-axis. Hence

$$\int_C \Phi(x, y)\, dx = \int_a^b \Phi\{x, \phi_1(x)\}\, dx + \int_b^a \Phi\{x, \phi_2(x)\}\, dx$$

$$= -\int_a^b (\Phi_2 - \Phi_1)\, dx,$$

which proves the lemma.

Similarly if C now denotes the boundary of one of the pieces into which the domain is dissected relative to the x-axis,

$$\int_C \Psi(x, y)\, dy = \int_\alpha^\beta (\Psi_2 - \Psi_1)\, dy.$$

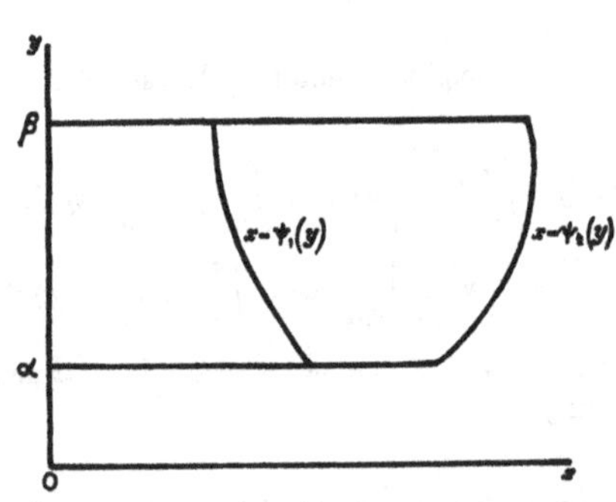

Fig. 26

THE TWO-DIMENSIONAL FORM OF GREEN'S THEOREM.

*Let D be a given domain, C its boundary, $P(x, y)$ and $Q(x, y)$ given integrands such that** (i) *C is a regular curve,* (ii) *D can be dissected relative to either axis,* (iii) *$P(x, y)$ and $Q(x, y)$ are continuous in both variables, and* (iv) $\dfrac{\partial Q}{\partial x}$ *and* $\dfrac{\partial P}{\partial y}$ *are also both continuous over D, then*

$$\int_C (P\,dx + Q\,dy) = \iint_D \left(\frac{\partial Q}{\partial x} - \frac{\partial P}{\partial y}\right) dx\,dy.$$

The theorem can be divided into two parts:

(1) $\displaystyle\int_C P\,dx = -\iint_D \frac{\partial P}{\partial y}\, dx\,dy,$ (2) $\displaystyle\int_C Q\,dy = \iint_D \frac{\partial Q}{\partial x}\, dx\,dy:$

* The conditions stated are the "standard conditions"; they are sufficient but not necessary.

consider the second of these. Since curvilinear and double integrals are both additive for ranges of integration, it is sufficient to prove (2) for one of the pieces into which D can be dissected relative to the axis of x. Consider such a piece as that shewn in Fig. 26.

Since $\frac{\partial Q}{\partial x}$ is continuous, the double integral in (2) may be evaluated as a repeated integral, and so

$$\iint_D \frac{\partial Q}{\partial x}\,dxdy = \int_\alpha^\beta dy \int_{\psi_1}^{\psi_2} \frac{\partial Q}{\partial x}\,dx = \int_\alpha^\beta (Q_2 - Q_1)\,dy$$

$$= \int_C Q\,dy$$

by the lemma.

Similarly the first part can be proved.

11·51. Deductions from Green's theorem.

We now consider two useful deductions from Green's theorem; the first concerns curvilinear integrals for area, and the second is Cauchy's fundamental theorem on regular functions of a complex variable.

I. *Curvilinear integrals for area.*

(i) If $Q = x$, $P = -y$ we get at once

$$\int_C (x\,dy - y\,dx) = 2\iint_D dx\,dy = 2A,$$

where A denotes the area of the domain D.

(ii) If we put either $Q = 0$, $P = -y$ or $Q = x$, $P = 0$ we get similarly

$$A = -\int_C y\,dx = \int_C x\,dy.$$

Hence the area of a closed domain D is given by any one of the three equivalent formulae

$$\int_C x\,dy, \quad -\int_C y\,dx, \quad \tfrac{1}{2}\int_C (x\,dy - y\,dx),$$

where C denotes the boundary of the closed domain D, to be described always in the positive (anticlockwise) sense.

II. CAUCHY'S THEOREM. *If C be a closed contour enclosing the domain D and if $f(z) = P(x, y) + iQ(x, y)$ be a regular function of $z\,(=x+iy)$, that is to say* (i) *$f(z)$ is single-valued, and* (ii) *$f'(z)$ exists uniquely everywhere in D, then*

$$\int_C f(z)\,dz = 0.$$

By definition
$$f'(z) = \lim_{\Delta z \to 0} \frac{f(z+\Delta z) - f(z)}{\Delta z},$$

and as is required in Cauchy's theory of regular functions, this limit must be the same by whatever path Δz approaches zero. By allowing Δz to approach zero by two different paths we deduce the Cauchy-Riemann differential equations which P and Q must satisfy if $f(z)$ is to be a regular function of z according to Cauchy's definition.

(i) Let Δz be taken along the axis of x so that $\Delta z = \Delta x$, then

$$f'(z) = \lim_{\Delta x \to 0} \frac{f(x+\Delta x) - f(x)}{\Delta x} = \frac{\partial P}{\partial x} + i\frac{\partial Q}{\partial x} \quad \text{..............(1).}$$

(ii) Let Δz be taken along the axis of y, so that $\Delta z = i\Delta y$, then similarly

$$f'(z) = \frac{1}{i}\left(\frac{\partial P}{\partial y} + i\frac{\partial Q}{\partial y}\right) \quad \text{..........................(2);}$$

and since (1) and (2) must be equal,

$$\frac{\partial P}{\partial x} = \frac{\partial Q}{\partial y}, \quad \frac{\partial Q}{\partial x} = -\frac{\partial P}{\partial y} \quad \text{..........................(3).}$$

Now

$$\int_C f(z)\,dz = \int_C (P + iQ)(dx + idy)$$

$$= \int_C (Pdx - Qdy) + i\int_C (Qdx + Pdy)$$

$$= \iint_D \left(-\frac{\partial Q}{\partial x} - \frac{\partial P}{\partial y}\right) dx\,dy + i\iint_D \left(\frac{\partial P}{\partial x} - \frac{\partial Q}{\partial y}\right) dx\,dy,$$

by Green's theorem. On using equations (3) we see that the integrand of each of the double integrals is zero, and Cauchy's theorem is proved.

Note on the assumptions required in the above proof.

The proof which has just been given is the simplest proof of Cauchy's theorem, but the assumptions required are more restricting than those required in the more difficult proofs. The preceding proof requires the assumption of the continuity (or something approaching it) of the function

$f'(z)$. The more difficult proof, depending on Goursat's Lemma*, assumes only the *existence* (not the continuity) of $f'(z)$. The fundamental part played by Cauchy's theorem in the Theory of Functions of a Complex Variable makes it important to be able to prove the theorem with the least possible number of restrictions on the functions concerned. The greater difficulty of the Goursat's Lemma proof is due to the reduction of the restrictions on the function $f'(z)$.

11·6. Transformations.

Many multiple integrals can be more easily evaluated by changing the variables and expressing the functions and the integrals concerned in terms of new coordinates. Before considering the change of variables in a double integral we shall consider the theory of transformation of spaces, dealing for simplicity mainly with two and three-dimensional spaces.

If the coordinates of a point (x, y, z) in three-dimensional space are related to the variables ξ, η, ζ by a set of equations of the form

$$x = \phi_1(\xi, \eta, \zeta), \quad y = \phi_2(\xi, \eta, \zeta), \quad z = \phi_3(\xi, \eta, \zeta) \quad \text{......(1)},$$

where the functions ϕ_1, ϕ_2, ϕ_3 are single-valued, we say that *(x, y, z)-space is a transformation of (ξ, η, ζ)-space*, and the transformation is defined by the equations (1). There is a single point (x, y, z) corresponding to a given point (ξ, η, ζ), but a point (x, y, z) may be the *image* of several points (ξ, η, ζ). If, given (x, y, z) there is only one point (ξ, η, ζ) of which the former point is the image, the transformation is said to be *reversible*. A sufficient condition for this is the non-vanishing of the Jacobian $\dfrac{\partial(x, y, z)}{\partial(\xi, \eta, \zeta)}$. This determinant is known as *the determinant of the transformation*, and, as we shall see, it plays a fundamental part in the theory of change of variable in a triple integral.

If $x = \phi_1(\xi, \eta)$, $y = \phi_2(\xi, \eta)$ the determinant of the transformation is of course $\dfrac{\partial(x, y)}{\partial(\xi, \eta)}$.

We shall speak of (ξ, η, ζ)-space as the *original space*, and (x, y, z)-space as the *image space*: properties of the original space may be said to be *reflected* in the image space.

* See Goursat, *Cours d'Analyse*, II, 76.

11·61. Curvilinear coordinates.

Suppose that we are given a transformation from (ξ, η, ζ)-space to (x, y, z)-space, and that attention is primarily directed to the latter space, but the formulae in which x, y, z figure are inconvenient, and we wish, if possible, to replace them by formulae in which ξ, η, ζ figure. Corresponding to any point P of (x, y, z)-space there are one or more points P_1 of the original space; the ξ, η, ζ coordinates of P_1 may be taken as coordinates, of a kind, of the point P; they are the *curvilinear coordinates* of P. They may be obtained in either of two ways, (i) by the solution of the equations of transformation

$$x = \phi_1(\xi, \eta, \zeta), \quad y = \phi_2(\xi, \eta, \zeta), \quad z = \phi_3(\xi, \eta, \zeta) \quad \ldots(1),$$

which in practical cases is usually difficult, or (ii) from a knowledge of the surfaces in the image space which correspond to constant values of the coordinates ξ, η, ζ respectively. The three sets of surfaces

$$\xi = \alpha, \quad \eta = \beta, \quad \zeta = \gamma,$$

for different values of the constants α, β, γ define a transformation just as well as the equations (1): they divide the image space into a network of cells, which is not of course rectangular.

Example. The equation

$$\frac{x^2}{a^2-\lambda} + \frac{y^2}{b^2-\lambda} + \frac{z^2}{c^2-\lambda} = 1 \quad \ldots\ldots\ldots\ldots\ldots\ldots\ldots\ldots(2)$$

defines, for varying values of λ, a set of confocal quadrics in the image space. It is known that through any point (x, y, z) there pass three of these quadrics, an ellipsoid, a hyperboloid of one sheet, and a hyperboloid of two sheets*. The corresponding values of λ, which are the three real roots of the cubic in λ given by equation (2) when x, y, z denote the coordinates of the particular point in question, are the so-called *confocal coordinates* of the point (x, y, z): we shall denote them by $\lambda_1, \lambda_2, \lambda_3$.

The transformation is only reversible as far as one-eighth of the image space is concerned. Corresponding to one point $(\lambda_1, \lambda_2, \lambda_3)$ there are, in all, eight corresponding points in the image space.

From equation (2) we have

$$F(\lambda) \equiv (a^2-\lambda)(b^2-\lambda)(c^2-\lambda) - x^2(b^2-\lambda)(c^2-\lambda)$$
$$- y^2(c^2-\lambda)(a^2-\lambda) - z^2(a^2-\lambda)(b^2-\lambda) = 0 \ldots\ldots(3);$$

* It is assumed that the reader's knowledge of analytical geometry of three dimensions is sufficient to enable the above result to be taken as known. The theory of confocal quadrics can be found in any text-book on Solid Analytical Geometry.

and to find where the roots lie, if we assume that $a^2 > b^2 > c^2$, the sign of $F(\lambda)$, for the important values of λ is given by the table

λ	$-\infty$	c^2	b^2	a^2	∞
$F(\lambda)$	+	−	+	−	−

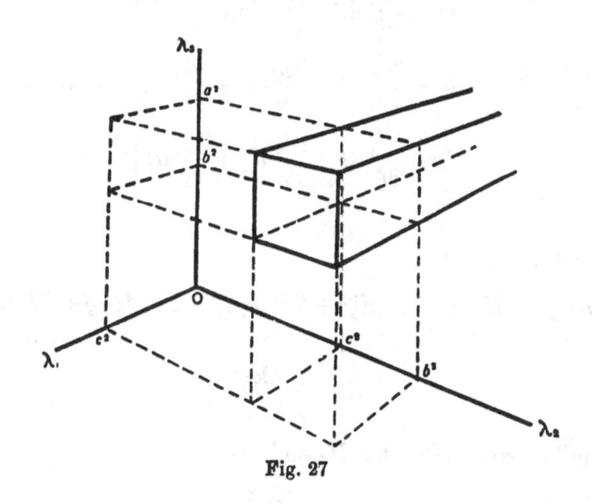

Fig. 27

From the table we see that the three roots of $F(\lambda)=0$ are such that

$$\lambda_1 < c^2, \quad c^2 < \lambda_2 < b^2, \quad b^2 < \lambda_3 < a^2;$$

and the positive octant of (x, y, z)-space is the image of an infinite slab of $(\lambda_1, \lambda_2, \lambda_3)$-space whose rectangular end is parallel to the $\lambda_2\lambda_3$-plane and whose infinite length is parallel to the axis of λ_1 for values of $\lambda_1 \leqslant c^2$. This is illustrated in Fig. 27.

In this particular case the surfaces

$$\lambda_1 = \alpha, \quad \lambda_2 = \beta, \quad \lambda_3 = \gamma,$$

where α, β, γ are constants, are respectively an ellipsoid, a hyperboloid of one sheet, and a hyperboloid of two sheets in the (x, y, z)-space, these quadrics being members of the confocal system defined by equation (2). To any given rectangular parallelepiped in the original space, there corresponds a curvilinear region, bounded by portions of the above-mentioned quadrics, in the image space.

It is a known property of confocal quadrics that the three quadrics which pass through any given point (x, y, z) cut orthogonally. The elliptic transformation is a case of an *orthogonal transformation*: in practice orthogonal transformations are always used because of their greater simplicity and convenience.

11·62. Elements of length and volume in a transformed space.

I. Let the transformation in question be defined by the equations

$$x = \phi_1(\xi, \eta, \zeta), \quad y = \phi_2(\xi, \eta, \zeta), \quad z = \phi_3(\xi, \eta, \zeta) \text{ ...(1)},$$

the functions concerned being single-valued and differentiable, then

$$\left.\begin{aligned} dx &= \frac{\partial\phi_1}{\partial\xi}d\xi + \frac{\partial\phi_1}{\partial\eta}d\eta + \frac{\partial\phi_1}{\partial\zeta}d\zeta \\ dy &= \frac{\partial\phi_2}{\partial\xi}d\xi + \frac{\partial\phi_2}{\partial\eta}d\eta + \frac{\partial\phi_2}{\partial\zeta}d\zeta \\ dz &= \frac{\partial\phi_3}{\partial\xi}d\xi + \frac{\partial\phi_3}{\partial\eta}d\eta + \frac{\partial\phi_3}{\partial\zeta}d\zeta \end{aligned}\right\} \text{ ....(2)},$$

so that

$$\begin{aligned} ds^2 &= dx^2 + dy^2 + dz^2 \\ &= H_1 d\xi^2 + H_2 d\eta^2 + H_3 d\zeta^2 + 2F_1 d\eta d\zeta + 2F_2 d\zeta d\xi + 2F_3 d\xi d\eta \end{aligned}$$
$$\text{.........(3)},$$

where
$$H_1 = \sum_r \left(\frac{\partial\phi_r}{\partial\xi}\right)^2,$$

with similar expressions for H_2 and H_3,

and
$$F_1 = \sum_r \frac{\partial\phi_r}{\partial\eta}\frac{\partial\phi_r}{\partial\zeta},$$

with similar expressions for F_2 and F_3.

The square root of the expression on the right-hand side of (3) is called the *element of length* in the transformed space.

II. Let P be any point in (x, y, z)-space, and P_1 the point in (ξ, η, ζ)-space of which P is the image. Let V_1 be the volume of an element of the original space, and suppose that V_1 surrounds the point P_1. If V be the corresponding volume in the image space, under appropriate conditions the ratio $V : V_1$ tends to a unique finite limit as the dimensions of V_1 are indefinitely decreased; this limit multiplied by $d\xi d\eta d\zeta$ is called the *element of volume* in the image space. We shall see later that its value is

$$|J|\, d\xi d\eta d\zeta,$$

where $J = \dfrac{\partial(x, y, z)}{\partial(\xi, \eta, \zeta)}$, the determinant of the transformation.

11·621. Elements of length and area in plane space.

If the plane transformation is

$$x = \phi_1(\xi, \eta), \quad y = \phi_2(\xi, \eta),$$

we have

$$ds^2 = Ed\xi^2 + 2Fd\xi d\eta + Gd\eta^2 \quad \text{..............(1)},$$

where

$$E = \left(\frac{\partial \phi_1}{\partial \xi}\right)^2 + \left(\frac{\partial \phi_2}{\partial \xi}\right)^2,$$

$$F = \frac{\partial \phi_1}{\partial \xi}\frac{\partial \phi_1}{\partial \eta} + \frac{\partial \phi_2}{\partial \xi}\frac{\partial \phi_2}{\partial \eta},$$

$$G = \left(\frac{\partial \phi_1}{\partial \eta}\right)^2 + \left(\frac{\partial \phi_2}{\partial \eta}\right)^2.$$

The square root of the expression on the right of equation (1) is the element of length, and similarly we get the element of area

$$|J|\, d\xi d\eta,$$

where

$$J = \frac{\partial(x, y)}{\partial(\xi, \eta)}.$$

In the next chapter, when areas of curved surfaces are considered, we define in a similar way the element of length and the element of surface-area.

11·63. Orthogonal transformations.

Any transformation for which the three sets of surfaces

$$\xi = \alpha, \quad \eta = \beta, \quad \zeta = \gamma$$

form a triply orthogonal system is said to be an *orthogonal transformation.*

THEOREM. *The necessary and sufficient condition for the transformation*

$$x = \phi_1(\xi, \eta, \zeta), \quad y = \phi_2(\xi, \eta, \zeta), \quad z = \phi_3(\xi, \eta, \zeta)$$

to be orthogonal is that $F_1 = F_2 = F_3 = 0$.

(1) It is necessary, for if the surfaces cut orthogonally at any point, the tangent lines at this point to the curves of intersection of the surfaces taken in pairs are a set of three mutually perpendicular lines. Consider the surfaces

$$\eta = \beta, \quad \zeta = \gamma;$$

then the curve of intersection of these surfaces has equations

$$x = \phi_1(\xi, \beta, \gamma), \quad y = \phi_2(\xi, \beta, \gamma), \quad z = \phi_3(\xi, \beta, \gamma),$$

and the direction cosines of the tangent line to this curve are proportional to

$$\frac{\partial \phi_1}{\partial \xi}, \frac{\partial \phi_2}{\partial \xi}, \frac{\partial \phi_3}{\partial \xi}.$$

Similarly on taking the surfaces

$$\zeta = \gamma, \quad \xi = \alpha,$$

the direction cosines of the tangent line to their curve of intersection are proportional to

$$\frac{\partial \phi_1}{\partial \eta}, \frac{\partial \phi_2}{\partial \eta}, \frac{\partial \phi_3}{\partial \eta}.$$

These two tangent lines are perpendicular if

$$\frac{\partial \phi_1}{\partial \xi}\frac{\partial \phi_1}{\partial \eta} + \frac{\partial \phi_2}{\partial \xi}\frac{\partial \phi_2}{\partial \eta} + \frac{\partial \phi_3}{\partial \xi}\frac{\partial \phi_3}{\partial \eta} = 0 \quad \text{...............(1)},$$

in other words if $F_3 = 0$. Similarly $F_1 = F_2 = 0$.

(2) It is also sufficient, for if $F_1 = F_2 = F_3 = 0$ the three equations of the type of (1) above express the condition that the above-mentioned set of tangent lines are mutually perpendicular, and so the surfaces

$$\xi = \alpha, \quad \eta = \beta, \quad \zeta = \gamma$$

cut orthogonally. The theorem is therefore proved.

For an orthogonal transformation the element of length takes the simpler form

$$\sqrt{(H_1 d\xi^2 + H_2 d\eta^2 + H_3 d\zeta^2)}.$$

It can be shown that the fundamental magnitudes of space transformation are connected by the relation

$$J^2 = \begin{vmatrix} H_1, & F_3, & F_2 \\ F_3, & H_2, & F_1 \\ F_2, & F_1, & H_3 \end{vmatrix}.$$

The proof is left as an exercise for the reader, since it depends only on the rule for squaring a determinant.

In the case of an *orthogonal* transformation the element of volume $|J| \, d\xi d\eta d\zeta$ becomes $\sqrt{(H_1 H_2 H_3)} \, d\xi d\eta d\zeta$.

11·64. Conformal transformations.

A transformation is said to be *conformal* if the magnitude of angles is preserved, that is to say the magnitude of the angle between any two lines in the original space is equal to that of the angle between the corresponding lines in the image space. If $d\sigma$ be the element of length in a given space, and ds the corresponding element in the image space, the necessary and sufficient condition that the transformation shall be conformal is that the ratio $d\sigma : ds$ shall be independent of direction.

Now

$$d\sigma^2 = d\xi^2 + d\eta^2 + d\zeta^2,$$

$$ds^2 = H_1 d\xi^2 + H_2 d\eta^2 + H_3 d\zeta^2 + 2F_1 d\eta\, d\zeta + 2F_2 d\zeta d\xi + 2F_3 d\xi d\eta,$$

and the condition therefore becomes

$$\frac{H_1}{1} = \frac{H_2}{1} = \frac{H_3}{1} = \frac{F_1}{0} = \frac{F_2}{0} = \frac{F_3}{0},$$

in other words, $H_1 = H_2 = H_3$; $F_1 = F_2 = F_3 = 0$.

Thus every three-dimensional conformal transformation is orthogonal*.

For a plane conformal transformation

$$d\sigma^2 = d\xi^2 + d\eta^2,$$

$$ds^2 = E d\xi^2 + 2F d\xi d\eta + G d\eta^2,$$

and so the conditions are

$$\frac{E}{1} = \frac{F}{0} = \frac{G}{1},$$

or $E = G$, $F = 0$.

It is easy to shew that $F = 0$ is the necessary and sufficient condition for the plane transformation to be orthogonal; hence a plane conformal transformation is also orthogonal*.

11·65. The polar transformations.

The polar transformations are so frequently required in practice for changing from Cartesian to polar coordinates that it is worth while making a special note of the fundamental magnitudes.

* This result is also obvious from the definition of a conformal transformation.

I. *The plane polar transformation.*

$$x = u\cos\phi, \quad y = u\sin\phi,$$
$$ds^2 = du^2 + u^2 d\phi^2,$$
$$|J| = u.$$

II. *The cylindrical polar transformation.*

$$x = u\cos\phi, \quad y = u\sin\phi, \quad z = z,$$
$$ds^2 = du^2 + u^2 d\phi^2 + dz^2,$$
$$|J| = u.$$

III. *The spherical polar transformation.*

$$x = r\sin\theta\cos\phi, \quad y = r\sin\theta\sin\phi, \quad z = r\cos\theta,$$
$$ds^2 = dr^2 + r^2 d\theta^2 + r^2\sin^2\theta\, d\phi^2,$$
$$|J| = r^2\sin\theta.$$

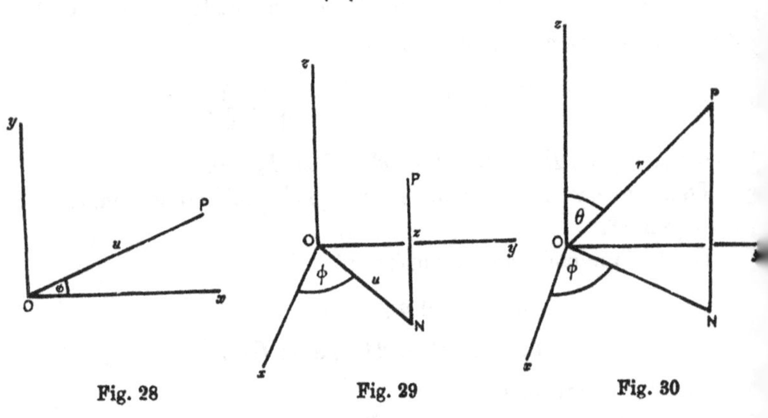

Fig. 28 Fig. 29 Fig. 30

11·66. Direct and inverse transformations of a domain.

Suppose that the domain Δ_1 of the (ξ, η)-plane transforms into the domain Δ of the (x, y)-plane. If C_1 is a closed curve in Δ_1 its image C is a closed curve in Δ, but the senses of C and C_1 may or may not be alike: if they are, the transformation is *direct*, if they are not, the transformation is *inverse.*

Thus, for example, the polar transformation

$$x = \xi\cos\eta, \quad y = \xi\sin\eta,$$

is *direct.* See Fig. 31.

The reflection transformation

$$x = -\xi, \quad y = \eta,$$

is *inverse.*

It is possible to determine analytically whether a transformation is direct or inverse: the criterion is the sign of the determinant J. In the next section it is shewn that, *provided that the Jacobian J*

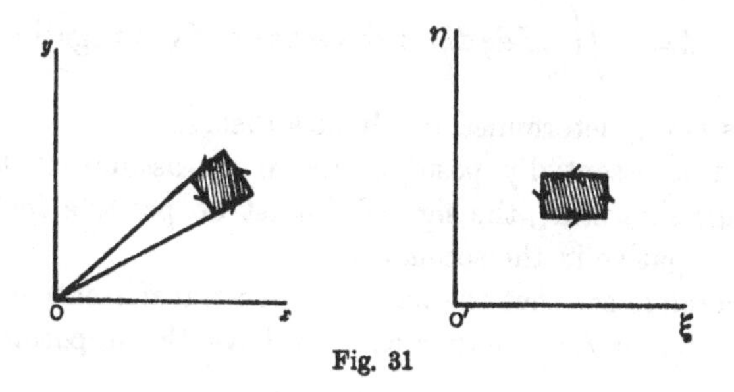

Fig. 31

does not change sign in Δ_1, the transformation is direct or inverse according as the sign of J is positive or negative.

11·7. Area in curvilinear coordinates.

Suppose that the domain D_1 is the interior of a closed curve C_1 within the region Δ_1, and let D be the image of D_1: then D will have as boundary a closed curve C within the region Δ. The area A of the domain D can be calculated by means of a curvilinear integral, and so

$$A = \int_C x\,dy,$$

where C is described in the positive (anticlockwise) sense. On changing the variables to ξ and η we get*

$$A = \int_{C_1} x\left\{\frac{\partial y}{\partial \xi}d\xi + \frac{\partial y}{\partial \eta}d\eta\right\},$$

and by applying Green's theorem

$$A = \pm\iint_{D_1}\left\{\frac{\partial}{\partial \xi}\left(x\frac{\partial y}{\partial \eta}\right) - \frac{\partial}{\partial \eta}\left(x\frac{\partial y}{\partial \xi}\right)\right\} d\xi\, d\eta,$$

the sign being + or − according as C_1 is described in the positive or negative sense. On evaluating the integrand we see that

$$A = \pm\iint_{D_1}\left(\frac{\partial x}{\partial \xi}\frac{\partial y}{\partial \eta} - \frac{\partial x}{\partial \eta}\frac{\partial y}{\partial \xi}\right) d\xi\, d\eta,$$

* The x is left unchanged.

and so

$$A = \iint_{D_1} J \, d\xi \, d\eta \quad \text{if the sense of } C_1 \text{ is positive,}$$

$$A = -\iint_{D_1} J \, d\xi \, d\eta \quad \text{if the sense of } C_1 \text{ is negative,}$$

the signs being determined by Green's theorem.

Now A is essentially positive, and if we assume that J does not change sign in Δ_1, the sign of J must be positive in the first case and negative in the second.

We therefore see that the nature of the transformation depends upon the sign of J. In either case we have the important result that

$$A = \iint_{D_1} |J| \, d\xi \, d\eta.$$

11·8. Change of variables in a double integral.

Before proving the fundamental theorem, it is convenient to state, in the form of a preliminary lemma, the result which is really *the generalised first mean-value theorem for double integrals.* The proof of the lemma follows the same lines as the proof of the generalised first mean-value theorem for integrals given in § 7·81.

LEMMA. *If $h(x, y)$ is a function which is always positive in a domain D, and if $m \leqslant g(x, y) \leqslant M$ in D, then*

$$m \iint_D h(x, y) \, dx \, dy \leqslant \iint_D h(x, y) \, g(x, y) \, dx \, dy \leqslant M \iint_D h(x, y) \, dx \, dy \quad \text{.........(1),}$$

the integrals involved being assumed existent.

THEOREM. *Let $f(x, y)$ be bounded and integrable in a domain D, then*

$$\iint_D f(x, y) \, dx \, dy = \iint_{D_1} f_1(\xi, \eta) |J| \, d\xi \, d\eta,$$

where $f_1(\xi, \eta)$ is the function obtained when $x = \phi_1(\xi, \eta)$, $y = \phi_2(\xi, \eta)$ are substituted for x and y in the function $f(x, y)$, and D_1 is the domain of which D is the image.

If the (ξ, η)-plane is covered with a rectangular network formed of lines drawn parallel to the axes of ξ and η respectively, the domain D_1 will be divided up into a finite number of sub-rectangles ρ_{rs}. In the (x, y)-plane the domain D will be divided up by the curves

$$\xi = \alpha, \quad \eta = \beta$$

into a curvilinear network. Let a_{rs} be the cell of this network which is the image of ρ_{rs}.

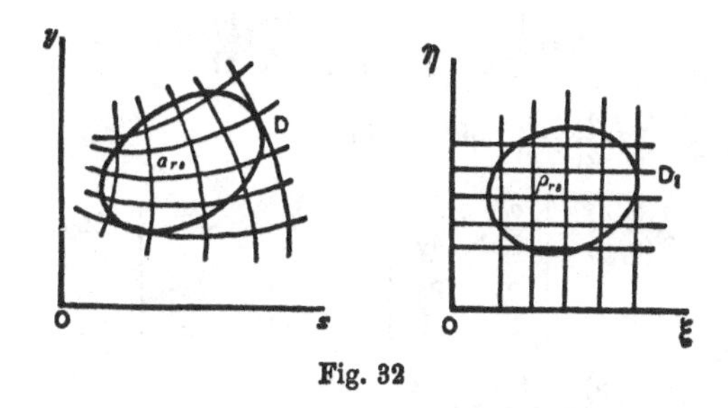

Fig. 32

Let m_{rs}, M_{rs} be the bounds of $f(x, y)$ in a_{rs}, and hence also the bounds of $f_1(\xi, \eta)$ in ρ_{rs}, then, by the lemma,

$$m_{rs}\iint_{\rho_{rs}} |J|\, d\xi\, d\eta \leqslant \iint_{\rho_{rs}} |J| f_1(\xi, \eta)\, d\xi\, d\eta \leqslant M_{rs}\iint_{\rho_{rs}} |J|\, d\xi\, d\eta \quad \text{.........(2).}$$

The extreme members of the above inequalities are respectively equal to $m_{rs}a_{rs}$ and $M_{rs}a_{rs}$, since, by § 11·7,

$$\iint_{\rho_{rs}} |J|\, d\xi\, d\eta = \iint_{a_{rs}} dx\, dy = a_{rs}.$$

On summing the inequalities (2) over the sub-rectangles, we get

$$\Sigma\Sigma m_{rs} a_{rs} \leqslant \iint_{D_1} |J| f_1(\xi, \eta)\, d\xi\, d\eta \leqslant \Sigma\Sigma M_{rs} a_{rs} \quad \text{......(3).}$$

Now $\iint_D f(x, y)\, dx\, dy$ exists, and the extreme members of the inequalities (3) are approximative sums s and S for this double integral: it follows that

$$\iint_{D_1} f_1(\xi, \eta) |J|\, d\xi\, d\eta$$

lies between two sums s and S which differ by as little as we please from

$$\iint_D f(x, y)\, dx\, dy;$$

the theorem is therefore established.

Example 1. *Shew that the area in the positive quadrant enclosed between the four curves* $a^2 y = x^3$, $b^2 y = x^3$, $p^2 x = y^3$, $q^2 x = y^3$ *is* $\frac{1}{2}(a-b)(p-q)$.

Let us make the transformation $\xi = x^3/y$, $\eta = y^3/x$, then

$$A = \iint_D dx\, dy = \iint_{D_1} \frac{\partial(x, y)}{\partial(\xi, \eta)}\, d\xi\, d\eta.$$

Now $\dfrac{\partial(x, y)}{\partial(\xi, \eta)} = 1 \Big/ \dfrac{\partial(\xi, \eta)}{\partial(x, y)}$, and so, since

$$\frac{\partial(\xi, \eta)}{\partial(x, y)} = \begin{vmatrix} \dfrac{\partial \xi}{\partial x}, & \dfrac{\partial \xi}{\partial y} \\ \dfrac{\partial \eta}{\partial x}, & \dfrac{\partial \eta}{\partial y} \end{vmatrix} = \begin{vmatrix} \dfrac{3x^2}{y}, & -\dfrac{x^3}{y^2} \\ -\dfrac{y^3}{x^2}, & \dfrac{3y^2}{x} \end{vmatrix} = 8xy,$$

$$\frac{\partial(x, y)}{\partial(\xi, \eta)} = \frac{1}{8xy} = \frac{1}{8\sqrt{(\xi\eta)}}.$$

Hence $$A = \iint_{D_1} \frac{1}{8\sqrt{(\xi\eta)}}\, d\xi\, d\eta = \frac{1}{8}\int_{a^2}^{b^2} \frac{d\xi}{\sqrt{\xi}} \int_{p^2}^{q^2} \frac{d\eta}{\sqrt{\eta}} = \tfrac{1}{2}(a-b)(p-q).$$

Example 2. *Evaluate the double integral*

$$A = \iint x^{\frac{1}{2}} y^{\frac{1}{3}} (1-x-y)^{\frac{2}{3}}\, dx\, dy$$

over the domain D bounded by the lines $x=0$, $y=0$, $x+y=1$.

The simplest method is to use the transformation $x+y=\xi$, $y=\xi\eta$, which may be written in the form

$$x = \xi(1-\eta), \quad y = \xi\eta.$$

Now

$$A = \iint_D x^{\frac{1}{2}} y^{\frac{1}{3}} (1-x-y)^{\frac{2}{3}}\, dx\, dy = \iint_{D_1} \xi^{\frac{1}{2}}(1-\eta)^{\frac{1}{2}} \,.\, \xi^{\frac{1}{3}}\eta^{\frac{1}{3}}(1-\xi)^{\frac{2}{3}} \frac{\partial(x, y)}{\partial(\xi, \eta)}\, d\xi\, d\eta,$$

where D_1 is the square $0 \leqslant \xi \leqslant 1$, $0 \leqslant \eta \leqslant 1$, and

$$\frac{\partial(x, y)}{\partial(\xi, \eta)} = \begin{vmatrix} 1-\eta, & -\xi \\ \eta, & \xi \end{vmatrix} = \xi.$$

Hence $$A = \int_0^1 \xi^{\frac{11}{6}}(1-\xi)^{\frac{2}{3}}\, d\xi \int_0^1 \eta^{\frac{1}{3}}(1-\eta)^{\frac{1}{2}}\, d\eta.$$

These integrals cannot be expressed in terms of ordinary simple functions, but they can be expressed as Beta functions (see § 8·21), and so we may write

$$A = \mathrm{B}\left(\tfrac{17}{6}, \tfrac{5}{3}\right) . \mathrm{B}\left(\tfrac{4}{3}, \tfrac{3}{2}\right).$$

Example 3. *Prove that* $B(m, n) = \dfrac{\Gamma(m)\Gamma(n)}{\Gamma(m+n)}$.

From the definition of the Gamma function, if $m > 0$ we have

$$\Gamma(m) = \int_0^\infty t^{m-1} e^{-t} dt,$$

and on writing $t = x^2$ we get

$$\Gamma(m) = \int_0^\infty x^{2m-2} e^{-x^2} 2x\,dx = 2\int_0^\infty x^{2m-1} e^{-x^2} dx.$$

Let us write $$\Gamma_a(m) \equiv 2\int_0^a x^{2m-1} e^{-x^2} dx,$$

whence $$\Gamma_a(m)\Gamma_a(n) = 4\int_0^a \int_0^a x^{2m-1} y^{2n-1} e^{-(x^2+y^2)} dx\,dy,$$

a repeated integral which has the same value as the double integral taken over the square $0 \leqslant x \leqslant a$, $0 \leqslant y \leqslant a$. On changing to polar coordinates we get

$$\Gamma_a(m)\Gamma_a(n) = 4\iint (r\cos\theta)^{2m-1} (r\sin\theta)^{2n-1} e^{-r^2} r\,dr\,d\theta$$

taken over the square.

Write $$I_a = 4\int_0^a \int_0^{\frac{1}{2}\pi} r^{2m+2n-1} e^{-r^2} \sin^{2n-1}\theta \cos^{2m-1}\theta\,dr\,d\theta,$$

and since this is a repeated integral equal to the double integral of the function over the positive quadrant of a circle of radius a,

$$I_a < \Gamma_a(m)\Gamma_a(n) < I_{\sqrt{2}a} \qquad \text{.........................(1).}$$

But $$I_a = 2\Gamma_a(m+n)\int_0^{\frac{1}{2}\pi} \sin^{2n-1}\theta \cos^{2m-1}\theta\,d\theta \qquad \text{..............(2).}$$

Now $$B(m, n) = \int_0^1 x^{m-1}(1-x)^{n-1} dx = 2\int_0^{\frac{1}{2}\pi} \sin^{2n-1}\theta \cos^{2m-1}\theta\,d\theta,$$

on writing $x = \cos^2\theta$.

From (1) and (2) we have

$$\Gamma_a(m+n)\,B(m, n) < \Gamma_a(m)\Gamma_a(n) < \Gamma_{\sqrt{2}a}(m+n)\,B(m, n)$$

and on making $a \to \infty$, the limit of the extreme members is $\Gamma(m+n)\,B(m, n)$ and the limit of the middle term is $\Gamma(m)\Gamma(n)$. The result is therefore established.

11·9. Formulae for calculating volumes.

(1) *Volume found by single integration.*

If the solid whose volume is sought is cut by a variable plane perpendicular to the line chosen as the axis of x, the area of the cross section of the solid made by any one of these planes is a function of x, say $A(x)$. Let a and b be the least and greatest values of x for points of the solid, then

$$V = \int_a^b A(x)\,dx.$$

For the ellipsoid $x^2/a^2+y^2/b^2+z^2/c^2=1$, the cross section by an arbitrary plane $x=x'$ is the ellipse

$$\frac{y^2}{b^2}+\frac{z^2}{c^2}=1-\frac{x'^2}{a^2},$$

the lengths of the semi-axes of which are $b\sqrt{\left(1-\frac{x'^2}{a^2}\right)}$ and $c\sqrt{\left(1-\frac{x'^2}{a^2}\right)}$, and so the area A in question is $\pi bc\left(1-\frac{x'^2}{a^2}\right)$, hence

$$V=\pi bc\int_{-a}^{a}\left(1-\frac{x^2}{a^2}\right)dx=\tfrac{4}{3}\pi abc.$$

(2) *Volume found by double integration.*

Suppose that the surface of the solid whose volume is sought is only cut in two points by a line parallel to the axis of z, and let z_1 and z_2 (which are functions of x and y) be the ordinates of these points. Let D denote the portion of the (x, y)-plane into which the solid projects, then

$$V=\iint_D dx\,dy\int_{z_1}^{z_2}dz=\iint_D (z_2-z_1)\,dx\,dy \text{ ....(1).}$$

If the solid has a plane base B in the (x, y)-plane itself, and is bounded laterally by a cylindrical surface with generators parallel to the axis of z, and above by the surface $z=f(x, y)$, then the domain D is the plane base B, $z_1=0$, $z_2=z$, and so

$$V=\iint_B z\,dx\,dy \text{ ....(2).}$$

It is sometimes convenient to change the variables in the double integrals (1) and (2) in order to simplify the calculations.

EXAMPLES XI.

1. (i) If S and s are the approximative sums as defined in § 11·2, prove that further subdivision of the rectangle R cannot increase S and cannot decrease s.

(ii) If S_1, s_1; S_2, s_2 refer to any two modes of subdivision of R, and Σ, σ refer to the mode consecutive to these, prove that

$$S_1\geqslant\Sigma\geqslant\sigma\geqslant s_2,$$

and deduce that $J\geqslant I$.

2. Prove, by evaluating the repeated integrals, that

$$\int_0^1 dx\int_0^1\frac{x^2-y^2}{(x^2+y^2)^2}dy \neq \int_0^1 dy\int_0^1\frac{x^2-y^2}{(x^2+y^2)^2}dx.$$

3. Evaluate $$\int_{1/a}^{3/a}\int_0^{y_1}\frac{(3y^2+1)}{a^2}\,dx\,dy,$$
where $y_1^2=a^2x^2-1$.

4. If $u+iv=(x+iy)^{\frac{1}{2}}$, shew that the curves $u=$constant, $v=$constant, are two sets of confocal parabolas.

If O be the common focus, and the two parabolas $u=u_1$, $v=v_1$ intersect at P, prove that
$$\iint\frac{dx\,dy}{\surd(x^2+y^2)}$$
extended over the area between the parabola $u=u_1$ and the lines OX and OP is $2u_1v_1$.

5. If $I=\int_0^a e^{-x^2}\,dx$, prove, by expressing I^2 as a double integral, that
$$\frac{\pi}{4}(1-e^{-a^2})<I^2<\frac{\pi}{4}(1-e^{-2a^2}),$$
and deduce that $$\int_0^\infty e^{-x^2}\,dx=\tfrac{1}{2}\surd\pi.$$

6. Prove that the volume of the wedge intercepted between the cylinder $x^2+y^2=2ax$ and the planes $z=mx$, $z=nx$ is $\pi(n-m)a^3$.

7. Evaluate

(i) $\iint r^2\sin\theta\,dr\,d\theta$ over the upper half of the circle $r=2a\cos\theta$;

(ii) $\iint x\,dx\,dy$ over the first quadrant of the ellipse $b^2x^2+a^2y^2=a^2b^2$;

and

(iii) find the integral of $x^2y\,dx-y^2x\,dy$ along the arc of the semicircle of $x^2+y^2=a^2$ which lies above the axis of x.

8. Find the area of either loop of the curve $x=a\sin 2\theta$, $y=b\cos\theta$.

9. Deduce from Green's theorem that if ϕ and ψ are functions of x and y
$$\iint_D(\phi\nabla^2\psi-\psi\nabla^2\phi)\,dx\,dy=\int_C\left\{\left(\phi\frac{\partial\psi}{\partial x}-\psi\frac{\partial\phi}{\partial x}\right)dy-\left(\phi\frac{\partial\psi}{\partial y}-\psi\frac{\partial\phi}{\partial y}\right)dx\right\},$$
where C is the closed curve which forms the boundary of the domain D. Specify what conditions have been imposed on ϕ and ψ.

10. Establish the formula
$$\Gamma(m)\,\Gamma(n)=\Gamma(m+n)\,\mathrm{B}(m,n)$$
by evaluating $$\iint e^{-x-y}\,x^{m-1}\,y^{n-1}\,dx\,dy$$
over the positive quadrant by means of the transformation
$$x+y=u,\quad y=uv.$$

11. Evaluate $$\iint_D x^2y^3\surd(1-x^3-y^3)\,dx\,dy,$$
where D is the domain defined by $x\geqslant 0$, $y\geqslant 0$, $x^3+y^3\leqslant 1$.

12. Find, in terms of Gamma functions, the area enclosed by the curve

$$\left(\frac{x}{a}\right)^4+\left(\frac{y}{b}\right)^{10}=1.$$

13. Shew that the area bounded by the four curves

$$y^3=ax^2,\quad y^3=bx^2,\quad x^4=cy^3,\quad x^4=dy^3,$$

is

$$\tfrac{6}{35}\,(b^{\frac{7}{6}}-a^{\frac{7}{6}})\,(d^{\frac{5}{6}}-c^{\frac{5}{6}}).$$

14. Find the volume enclosed by the cylinder $x^2+y^2=2ax$, the paraboloid $4az=x^2+y^2$ and the plane $z=0$.

15 Find the value of

$$\int_0^2\int_0^x\{(x-y)^2+2\,(x+y)+1\}^{-\frac{1}{2}}\,dx\,dy$$

by means of the change of variables defined by the transformation

$$x=u\,(1+v),\quad y=v\,(1+u).$$

16. Shew that the volume common to the sphere $x^2+y^2+z^2=a^2$ and the cylinder $x^2+y^2=ax$ is

$$a^3\left(\frac{2\pi}{3}-\frac{8}{9}\right)$$

17. Find the volume between the $(x,\,y)$-plane, the elliptic paraboloid $z=px^2+qy^2$ and the cylinder $x^2/a^2+y^2/b^2=1$.

18. Prove that

$$\int_0^a dx\int_0^x f(x,\,y)\,dy=\int_0^a dy\int_y^a f(x,\,y)\,dx$$

Deduce Dirichlet's formula

$$\int_0^t dx\int_0^x f(y)\,dy=\int_0^t (t-y)\,f(y)\,dy,$$

and, by applying this formula successively, establish the result

$$\int_0^x dx\int_0^x dx\,\ldots\int_0^x f(x)\,dx=\frac{1}{(n-1)\,!}\int_0^x (x-y)^{n-1}\,f(y)\,dy.$$

19. Find the centroid of a plane lamina in the form of a quadrant of an ellipse of semi-axes a and b (i) when the surface density is uniform, (ii) when it varies as the product xy.

Find also the moment of inertia of this lamina (i) about the minor axis, and (ii) about an axis through the centre of the ellipse perpendicular to its plane, in each of the above cases.

20. A system of curvilinear coordinates in the first quadrant is given by the two families of confocal parabolas

$$y^2=-2ux+u^2,\quad y^2=2vx+v^2\,;$$

find the moment of inertia about an axis through the origin perpendicular to its plane of the region S bounded by two parabolas from each family.

CHAPTER XII

TRIPLE AND SURFACE INTEGRALS

12·1. Introduction.

Although in the preceding chapter double integrals were the main topic for consideration, it may be remarked that most of the theory of multiple integrals is illustrated by the two-dimensional case. In the last chapter, where the extensions were immediate, the corresponding results about triple integrals were given immediately after the theorems proved about double integrals. As we shall see, the formula for the change of variables in a triple integral is the same in form as that already proved for a double integral, and it is proved in a similar way by an appeal to Green's theorem. Just as Green's theorem in two dimensions establishes a relation between curvilinear and double integrals, the three-dimensional form of this theorem is a relation between triple integrals and surface integrals. It is therefore necessary first to consider surface integrals and the theory of area of curved surfaces.

12·2. Surfaces.

The equations

$$x=f(u, v), \quad y=g(u, v), \quad z=h(u, v) \quad \text{.........(1)},$$

where u and v are variable parameters, define a surface. Since a surface requires only *two* parameters to define it, a portion of any given surface may be regarded as the image of a plane domain in the (u, v)-plane. When the point (u, v) moves so as to trace out a plane domain D, the point (x, y, z) traces out a portion of the surface S. We are assuming that the functions concerned in the equations (1) are single-valued and continuous.

Example. If

$$x=a \sin\theta \cos\phi, \quad y=a \sin\theta \sin\phi, \quad z=a \cos\theta,$$

then as θ goes from 0 to $\frac{1}{2}\pi$ and ϕ goes from 0 to $\frac{1}{2}\pi$, the point (x, y, z) traces out the surface of an octant of a sphere whose centre is at the origin and whose radius is a.

The parameters u and v may be regarded as curvilinear coordinates: if we divide the (u, v)-plane into a network of meshes by lines parallel to the axes, the curves

$$u=\alpha, \quad v=\beta,$$

where α and β are constants, divide up the surface into a network: this curvilinear network on the surface S in the (x, y, z)-plane corresponds to the network of rectangles in the (u, v)-plane.

In the above example if θ is constant the curves on the surface of the sphere are lines of latitude, and if ϕ is constant the curves are lines of longitude.

12·21. Regular surfaces.

The surface defined by the equations

$$x=f(u, v), \quad y=g(u, v), \quad z=h(u, v) \quad \text{.........(1)}$$

is said to be a *regular surface* if the domain in the (u, v)-plane to which it corresponds can be divided into a finite number of pieces within each of which none of the determinants

$$\frac{\partial(y, z)}{\partial(u, v)}, \quad \frac{\partial(z, x)}{\partial(u, v)}, \quad \frac{\partial(x, y)}{\partial(u, v)} \quad \text{..............(2)}$$

vanishes. These three Jacobians are among the fundamental magnitudes of surface transformations, and they are usually denoted by

$$A, \quad B, \quad C.$$

Over any elementary portion of a regular surface we can make the dependent variables independent. This is an immediate consequence of the fact that over an elementary portion of such a surface the equations defining the surface can be solved.

If $A \neq 0$, this means that the equations

$$y=g(u, v), \quad z=h(u, v)$$

can be solved for u and v in terms of y and z: on substituting these values in the equation $x=f(u, v)$ we get the equation of the surface in the form

$$x=F(y, z).$$

Similar remarks apply if $B \neq 0$ or $C \neq 0$.

The reader should observe that the usual form of the equation of a surface $z=\phi(x, y)$ implies that the Jacobian

$$C=\frac{\partial(x, y)}{\partial(u, v)} \neq 0.$$

12·22. Elements of length and area on a surface.

Let the surface be defined by the equations (1) above, then since

$$dx = \frac{\partial f}{\partial u}\,du + \frac{\partial f}{\partial v}\,dv,$$

$$dy = \frac{\partial g}{\partial u}\,du + \frac{\partial g}{\partial v}\,dv,$$

$$dz = \frac{\partial h}{\partial u}\,du + \frac{\partial h}{\partial v}\,dv,$$

we get

$$ds^2 = E\,du^2 + 2F\,du\,dv + G\,dv^2 \quad \text{..............(3)},$$

where

$$E \equiv \left(\frac{\partial f}{\partial u}\right)^2 + \left(\frac{\partial g}{\partial u}\right)^2 + \left(\frac{\partial h}{\partial u}\right)^2, \quad F \equiv \frac{\partial f}{\partial u}\frac{\partial f}{\partial v} + \frac{\partial g}{\partial u}\frac{\partial g}{\partial v} + \frac{\partial h}{\partial u}\frac{\partial h}{\partial v},$$

$$G \equiv \left(\frac{\partial f}{\partial v}\right)^2 + \left(\frac{\partial g}{\partial v}\right)^2 + \left(\frac{\partial h}{\partial v}\right)^2.$$

The square-root of the expression on the right of (3) is called the *element of length* on the given surface.

Let P be a point on the surface, and P_1 the point in the (u, v)-plane of which P is the image: if Δ_1 be any plane element of area enclosing the point P_1, and Δ the corresponding element of area on the surface, under the appropriate conditions the ratio $\Delta : \Delta_1$ tends to a unique finite limit as the dimensions of Δ_1 tend to zero. This limit, multiplied by $du\,dv$, is the *element of area* on the surface: it is found to be

$$\sqrt{(EG - F^2)}\,du\,dv.$$

12·3. The area of a curved surface.

As the length of a curve can be obtained by means of inscribed polygons, it is not unreasonable to expect that the area of a curved surface might be obtained in a similar way by means of inscribed polyhedra. For a long time this was the method generally employed, and standard treatises contained demonstrations to this effect. It was pointed out by Schwarz* that, even for such a simple surface as the cylinder, this method gives us no definite result, the areas of the polyhedra varying over a wide range with the shape of the triangles which form their faces.

* *Werke*, II, 309. See also Pierpont, *Theory of Functions of a Real Variable*, II, §§ 603, 604.

Satisfactory definitions of the area of a curved surface have only been given within recent years, but unfortunately the preliminary ideas necessary for enabling an account of these theories to be given are outside the scope of this book*. In order to shew the reader something of the difficulties involved, W. H. Young's definition of the area of a curved surface is given in the next section, but, for the reason just stated, it cannot be fully treated, and further details may be found in the original memoir referred to in the footnote.

12·31. Young's definition of the area of a curved surface.

The fundamental idea is obtained in the following way. Consider an element of the surface in question, and inscribe, in the skew curve which forms the contour of this surface element, a polygonal figure the lengths of whose sides shall be all less than a certain assigned number. Suppose that there are, acting along the sides of this skew polygon, vectors represented by the sides in magnitude, line of action and sense; these vectors are equivalent to a couple. If we suppose that the magnitude of this couple tends to a unique finite limit as the lengths of the sides of the polygon tend to zero, this limit is defined to be the area of the skew curve.

Let C be a circuit in space described in a definite sense, and let C_1, C_2, C_3 be the closed plane curves which are respectively the projections of the circuit C on the three coordinate planes; and suppose that each of the three plane curves is described in a definite sense. Let the areas of the three closed curves C_1, C_2, C_3 be a_1, a_2, a_3, and express them by the formulae for area in terms of curvilinear integrals†,

$$a_1 = \tfrac{1}{2}\int_{C_1}(y\,dz - z\,dy),\quad a_2 = \tfrac{1}{2}\int_{C_2}(z\,dx - x\,dz),\quad a_3 = \tfrac{1}{2}\int_{C_3}(x\,dy - y\,dx).$$

The vector $\mathbf{a}$, whose coordinates are (a_1, a_2, a_3), is taken as measuring the *vector-area* of C. The modulus of the vector-area,

$$\sqrt{(a_1^2 + a_2^2 + a_3^2)},$$

is taken to be the *absolute area* of C.

Let S be any curved surface in (x, y, z)-space, and Δ the domain in the (u, v)-plane of which S is the image: corresponding to a division of Δ into a network of rectangular meshes ρ_{rs}, we have a division of S into curvilinear meshes β_{rs}. Let b_{rs} be the absolute area of the perimeter of the circuit β_{rs}, and write

$$\sigma = \Sigma\Sigma\, b_{rs}.$$

* See W. H. Young, *Proc. Royal Soc.* XCVI A (1920), 72; and J. C. Burkill, *Proc. London Math. Soc.* XXII (1923), 311.

† See § 11·51, I.

The number σ is an approximation to the area of S, and the upper bound of the sums σ is defined to be the area of S.

If d_{rs} is the projection of β_{rs} on the (yz)-plane, then the area of the closed plane curve C_1 arising from d_{rs} is

$$\iint_{d_{rs}} dy\,dz = \iint_{\rho_{rs}} A\,du\,dv,$$

where $A = \dfrac{\partial(y, z)}{\partial(u, v)}$ as previously defined.

The area a required is therefore, by definition,

$$\left\{\left(\iint_{\rho_{rs}} A\,du\,dv\right)^2 + \left(\iint_{\rho_{rs}} B\,du\,dv\right)^2 + \left(\iint_{\rho_{rs}} C\,du\,dv\right)^2\right\}^{\frac{1}{2}}.$$

By assuming for the functions

$$x = f(u, v), \quad y = g(u, v), \quad z = h(u, v)$$

absolute continuity, it can be proved that

$$\lim \Sigma a \geqslant \iint_\Delta \sqrt{(A^2 + B^2 + C^2)}\,du\,dv \quad\text{.....................(1)};$$

and, on using Minkowski's inequality for integrals, that

$$\Sigma a \leqslant \iint_\Delta \sqrt{(A^2 + B^2 + C^2)}\,du\,dv \quad\text{.....................(2)};$$

hence

$$\lim \Sigma a = \iint_\Delta \sqrt{(A^2 + B^2 + C^2)}\,du\,dv \quad\text{.....................(3)}.$$

It is not within the scope of this book to discuss the concept of absolute continuity, and the proofs of the inequalities (1) and (2) are not therefore given. Further details can be found in the original paper to which reference has already been made.

Enough has been said to shew the reader that the area of a curved surface has to be defined quite differently from the way in which either the length of a curve or the area of a plane domain has been defined.

12·32. Note on other methods of defining surface-area.

In view of what has already been said, the author is of opinion that the best *elementary* approach is on the lines of de la Vallée Poussin*, who defines the area of the surface *a priori* by the integral (3) above. This method is described in the next section.

It may be remarked that attempts to define the area of a surface as the limit of the areas of approximative polyhedra have generally been successful only if the surface has a continuously varying

* *Cours d'Analyse Infinitésimale*, I (1921), § 287.

tangent plane. Under suitable restrictions a surface can be proved to have the area given by the integral (3) above, by projecting an arbitrary element of the surface on to one of the tangent planes belonging to that element. Provided that the normals to the tangent planes only cut the surface once, and the first partial derivatives of x, y, z with respect to u and v are all continuous, the limiting value of the sum of all the plane areas obtained by the above projection is the value of the area of the surface*.

Instead of projecting the element of surface on to a tangent plane, another method is to project it on to a plane for which the projection is a maximum. Provided that the surface S has a continuously turning tangent plane nowhere parallel to the axis of z, this definition leads to the same value for the area S as before.

12·33. De la Vallée Poussin's definition of surface-area.

Let Δ be the domain in the (u, v)-plane of which a curved surface S in the (x, y, z)-plane is the image. Suppose that the functions

$$x = f(u, v), \quad y = g(u, v), \quad z = h(u, v),$$

which define the surface, together with their first partial derivatives, are single-valued and continuous in the regions concerned, and that the three determinants A, B, C are not simultaneously zero in Δ. We *define* the area of the surface S *a priori* to be

$$\iint_\Delta \sqrt{(EG - F^2)}\, du\, dv \quad \text{..................(1)}.$$

It is easy to establish the identity

$$A^2 + B^2 + C^2 \equiv EG - F^2,$$

and so the integral (1) is the same as

$$\iint_\Delta \sqrt{(A^2 + B^2 + C^2)}\, du\, dv \quad \text{..............(2)}.$$

The element of area on the surface may therefore be expressed in either of the equivalent forms

$$\sqrt{(A^2 + B^2 + C^2)}\, du\, dv, \quad \sqrt{(EG - F^2)}\, du\, dv \quad \text{......(3)}.$$

* See Goursat, *Cours d'Analyse*, I (1917), § 131.

The latter expression shews that the element of area is invariant for changes of rectangular axes, since it depends only on

$$ds^2 = dx^2 + dy^2 + dz^2$$

and this is certainly invariant. Hence the integral (1) is independent of the choice of the rectangular axes of reference Ox, Oy, Oz.

Let us consider the geometrical interpretation of the integral (2); since the direction-cosines of the normal to the surface at any point are proportional to A, B, C, if these direction-cosines are denoted by l, m, n, we have

$$\frac{l}{A} = \frac{m}{B} = \frac{n}{C} = \frac{1}{\sqrt{(A^2 + B^2 + C^2)}} \quad \text{..............(4).}$$

If we suppose for definiteness that $C \neq 0$ in Δ, the geometrical interpretation of this is that the normal is not perpendicular to the axis of z, and it can be drawn so as to make an acute angle with this axis, so that, since n is positive,

$$n = \frac{|C|}{\sqrt{(A^2 + B^2 + C^2)}},$$

and the integral (2) becomes

$$\iint_\Delta \frac{|C|}{n}\, du\, dv \quad \text{......................(5).}$$

If D denotes the projection of the surface S on the (x, y)-plane, then on transferring the integral (5) to an integral over the plane domain D, since $J = \dfrac{\partial(u, v)}{\partial(x, y)} = \dfrac{1}{C}$, we get

$$\iint_D \frac{dx\,dy}{n} \quad \text{..........................(6).}$$

Similar integrals are obtainable by assuming that $A \neq 0$, $B \neq 0$. The integral (6), and those analogous to it, shew that the value of the integral (2) does not depend upon the choice of the parameters u and v. It must therefore depend only upon the shape of the surface S.

If the surface S does not satisfy the condition that the three determinants A, B, C must not be simultaneously zero in the domain Δ, but if we can divide up the surface S into a finite number of pieces S_r $(r = 1, 2, \ldots, n)$, such that the three determinants A, B, C are not simultaneously zero in each of the domains Δ_r $(r = 1, 2, \ldots, n)$, the above argument is applied to each piece S_r separately, and the results are added.

The condition under which formula (6) was established, namely that $C \neq 0$ in Δ, is equivalent to the geometrical condition that the surface S, which is the image of the domain Δ, has no tangent plane anywhere perpendicular to the (x, y)-plane. When $C \neq 0$ we know that the equation of the surface assumes the form

$$z = \phi(x, y);$$

now

$$n = 1/\sqrt{(1 + p^2 + q^2)},$$

and so the area of the curved surface S is given by

$$\iint_D \sqrt{(1 + p^2 + q^2)}\, dx\, dy \quad \text{..................(7)}.$$

The reader is advised to verify that the above integrals give the right results for the area of elementary curved surfaces. For example, in the case of the sphere

$$x = a \sin\theta \cos\phi, \quad y = a \sin\theta \sin\phi, \quad z = a \cos\theta,$$

it is easily verified that

$$E = a^2 \sin^2\theta, \quad F = 0, \quad G = a^2$$

and so the area of the surface of the sphere is given by

$$A = \iint_\Delta \sqrt{(EG - F^2)}\, du\, dv = \iint_\Delta a^2 \sin\theta\, d\theta\, d\phi.$$

If θ ranges from 0 to π and ϕ from 0 to 2π the point (x, y, z) traces out the whole surface of the sphere, and so

$$A = a^2 \int_0^\pi \sin\theta\, d\theta \int_0^{2\pi} d\phi = 4\pi a^2.$$

The calculation of the surface-area by means of the integral (7) is left as an exercise for the reader.

12·4. The "sides" of a curved surface.

Let S be a portion of a surface which is bounded by a definite contour. The surfaces with which we shall be mainly concerned must possess two distinct *sides*: by this we mean that if a moving point traces out a path in any manner on one side of the surface it cannot come to the other side by any means save by crossing the contour which forms the boundary of the surface.

An example of a surface which possesses two distinct "sides" is a spherical cap. Consider the portion of the surface of the sphere $x^2 + y^2 + z^2 = a^2$ which lies between the planes $z = c$, $z = a$, where $c < a$; it is clear that on whatever "side" the moving point starts it cannot get to the other "side" without crossing the boundary, which in this case is the circle $z = c$, $x^2 + y^2 = a^2 - c^2$.

It is easy to construct a surface which does not possess two distinct "sides." Cut out a rectangular strip of paper, and on one face of it mark the letters A, B, C, D as shewn in Fig. 33.

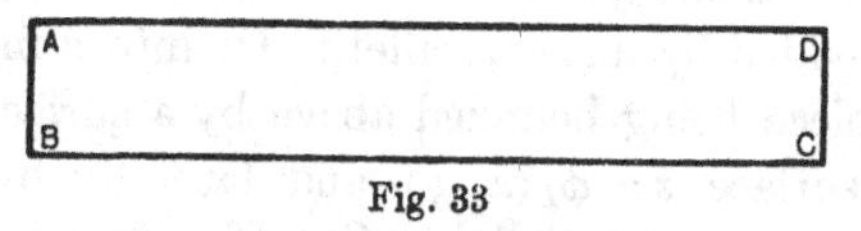

Fig. 33

Give the strip of paper a twist in such a way that the two narrow edges come into contact with the letter C falling on A, and D on B. It will be found that the surface so formed does not possess *distinct sides*, for a point may start anywhere on the surface and move without ever crossing the boundary and yet return to its starting-point on the opposite side of the paper from the one on which it started.

Suppose that we have a portion of a surface defined by the equation

$$z = \phi(x, y),$$

and let the axis of z be vertically upwards as illustrated in Fig. 34.

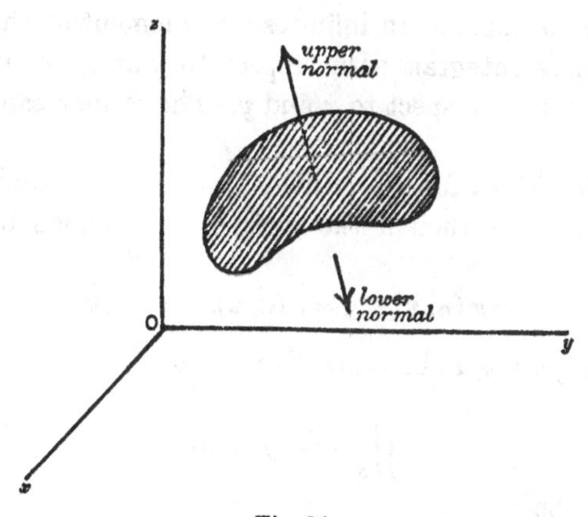

Fig. 34

The side of this surface from which the outward-drawn normal makes an acute angle with Oz is called the *upper side*, and the other side, which faces the (x, y)-plane, is called the *lower side*. The half-line drawn perpendicular to the tangent plane at any point on the upper side in a direction outwards above the upper side may be called the *upper normal*: the other half-line drawn in a direction downwards below the lower side is the *lower normal*.

12·41. Dissection of a three-dimensional domain.

Suppose that we have a domain T in (x, y, z)-space bounded by a surface S: the domain is said to be *dissected relative to the axis of* z, if it is divided by lines parallel to Oz into a finite number of pieces, each piece being bounded above by a surface $z=\phi_2(x, y)$, below by a surface $z=\phi_1(x, y)$, and laterally by a cylindrical surface with generators parallel to Oz. If a domain T is dissected relative to the axis of z we shall suppose that for any piece T_r the two surfaces $z=\phi_2(x, y)$, $z=\phi_1(x, y)$ are such that the tangent plane at any point is nowhere perpendicular to the (x, y)-plane, except perhaps on the curves which form the rims of these two surfaces. Such a dissected piece T_r is bounded by surfaces which are only met by a line parallel to any one of the axes in two points.

Similarly we define dissection relative to either of the other axes.

12·42. Surface integrals.

There are two kinds of surface integrals, (i) integrals with respect to σ, where $d\sigma$ denotes an infinitesimal element of the surface S, and (ii) surface integrals with respect to y and z, or with respect to z and x or with respect to x and y. The former can be defined at once.

Let $f(x, y, z)$ be a function which is single-valued and continuous at all points of S, then if the surface S is defined by the usual equations

$$x=x(u, v), \quad y=y(u, v), \quad z=z(u, v),$$

the function $f(x, y, z)$ becomes $F(u, v)$ and

$$\iint_S f(x, y, z)\, d\sigma$$

is *defined* to be

$$\iint_\Delta F(u, v)\sqrt{(EG-F^2)}\, du\, dv,$$

where Δ is the domain in the (u, v)-plane of which S is the image.

The latter type is defined by first dividing up the surface S, which is assumed to be regular, into pieces each of which has an equation of a particular form. Take, for example, the case of integrals with respect to x and y. Suppose that S is divided into a finite number of pieces S_r each of which has an equation of the form $z=\nu(x,y)$ and no tangent plane is perpendicular to the (x,y)-plane; then if Σ_r is the projection of S_r on the (x,y)-plane,

$$\iint_{S_r} f(x,y,z)\,dx\,dy = \pm\iint_{\Sigma_r} f\{x,y,\nu(x,y)\}\,dx\,dy,$$

the positive sign being taken for the upper side of S_r, the negative sign for the lower side of S_r. By adding the results for all the pieces S_r we obtain

$$\iint_S f(x,y,z)\,dx\,dy.$$

If any of the pieces into which S is divided are portions of cylindrical surfaces with generators parallel to Oz, these portions give no contribution to integrals with respect to x and y.

Integrals of the second type can be transformed into double integrals over the plane domain Δ in the (u, v)-plane. By the formula for change of variables in a double integral

$$\iint_{\Sigma_r} f\{x,y,\nu(x,y)\}\,dx\,dy = \iint_{\Delta_r} F(u,v)\left|\frac{\partial(x,y)}{\partial(u,v)}\right| du\,dv$$

$$= \iint_{\Delta_r} F(u,v)\,|C|\,du\,dv,$$

and so by addition we get

$$\iint_S f(x,y,z)\,dx\,dy = \iint_\Delta F(u,v)\,C\,du\,dv,$$

the side of S which is taken being determined by the sign of C.

Similar results hold for integrals with respect to z and x, and with respect to y and z.

12·43. Relation between the two kinds of surface integral.

We now shew that

$$\iint_S lf(x,y,z)\,d\sigma = \iint_S f(x,y,z)\,dy\,dz \quad\text{.........(1)},$$

where (l, m, n) are the direction-cosines of the normal to the surface S drawn outwards from the side of S in question.

Now
$$\frac{l}{A}=\frac{m}{B}=\frac{n}{C}=\frac{1}{\sqrt{(A^2+B^2+C^2)}},$$

and so
$$l=\frac{A}{\sqrt{(EG-F^2)}},$$

hence the left-hand side of (1) above reduces to

$$\iint_{\Delta} \frac{A}{\sqrt{(EG-F^2)}} F(u,v)\sqrt{(EG-F^2)}\,du\,dv = \iint_{\Delta} F(u,v)\,A\,du\,dv,$$

which is identical with the right-hand side of (1).

12·44. Integrals over closed surfaces.

If the closed surface S is of the simple oval type, such as an ellipsoid, and the axis of z is vertically upwards, the points at which the tangent planes to S are perpendicular to the (x, y)-plane lie on a certain twisted curve C drawn on the surface S. Each of the two parts into which S is divided by this curve C has two distinct sides. The *outer side* of the given closed surface S is the upper side of that portion of the surface which lies above the curve C and the lower side of the portion which lies below C.

Green's theorem (see § 12·5) connects a triple integral through a domain T with a surface integral taken over the closed boundary surface S of the domain T.

In order to prove this theorem we divide up the domain T into a finite number of pieces by dissection relative to one or more of the axes Ox, Oy, Oz. For definiteness, suppose that the domain T is dissected relative to Oz. In this case we divide up the domain into pieces T_r bounded by surfaces S_r, each of which consists of portions having an equation of the form $z = \nu(x, y)$ and cylindrical portions with generators parallel to Oz. The only curves on S along which the tangent planes are perpendicular to the (x, y)-plane will be some of the boundary curves of certain portions S_r. Since surface integrals with respect to x and y are zero over any cylindrical portion with generators parallel to Oz, the sum of the surface integrals over all the dissected pieces S_r is clearly equal to the surface integral over the closed boundary surface S of the domain T. Similar remarks apply to dissection relative to Ox and Oy.

12·45. Change of variables in a surface integral.

Let the surface S in (x, y, z)-space be the image of a surface S_1 in (ξ, η, ζ)-space, and let Δ be the domain in the (u, v)-plane to which both these surfaces correspond. Suppose that

$$f(x, y, z) = f_1(\xi, \eta, \zeta) = F(u, v).$$

We have already seen that

$$\iint_S f(x, y, z)\, dy\, dz = \iint_\Delta F(u, v) \frac{\partial(y, z)}{\partial(u, v)} du\, dv \;\ldots\ldots(1),$$

but*

$$\frac{\partial(y, z)}{\partial(u, v)} = \frac{\partial(y, z)}{\partial(\xi, \eta)} \frac{\partial(\xi, \eta)}{\partial(u, v)} + \frac{\partial(y, z)}{\partial(\eta, \zeta)} \frac{\partial(\eta, \zeta)}{\partial(u, v)} + \frac{\partial(y, z)}{\partial(\zeta, \xi)} \frac{\partial(\zeta, \xi)}{\partial(u, v)},$$

and on substituting this in the right-hand side of (1), since

$$\frac{\partial(\xi, \eta)}{\partial(u, v)} du\, dv = d\xi\, d\eta,$$

we get

$$\iint_S f(x, y, z)\, dy\, dz$$

$$= \iint_{S_1} f_1(\xi, \eta, \zeta) \left\{ \frac{\partial(y, z)}{\partial(\xi, \eta)} d\xi\, d\eta + \frac{\partial(y, z)}{\partial(\eta, \zeta)} d\eta\, d\zeta + \frac{\partial(y, z)}{\partial(\zeta, \xi)} d\zeta\, d\xi \right\}.$$

12·5. Green's theorem.

The three-dimensional form of Green's theorem may be stated as follows.

Let $P(x, y, z)$, $Q(x, y, z)$, $R(x, y, z)$ be three given integrands, T a three-dimensional domain; then if the functions

$$P,\ Q,\ R,\ \frac{\partial P}{\partial x},\ \frac{\partial Q}{\partial y},\ \frac{\partial R}{\partial z}$$

are all continuous throughout T and on S, its closed boundary surface, and, if the surface integral be taken over the outer side of S,

$$\iint_S (P\, dy\, dz + Q\, dz\, dx + R\, dx\, dy) = \iiint_T \left(\frac{\partial P}{\partial x} + \frac{\partial Q}{\partial y} + \frac{\partial R}{\partial z} \right) dx\, dy\, dz$$

$$\ldots\ldots\ldots(1).$$

To prove the theorem it is sufficient to consider one of the three parts into which the formula (1) may be divided, say

$$\iint_{S_r} R\, dx\, dy = \iiint_{T_r} \frac{\partial R}{\partial z} dx\, dy\, dz \;\ldots\ldots\ldots\ldots(2),$$

where S_r and T_r refer to one of the pieces into which the domain T has been dissected relative to the axis of z.

* See § 10·64, III.

† These are the "standard conditions," and they are sufficient but not necessary.

Now the right-hand side of (2) becomes

$$\iint_{\Sigma_r} dx\,dy \int_{z_1}^{z_2} \frac{\partial R}{\partial z}\,dz \quad \text{.....................(3)},$$

where Σ_r is the projection of the dissected domain T_r on the (x, y)-plane and $z_1 = \phi_1(x, y)$, $z_2 = \phi_2(x, y)$, where $z_2 > z_1$. The integral (3) therefore becomes

$$\iint_{\Sigma_r} \{R(x, y, z_2) - R(x, y, z_1)\}\,dx\,dy \quad \text{...........(4)}$$

The integral (4) is the difference between two double integrals which are respectively the same as two surface integrals, the former over the upper side of the surface $z = \phi_2(x, y)$ and the latter over the lower side of the surface $z = \phi_1(x, y)$. In each case it is the outer side of those portions of the surface S which belong to the dissected piece T_r under consideration. Since integrals with respect to x and y over the cylindrical surface with generators parallel to the axis of z are zero, it follows that the integral (4) is the same as

$$\iint_{S_r} R\,dx\,dy,$$

hence (2) is established.

Similarly we prove the other two parts, and the result for the whole domain T follows by addition.

12·51. Volume in terms of surface integrals.

In Green's theorem, if we write $P = x$, $Q = 0$, $R = 0$, we get

$$\iint_S x\,dy\,dz = \iiint_T dx\,dy\,dz$$

and so it follows that the volume of the domain T can be found by evaluating any one of the three surface integrals

$$\iint_S x\,dy\,dz, \quad \iint_S y\,dz\,dx, \quad \iint_S z\,dx\,dy,$$

the integral being taken over the outer side of S in each case.

12·52. Volume in curvilinear coordinates.

Just as in the two-dimensional case, Green's theorem is used to obtain the required formula.

THEOREM. *If the transformation is defined by the equations*

$$x = \phi_1(\xi, \eta, \zeta), \quad y = \phi_2(\xi, \eta, \zeta), \quad z = \phi_3(\xi, \eta, \zeta),$$

then the integral for volume in the coordinates ξ, η, ζ *is*

$$\iiint_{T_1} |J|\, d\xi\, d\eta\, d\zeta,$$

where $J \equiv \dfrac{\partial(x, y, z)}{\partial(\xi, \eta, \zeta)}$ *and* T_1 *is the domain in* (ξ, η, ζ)*-space of which the domain* T *whose volume is required is the image.*

Let S and S_1 be the closed surfaces bounding the domains T and T_1, then, if V be the volume required,

$$V = \iint_S x\, dy\, dz$$

$$= \iint_{S_1} x \left\{ \frac{\partial(y, z)}{\partial(\eta, \zeta)} d\eta\, d\zeta + \frac{\partial(y, z)}{\partial(\zeta, \xi)} d\zeta\, d\xi + \frac{\partial(y, z)}{\partial(\xi, \eta)} d\xi\, d\eta \right\}.$$

On applying Green's theorem we get

$$V = \pm \iiint_{T_1} \left[\frac{\partial}{\partial \xi} \left\{ x \frac{\partial(y, z)}{\partial(\eta, \zeta)} \right\} + \frac{\partial}{\partial \eta} \left\{ x \frac{\partial(y, z)}{\partial(\zeta, \xi)} \right\} + \frac{\partial}{\partial \zeta} \left\{ x \frac{\partial(y, z)}{\partial(\xi, \eta)} \right\} \right] d\xi\, d\eta\, d\zeta,$$

and the integrand reduces to

$$\frac{\partial x}{\partial \xi} \frac{\partial(y, z)}{\partial(\eta, \zeta)} + \frac{\partial x}{\partial \eta} \frac{\partial(y, z)}{\partial(\zeta, \xi)} + \frac{\partial x}{\partial \zeta} \frac{\partial(y, z)}{\partial(\xi, \eta)} = \frac{\partial(x, y, z)}{\partial(\xi, \eta, \zeta)} \quad \text{......(1)},$$

since the second order derivatives are annulled. t follows that

$$V = \pm \iiint_{T_1} J\, d\xi\, d\eta\, d\zeta.$$

Since V is essentially positive we shew as before (§ 11·7) that the sign of J determines the nature of the transformation: if J is positive the outside of S corresponds to the outside of S_1, while if J is negative the outside of S corresponds to the inside of S_1. In either case

$$V = \iiint_{T_1} |J|\, d\xi\, d\eta\, d\zeta.$$

12·6. Change of variables in a triple integral.

Without giving the detailed proof, we now state that if the domain T be divided into cells by the surfaces

$$\xi=\alpha, \quad \eta=\beta, \quad \zeta=\gamma,$$

and we apply the formula for the volume of a cell and the generalised first mean-value theorem for a triple integral, then

$$\iiint_T f(x, y, z)\, dx\, dy\, dz=\iiint_{T_1} f_1(\xi, \eta, \zeta)\, |J|\, d\xi\, d\eta\, d\zeta.$$

The reader is advised to write out the formal proof as an exercise. The proof is on exactly the same lines as the one given for double integrals in § 11·8.

Example. Calculate Dirichlet's integral

$$I \equiv \iiint x^{\alpha} y^{\beta} z^{\gamma} (1-x-y-z)^{\lambda}\, dx\, dy\, dz$$

over the interior of the tetrahedron formed by the coordinate planes and the plane $x+y+z=1$.

Write $$x+y+z=\xi, \quad y+z=\xi\eta, \quad z=\xi\eta\zeta \quad \text{..................(1)},$$

and use the formula for change of variables above.

The equations (1) may be written

$$x=\xi(1-\eta), \quad y=\xi\eta(1-\zeta), \quad z=\xi\eta\zeta \quad \text{..................(2)},$$

and the tetrahedron in (x, y, z)-space is seen to be the image of a unit cube in (ξ, η, ζ)-space. The determinant

$$J=\frac{\partial(x, y, z)}{\partial(\xi, \eta, \zeta)}$$

may be calculated directly from the equations (2), but if we write

$$u=\xi, \quad v=\xi\eta, \quad w=\xi\eta\zeta \quad \text{..................(3)},$$

so that $$x=u-v, \quad y=v-w, \quad z=w \quad \text{..................(4)},$$

then
$$\frac{\partial(x, y, z)}{\partial(\xi, \eta, \zeta)}=\frac{\partial(x, y, z)}{\partial(u, v, w)} \cdot \frac{\partial(u, v, w)}{\partial(\xi, \eta, \zeta)}$$

$$=\begin{vmatrix} 1, & -1, & 0 \\ 0, & 1, & -1 \\ 0, & 0, & 1 \end{vmatrix} \times \begin{vmatrix} 1, & \eta, & \eta\zeta \\ 0, & \xi, & \zeta\xi \\ 0, & 0, & \xi\eta \end{vmatrix},$$

$$=\xi^2\eta.$$

The triple integral in (ξ, η, ζ)-coordinates can be calculated as a repeated integral, and so

$$I=\int_0^1 \xi^{\alpha+\beta+\gamma+2}(1-\xi)^{\lambda}\, d\xi \cdot \int_0^1 \eta^{\beta+\gamma+1}(1-\eta)^{\alpha}\, d\eta \cdot \int_0^1 \zeta^{\gamma}(1-\zeta)^{\beta}\, d\zeta$$

$$=\mathrm{B}(\alpha+\beta+\gamma+3, \lambda+1) \cdot \mathrm{B}(\beta+\gamma+2, \alpha+1) \cdot \mathrm{B}(\gamma+1, \beta+1)$$

$$=\frac{\Gamma(\alpha+1)\Gamma(\beta+1)\Gamma(\gamma+1)\Gamma(\lambda+1)}{\Gamma(\alpha+\beta+\gamma+\lambda+4)}.$$

12·61. Laplace's equation in orthogonal curvilinear coordinates.

Laplace's equation

$$\nabla^2 V \equiv \frac{\partial^2 V}{\partial x^2} + \frac{\partial^2 V}{\partial y^2} + \frac{\partial^2 V}{\partial z^2} = 0$$

is one of the fundamental differential equations of Applied Mathematics and Physics, and it is frequently necessary, for the application of this equation to many of the problems which involve its use, to be able to express it in terms of other systems of coordinates. The direct method of transforming $\nabla^2 V$ into spherical polar coordinates was given in § 10·4. The method which is now given enables us to express Laplace's equation in *any* system of orthogonal curvilinear coordinates.

If (l, m, n) are the direction-cosines of the normal drawn outwards from the surface S which bounds a given domain T, Green's theorem may be written

$$\iint_S (lP + mQ + nR)\, d\sigma = \iiint_T \left(\frac{\partial P}{\partial x} + \frac{\partial Q}{\partial y} + \frac{\partial R}{\partial z}\right) dx\,dy\,dz \quad (1);$$

and on writing $P = \dfrac{\partial V}{\partial x}$, $Q = \dfrac{\partial V}{\partial y}$, $R = \dfrac{\partial V}{\partial z}$, equation (1) becomes

$$\iint_S \frac{\partial V}{\partial \nu}\, d\sigma = \iiint_T \nabla^2 V\, dx\,dy\,dz \quad \text{..............(2)},$$

where $\dfrac{\partial}{\partial \nu} \equiv l\dfrac{\partial}{\partial x} + m\dfrac{\partial}{\partial y} + n\dfrac{\partial}{\partial z}$ and denotes the derivative in the direction of the normal to the surface element in question.

Let u, v, w be any system of orthogonal curvilinear coordinates, and let T_1 be the rectangular parallelepiped bounded by the planes*

$$u = u_1,\ v = v_1,\ w = w_1; \quad u = u_2,\ v = v_2,\ w = w_2,$$

then T (the image of T_1) in (x, y, z)-space is bounded by surfaces which will be denoted by

$$U_1,\ V_1,\ W_1; \quad U_2,\ V_2,\ W_2 \quad \text{..............(3)}.$$

On writing $H_1 = 1/h_1^2$, $H_2 = 1/h_2^2$, $H_3 = 1/h_3^2$, the element of length is given by

$$ds^2 = \frac{du^2}{h_1^2} + \frac{dv^2}{h_2^2} + \frac{dw^2}{h_3^2}.$$

* We are assuming that $u_2 > u_1$, $v_2 > v_1$, $w_2 > w_1$.

and $|J| = \sqrt{(H_1 H_2 H_3)} = 1/h_1 h_2 h_3$; hence

$$\iint_S \frac{\partial V}{\partial \nu} d\sigma = \iiint_T \nabla^2 V \, dx\,dy\,dz = \iiint_{T_1} \frac{\nabla^2 V}{h_1 h_2 h_3} du\,dv\,dw \;\ldots(4).$$

The first integral in (4) may be calculated in another way. The surface S consists of the six surfaces (3), and by considering U_1 only we get

$$\iint_{U_1} \frac{\partial V}{\partial \nu} d\sigma = \iint_{u_1} \left(-h_1 \frac{\partial V}{\partial u}\right)\left(-\frac{1}{h_2 h_3} dv\,dw\right) \quad \ldots\ldots(5),$$

for $\dfrac{\partial}{\partial \nu}$ is a derivation along the normal to the plane $u = u_1$ in the direction of u decreasing, so that $\dfrac{\partial}{\partial \nu} = -h_1 \dfrac{\partial}{\partial u}$, and the second term in the last integral has a negative sign because the integral on the left-hand side of (5) is taken over the outer, which is the lower, side of U_1. Hence, on assuming that the transformation is direct, we have

$$\iint_{U_1} \frac{\partial V}{\partial \nu} d\sigma = \iint_{u_1} \frac{h_1}{h_2 h_3} \frac{\partial V}{\partial u} dv\,dw.$$

Similarly for integration over the surface U_2, and since integrals with respect to v and w are zero over the remainder of S,

$$\iint_{U_1 + U_2} \frac{\partial V}{\partial \nu} d\sigma = \iint_{S_1} \frac{h_1}{h_2 h_3} \frac{\partial V}{\partial u} dv\,dw.$$

On treating the other four pieces of the bounding surface S in a similar way we get

$$\iint_S \frac{\partial V}{\partial \nu} d\sigma = \iint_{S_1} \left(\frac{h_1}{h_2 h_3} \frac{\partial V}{\partial u} dv\,dw + \frac{h_2}{h_3 h_1} \frac{\partial V}{\partial v} dw\,du + \frac{h_3}{h_1 h_2} \frac{\partial V}{\partial w} du\,dv\right),$$

and by Green's theorem the last integral transforms into

$$\iiint_{T_1} \left\{\frac{\partial}{\partial u}\left(\frac{h_1}{h_2 h_3} \frac{\partial V}{\partial u}\right) + \frac{\partial}{\partial v}\left(\frac{h_2}{h_3 h_1} \frac{\partial V}{\partial v}\right) + \frac{\partial}{\partial w}\left(\frac{h_3}{h_1 h_2} \frac{\partial V}{\partial w}\right)\right\} du\,dv\,dw \quad \ldots\ldots\ldots(6).$$

By comparing (4) and (6) we get

$$\nabla^2 V = h_1 h_2 h_3 \left\{\frac{\partial}{\partial u}\left(\frac{h_1}{h_2 h_3} \frac{\partial V}{\partial u}\right) + \frac{\partial}{\partial v}\left(\frac{h_2}{h_3 h_1} \frac{\partial V}{\partial v}\right) + \frac{\partial}{\partial w}\left(\frac{h_3}{h_1 h_2} \frac{\partial V}{\partial w}\right)\right\}.$$

For the polar transformation we have

$$ds^2 = dr^2 + r^2 d\theta^2 + r^2 \sin^2 \theta \, d\phi^2,$$

and so $\quad 1/h_1^2 = 1, \quad 1/h_2^2 = r^2, \quad 1/h_3^2 = r^2 \sin^2 \theta;$

hence

$$\nabla^2 V = \frac{1}{r^2 \sin\theta}\left\{\frac{\partial}{\partial r}\left(r^2 \sin\theta \frac{\partial V}{\partial r}\right) + \frac{\partial}{\partial \theta}\left(\sin\theta \frac{\partial V}{\partial \theta}\right) + \frac{\partial}{\partial \phi}\left(\frac{1}{\sin\theta}\frac{\partial V}{\partial \phi}\right)\right\}$$

$$= \frac{1}{r^2}\frac{\partial}{\partial r}\left(r^2 \frac{\partial V}{\partial r}\right) + \frac{1}{r^2}\frac{\partial^2 V}{\partial \theta^2} + \frac{\cot\theta}{r^2}\frac{\partial V}{\partial \theta} + \frac{1}{r^2 \sin^2\theta}\frac{\partial^2 V}{\partial \phi^2}.$$

12·7. Curvilinear integrals in three-dimensional space.

The notion of a curvilinear integral can be extended at once to three-dimensional space. Consider an arc AB of a twisted curve on which x varies in such a way that it is either always increasing or always decreasing from $x = a$ to $x = b$. If the equations defining the curve are $x = f(t)$, $y = \phi(t)$, $z = \psi(t)$, by the inverse function theorem t is a monotonic function of x, and so y and z can be expressed as functions of x. If $P(x, y, z)$ is a continuous function of x, y and z, and y and z are themselves continuous functions of x, we define the curvilinear integral along the curve C,

$$\int_C P(x, y, z)\, dx \quad \text{.......................(1)},$$

to be the ordinary Riemann integral

$$\int_a^b P\{x, y(x), z(x)\}\, dx = \int_a^b \Phi(x)\, dx \quad \text{...........(2)}.$$

By an argument similar to that used in § 8·7, it can be shewn that the integral (2) certainly exists under the conditions stated above.

If a given curve L does not satisfy the condition that the equations $x = f(t)$, $y = \phi(t)$, $z = \psi(t)$ are monotonic functions of t, the definition can be extended to such a curve provided that it is possible to divide up the curve L into a finite number of curves C_r on each of which the coordinate x is always increasing or always decreasing. In this case

$$\int_L P(x, y, z)\, dx = \sum_{r=1}^{n} \int_{C_r} P(x, y, z)\, dx.$$

In the same way we can define curvilinear integrals with respect to y and with respect to z, and the most general expression of a curvilinear integral in three dimensions is

$$\int_L (P\,dx + Q\,dy + R\,dz),$$

where P, Q, R are continuous functions of the three variables x, y, z.

12·8. Stokes's theorem.

In certain problems which occur in Applied Mathematics and Physics, it is useful to be able to express a combination of surface integrals over a given surface in terms of curvilinear integrals round the circuit in space which forms the rim of the given surface. This is effected by means of Stokes's theorem, of which a proof is now given.

Stokes's theorem is proved here mainly because of its importance in the applications of mathematics to physical problems. It occupies a very different position from Green's theorem, which, although it has important practical applications, is one of the fundamental theorems in the theory of multiple integrals.

STOKES'S THEOREM. *If C be the circuit in space which forms the rim of a given surface S, then*

$$\int_C (P\,dx + Q\,dy + R\,dz) = \iint_S \left\{\left(\frac{\partial R}{\partial y} - \frac{\partial Q}{\partial z}\right) dy\,dz + \left(\frac{\partial P}{\partial z} - \frac{\partial R}{\partial x}\right) dz\,dx + \left(\frac{\partial Q}{\partial x} - \frac{\partial P}{\partial y}\right) dx\,dy\right\} \quad \text{......(1)},$$

where P, Q, R are continuous functions of x, y, z, and the sense of description of C is so related to the side of S that a person traversing C and keeping on the correct side of S has that side of the surface on his left.

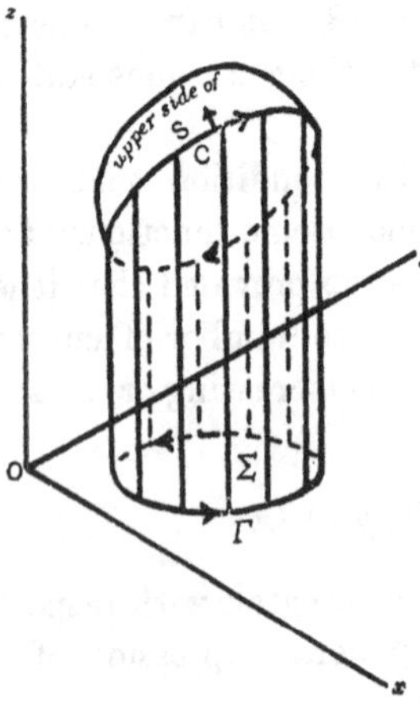

Fig. 35

The result may be divided into three parts: consider one of these parts,

$$\int P\,dx = \iint_S \left(\frac{\partial P}{\partial z}\,dz\,dx - \frac{\partial P}{\partial y}\,dx\,dy\right) \quad \text{...(2)}.$$

We are assuming that the surface S is one whose equation can be expressed in the form $z = \nu(x, y)$.

If this is not the case, then so long as the surface possesses two distinct sides we divide it up into suitable pieces by dissection parallel to the axis of z so that on each of the dissected pieces the tangent plane is nowhere perpendicular to the (x, y)-plane. If the theorem is proved for any one of such pieces, it follows for the whole surface S by addition.

Let Σ be the plane domain in the (x, y)-plane which is bounded by the closed curve Γ, the projection of the given circuit C on the (x, y)-plane. If (l, m, n) are the direction-cosines of the upper normal to S,

$$\frac{l}{\frac{\partial z}{\partial x}} = \frac{m}{\frac{\partial z}{\partial y}} = \frac{n}{-1};$$

and since $dz\,dx = m\,d\sigma$, $dx\,dy = n\,d\sigma$, the right-hand side of (2) becomes

$$\iint_S \left(\frac{\partial P}{\partial z}\frac{m}{n} - \frac{\partial P}{\partial y}\right) dx\,dy \quad \text{..................(3).}$$

The integrand of (3) is

$$-\left(\frac{\partial P}{\partial z}\frac{\partial z}{\partial y} + \frac{\partial P}{\partial y}\right) = -\frac{\partial P_1}{\partial y} \quad \text{..................(4),}$$

if
$$P_1(x, y) = P\{x, y, \nu(x, y)\}.$$

The integral (3) therefore reduces to a double integral over the plane domain Σ

$$-\iint_\Sigma \frac{\partial P_1}{\partial y}\,dx\,dy.$$

Now
$$\int_C P\,dx = \int_\Gamma P_1\,dx,$$

and so the required result is established if we prove that

$$-\iint_\Sigma \frac{\partial P_1}{\partial y}\,dx\,dy = \int_\Gamma P_1\,dx,$$

which is part of the two-dimensional form of Green's theorem.

Thus equation (2) is proved, and the other two parts of the theorem may be similarly established.

12·9. Multiple integrals.

Although there is no geometrical analogy for the case of more than three variables, the analysis of most of the work of this chapter is easily extended to n variables.

If $f(x_1, x_2, \ldots, x_n)$ is an integrable function of the n independent variables $x_1, x_2, \ldots, x_n$ we can easily define the n-ple integral

$$\iint \ldots \int_R f(x_1, x_2, \ldots, x_n)\,dx_1\,dx_2 \ldots dx_n,$$

where R is a hyper-rectangle in n-dimensional space.

It is unnecessary to restate theorems which readily extend from three to n dimensions, and one example is sufficient by way of illustration. The formula for change of variables remains the same for n-ple integrals, for let

$$x_r = \phi_r(x_1', x_2', \ldots, x_n') \quad (r = 1, 2, \ldots, n)$$

be the equations which define the transformation, and let

$$J \equiv \frac{\partial(x_1, x_2, \ldots, x_n)}{\partial(x_1', x_2', \ldots, x_n')},$$

then

$$\iint \ldots \int_D f(x_1, x_2, \ldots, x_n)\, dx_1 dx_2 \ldots dx_n$$
$$= \iint \ldots \int_{D'} f\{\phi_1, \phi_2, \ldots, \phi_n\} \,|J|\, dx_1' dx_2' \ldots dx_n',$$

where D and D' are the domains in the respective spaces which correspond by the given transformation.

EXAMPLES XII.

1. Prove that the volume enclosed by the cylinders

$$x^2 + y^2 = 2ax, \quad z^2 = 2ax$$

is $128a^3/15$.

2. A solid is bounded by the cone $z^2 = 2xy$ and the cylinder $\sqrt{x} + \sqrt{y} = 1$. Prove that the volume of the solid is $2\sqrt{2}/45$, and find the area of the conical surface bounding it.

3. Prove that

$$\iiint_T x^{p-1} y^{q-1} z^{r-1}\, dx\, dy\, dz = a^p b^q c^r \frac{\Gamma(p)\Gamma(q)\Gamma(r)}{\Gamma(p+q+r+1)},$$

where T is the tetrahedron formed by the coordinate planes and the plane $x/a + y/b + z/c = 1$.

Verify that the moments and products of inertia of the tetrahedron about the axes are the same as those of four particles, each $\frac{1}{20}V$, at the corners of the tetrahedron together with a fifth particle, $\frac{4}{5}V$, at the centroid, where $V = \frac{1}{6}abc$.

4. If T denote the domain in which x, y, z take all positive values such that $(x/a)^p + (y/b)^q + (z/c)^r \leqslant 1$, shew that the triple integral

$$\iiint_T F\left\{\left(\frac{x}{a}\right)^p + \left(\frac{y}{b}\right)^q + \left(\frac{z}{c}\right)^r\right\} x^{\alpha-1} y^{\beta-1} z^{\gamma-1}\, dx\, dy\, dz$$

can be reduced to a constant multiple of the single integral

$$\int_0^1 F(t)\, t^\lambda\, dt,$$

where $$\lambda = \alpha/p + \beta/q + \gamma/r - 1.$$

Deduce that if T is the unit sphere with its centre at the origin,

$$\iiint_T \frac{dx\,dy\,dz}{(1-x^2-y^2-z^2)^{\frac{1}{2}}} = \pi^2.$$

5. Prove that the area of the surface $z^2 = 2xy$ included between the planes $x=0$, $x=a$, $y=0$, $y=b$ is $4\sqrt{(ab)}\,(a+b)/3\sqrt{2}$.

6. Prove that the area of the portion of the sphere $x^2+y^2+(z-c)^2=c^2$ contained within the paraboloid $2z = x^2/a + y^2/b$ is $4\pi c\sqrt{(ab)}$.

7. Shew that, if the x and y coordinates of any point on the paraboloid of Question 6 are expressed in the form

$$x = a\tan\theta\cos\phi, \quad y = b\tan\theta\sin\phi,$$

the angle θ is the inclination of the normal at any point to the axis of z.

Prove that the area of the cap of the surface cut off by the curve $\theta=\lambda$ is $\frac{2}{3}\pi ab(\sec^3\lambda - 1)$.

8. (i) Find the area of the surface of the spheroid

$$(x^2+y^2)/a^2 + z^2/c^2 = 1.$$

(ii) A nearly spherical ellipsoid is defined by the equations

$$x = a\sin\theta\cos\phi, \quad y = (a+\omega)\sin\theta\sin\phi, \quad z = (a+\epsilon)\cos\theta,$$

where ϵ and ω are small compared with a. Prove that, if squares of ϵ and ω are neglected, the area of the surface is approximately $4\pi a^2 + \frac{8}{3}\pi a(\omega+\epsilon)$.

9. If C is the curve defined by $x^2+y^2+z^2-2bx-2by=0$, $x+y=2b$, prove that

$$\int_C (y\,dx + z\,dy + x\,dz) = -2\sqrt{2}\pi b^2,$$

the path beginning at the point $(2b, 0, 0)$ and lying at first in the portion of space for which z is negative.

10. Prove that the volume of a cone which extends from the origin to the surface $x=f(u,v)$, $y=g(u,v)$, $z=h(u,v)$ is given by

$$\tfrac{1}{3}\iint \Delta\, du\, dv,$$

where $$\Delta = \begin{vmatrix} x, & y, & z \\ x_u, & y_u, & z_u \\ x_v, & y_v, & z_v \end{vmatrix},$$

and suffixes denote partial derivation, $x_u = \dfrac{\partial x}{\partial u}$, ...

Hence prove that the volume of one octant of the cone $x^4+y^4=z^4\tan^2\alpha$ between its vertex and the surface $x^4+y^4+z^4=1$ is

$$\frac{\{\Gamma(\frac{1}{4})\}^2}{24\,\Gamma(\frac{1}{2})}\int_0^\alpha \frac{du}{\sqrt{(\cos u)}}.$$

11. Find the position of the centroid of the spherical sector formed by the revolution of a sector of a circle of angle 2α about its axis of symmetry (i) when the density is uniform, (ii) when it is proportional to the distance r from the centre.

12. Find the moment of inertia of an octant of a sphere about one of its straight edges, by direct evaluation of the triple integral.

13. By choosing as axes of coordinates three coterminous edges of a cube, prove that the coordinates of the centroid of that portion of the cube which remains when the corner bounded by the plane $x+y+z=2a$ has been removed are given by

$$\bar{x}=\bar{y}=\bar{z}=9a/20,$$

where a is the length of an edge of the cube.

14. Find the position of the centroid of the solid formed by rotating a plane semicircle of radius a about an axis in its plane perpendicular to the base of the semicircle when the distance of the axis of rotation from the centre of the semicircle is b, $(b > a)$.

[The solid is one-half of an anchor-ring.]

15. Deduce from Green's theorem that

$$\iiint_T (u\nabla^2 v - v\nabla^2 u)\, dx\, dy\, dz = \iint_S \left(u\frac{\partial v}{\partial n} - v\frac{\partial u}{\partial n}\right) d\sigma,$$

where $d\sigma$ denotes an element of the surface S which bounds the domain T, and $\dfrac{\partial}{\partial n}$ denotes derivation along the outward-drawn normal.

Deduce also the "generalisation" of the formula for integration by parts

$$\iiint_T \left(P_1\frac{\partial P_2}{\partial x} + Q_1\frac{\partial Q_2}{\partial y} + R_1\frac{\partial R_2}{\partial z}\right) dx\, dy\, dz$$

$$= \iint_S (P_1P_2\, dy\, dz + Q_1Q_2\, dz\, dx + R_1R_2\, dx\, dy)$$

$$- \iiint_T \left(P_2\frac{\partial P_1}{\partial x} + Q_2\frac{\partial Q_1}{\partial y} + R_2\frac{\partial R_1}{\partial z}\right) dx\, dy\, dz.$$

16. If by the inversion transformation $x=k^2\xi/\rho^2$, $y=k^2\eta/\rho^2$, $z=k^2\zeta/\rho^2$, where $r\rho=k^2$, $r^2=x^2+y^2+z^2$, $\rho^2=\xi^2+\eta^2+\zeta^2$, the twice differentiable function $V(x, y, z)$ becomes $V_1(\xi, \eta, \zeta)$, prove that, if $\nabla^2 V_1=0$, then $\nabla^2(V/r)=0$.

17. Prove that

$$\iint\cdots\int \frac{dx_1\, dx_2 \ldots dx_n}{(1-x_1^2-x_2^2-\ldots-x_n^2)^{\frac{1}{2}}} = \frac{\pi^{\frac{1}{2}n+\frac{1}{2}}}{2^n\,\Gamma(\frac{1}{2}n+\frac{1}{2})},$$

the integration extending over all positive values of $x_1, x_2, \ldots, x_n$ which satisfy the inequality $x_1^2+x_2^2+\ldots+x_n^2 \leqslant 1$.

18. Prove that $$\iiint dx\, dy\, dz = \frac{\frac{4}{3}\pi}{(1-m)\sqrt{(1+2m)}},$$

where the integration extends over all values of x, y, z such that $x^2+y^2+z^2+2m(xy+yz+zx)$ does not exceed unity, and $-\frac{1}{2}<m<1$.

CHAPTER XIII

POWER SERIES

13·1. The definition of a power series.

The series

$$a_0 + a_1(z-\alpha) + a_2(z-\alpha)^2 + \dots + a_n(z-\alpha)^n + \dots \quad \dots(1),$$

in which z, α and the coefficients a_n are in general complex numbers, is called a *power series about the point* $z=\alpha$ *as base.* The Taylor expansion of $f(x)$ as a series of ascending powers of $x-a$,

$$f(x) = a_0 + a_1(x-a) + a_2(x-a)^2 + \dots \quad \dots\dots\dots(2),$$

where

$$a_0 = f(a), \quad a_1 = f'(a), \quad a_2 = \frac{f''(a)}{2!}, \dots, \quad a_n = \frac{f^{(n)}(a)}{n!}, \dots,$$

is an example of a power series with which the reader is already familiar; but in our previous discussion of these series we have only considered the case in which all the numbers involved are real. The Maclaurin series is a power series about the origin, and it corresponds to the series (1) above when $\alpha = 0$.

Power series are some of the simplest as well as the most important series which occur in Analysis. We have already considered the exponential and logarithmic series, and, as we shall see, one of the rigorous methods of introducing the circular functions $\sin x$ and $\cos x$ is by defining them as the sum-functions of certain power series.

13·2. The concept of uniform convergence.

In § 3·431 we have already mentioned the concept of uniformity with special reference to uniform continuity, and it was stated there that although uniformity of continuity was not an important concept, uniformity of convergence was of very great importance in Analysis. Before discussing the theory of power series it is convenient to establish some fundamental results about operations on uniformly convergent series. When this has been done it is an easy matter to deduce from them their application to power series. This procedure is preferable to proving these properties only for the special case of power series.

Consider the series

$$u_1(x) + u_2(x) + \ldots + u_n(x) + \ldots \quad \text{............(1)},$$

and write $$s_n(x) = \sum_{r=1}^{n} u_r(x).$$

If $s_n(x)$ tends to the limit $s(x)$ as $n \to \infty$, then $\{s_n(x)\}$ is said to converge to the sum $s(x)$, and $s(x)$ is the "sum-function" of the series (1). As we have already seen in the case of a sequence $\{s_n\}$ which does not depend on x, the number ν involved in the definition of a limit in § 2·53 necessarily depends on ϵ. When the sequence $\{s_n(x)\}$ is considered, the number ν is, in general, a function of ϵ AND OF x. It is precisely here that the concept of uniformity becomes important. *If, given ϵ, a number $m(\epsilon)$ can be found* WHICH DOES NOT DEPEND ON x *such that* $|s_n(x) - s(x)| < \epsilon$ *for all values of* $n \geqslant m(\epsilon)$ *and for all values of x in* $a \leqslant x \leqslant b$, *then* $s_n(x)$ *converges* UNIFORMLY *to* $s(x)$ *in* (a, b).

Let $m(\epsilon, x)$ denote the *least* m for which the above holds, then the question as to whether a sequence $\{s_n(x)\}$ converges uniformly to $s(x)$ in (a, b) is precisely the same question as the following. *For a fixed value of ϵ, is the function $m(\epsilon, x)$ bounded for all values of x in $a \leqslant x \leqslant b$?*

For if $m(\epsilon, x) \leqslant \mu(\epsilon)$ for all values of x in (a, b), then the number μ may be chosen instead of m, and since μ *does not depend on x* the condition for uniformity of convergence is satisfied.

Example 1. *Let* $s_n(x) = x^n$ *in the range* $0 \leqslant x \leqslant 1$.

Clearly here $s(x) = 0$ in $0 \leqslant x < 1$, and $s(1) = 1$.

(*a*) If $0 \leqslant x < 1$, then $|s_n(x) - s(x)| = x^n < \epsilon$ if, and only if,

$$n > \frac{\log(1/\epsilon)}{\log(1/x)},$$

where $\epsilon < 1$: and so

$$m(\epsilon, x) = \left[\frac{\log(1/\epsilon)}{\log(1/x)}\right]^{*} \quad \text{............................(2)}.$$

* The symbol $[x]$ is used to denote the greatest integer contained in x. The reader should realise that m must be an integer.

(b) If $x=1$ then $|s_n(x)-s(x)|=0$ and so we have $m(\epsilon, x)=0$.

We therefore see that $m(\epsilon, x)$ exists for every value of x in $0 \leqslant x \leqslant 1$, but it is not bounded in this interval, for as x approaches the value 1 from below the value of $m(\epsilon, x)$ given by (2) increases indefinitely. We therefore see that in the interval $0 \leqslant x \leqslant 1$ this sequence $\{s_n(x)\}$ is non-uniformly convergent, and further that the non-uniformity is due to the inclusion of the point $x=1$ in the range. The sequence considered is uniformly convergent in the interval $0 \leqslant x \leqslant 1-\delta$, where $\delta > 0$ and as small as we please.

Note. The reader will observe that uniform convergence is essentially the property of an interval, and the definition of uniform convergence presupposes a specified range of values of x, usually some given interval (a, b), which is always a *closed* interval.

The concept of *uniform convergence at a given point* has however a good deal of importance, and for our present purpose the following definition may be given. *The sequence $\{s_n(x)\}$ is said to converge uniformly at the point x_0, if, given ϵ, we can find a number m, and an interval surrounding the point x_0, such that throughout it, and from and after $n=m$,*

$$|s_n(x)-s(x)|<\epsilon$$

If the interval is $(x_0-\delta, x_0+\delta)$, then both δ and m are functions both of x_0 and of ϵ.

This may be contrasted with the concept of continuity, where *continuity at a point* is the essential original concept, and "continuity in an interval" is derived from it.

13·21. Discontinuity of the sum-function of a power series.

Let us consider the series

$$x^2+\frac{x^2}{1+x^2}+\frac{x^2}{(1+x^2)^2}+\ldots+\frac{x^2}{(1+x^2)^n}+\ldots.$$

Since this series is a geometrical progression with common ratio $1/(1+x^2)$ which is always less than unity, save when $x=0$, the series is absolutely convergent for all real values of x (each term is in fact essentially positive).

Now $$s_n(x)=1+x^2-1/(1+x^2)^{n-1},$$

and so $$s(x)=\lim_{n\to\infty} s_n(x)=1+x^2, \quad \text{when } x \neq 0,$$

and clearly from the series $s(0)=0$.

Thus, although the series is an absolutely convergent series of continuous functions, its sum-function $s(x)$ has a discontinuity at the point $x=0$. The fact that a convergent series of continuous functions can have a discontinuous sum-function was first pointed out by Abel*, and, as we shall see, this can only occur when the series in question is non-uniformly convergent.

13·22. The uniform convergence of series.

When the definition of uniformity of convergence for a sequence $\{s_n(x)\}$ has been given, the case of uniform convergence of a series follows at once by writing $s_n(x) = \sum_{r=1}^{n} u_r(x)$.

THE GENERAL PRINCIPLE OF UNIFORM CONVERGENCE. *The necessary and sufficient condition that a sequence $\{s_n(x)\}$ should converge uniformly to $s(x)$ in a given range $a \leqslant x \leqslant b$, is that, given ϵ, there is a number $m(\epsilon)$* INDEPENDENT OF x, *such that*

$$|s_n(x) - s_{n'}(x)| < \epsilon,$$

for all values of $n' > n$ where $n \geqslant m$, and for all values of x in (a, b).

By recalling the general principle of ordinary convergence, § 2·53, we see that the necessity is trivial. To prove the sufficiency we assume that the condition holds, and then prove (i) that $s_n(x)$ tends to a limit for each value of x, and (ii) that the convergence of $s_n(x)$ to this limit $s(x)$ is uniform. Result (i) follows at once from the general principle of convergence. To prove result (ii) we observe that the above inequality holds for all values of n': make $n' \to \infty$, then

$$|s_n(x) - s_{n'}(x)| \to |s_n(x) - s(x)|,$$

and so

$$|s_n(x) - s(x)| \leqslant \epsilon,$$

for $n \geqslant m$, and for all values of x in (a, b). Hence $s_n(x)$ tends uniformly to $s(x)$.

COROLLARY. *The necessary and sufficient condition that the series $\Sigma u_n(x)$ should converge uniformly in a given interval $a \leqslant x \leqslant b$ is that, given ϵ, it must be possible to choose a number m, independent of x (but depending on ϵ), such that*

$$|u_{m+1}(x) + u_{m+2}(x) + \ldots + u_{m+p}(x)| < \epsilon \quad \ldots\ldots\ldots(1)$$

for all positive integral values of p.

* *J. Math.* (Berlin, 1826). Cauchy in 1821 stated definitely that a function defined by a convergent series of functions which are continuous in an interval must itself be continuous in that interval. See *Cours d'Analyse*, I, 131 (1821). Compare with Theorem 1 below.

The proof follows at once from the theorem, for we have only to write $n = m$, $n' = m + p$ in the statement of the general principle above.

It is convenient to write

$$R_{n,p}(x) \equiv u_{n+1}(x) + u_{n+2}(x) + \ldots + u_{n+p}(x),$$

and then the inequality (1) above becomes simply

$$|R_{m,p}(x)| < \epsilon.$$

13·3. General properties of uniformly convergent series.

THEOREM 1. *If every term of the series $\Sigma u_n(x)$ is continuous in the interval $a \leqslant x \leqslant b$, and if further the series converges uniformly to $s(x)$, then the sum-function $s(x)$ is also continuous in $a \leqslant x \leqslant b$.*

Write $$s_n(x) = s(x) - R_n(x),$$

so that, on the hypothesis of uniformity of convergence of the series, we have

$$|R_n(x)| < \epsilon$$

for all values of $n \geqslant m(\epsilon)$ independent of x.

Let c be any point of (a, b), and, to fix the ideas, suppose that it is an internal point, so that $a < c < b$.

Now

$$\begin{aligned} |s(x) - s(c)| &= |\{s_m(x) - s_m(c)\} + \{s(x) - s_m(x)\} - \{s(c) - s_m(c)\}| \\ &\leqslant |s_m(x) - s_m(c)| + |R_m(x)| + |R_m(c)| \\ &\leqslant |s_m(x) - s_m(c)| + 2\epsilon; \end{aligned}$$

but, since $s_m(x)$ is continuous at the point $x = c$, there is a number $\delta(\epsilon, m)$ such that, whenever $|x - c| \leqslant \delta$,

$$|s_m(x) - s_m(c)| < \epsilon;$$

and so $$|s(x) - s(c)| < 3\epsilon,$$

under the sole restriction that $|x - c| \leqslant \delta$. Since further m is a function of ϵ only, δ is a function of ϵ only, which proves that $s(x)$ is continuous at $x = c$.

A slight modification of the argument covers the case when c coincides with either a or b. This is left as an exercise for the reader.

The above theorem still holds if x is a complex variable.

Note on the theorem.

The reader should observe that discontinuity in the sum-function involves non-uniformity of convergence of the series, but the following example illustrates the important fact that series exist which have a continuous sum-function but which are nevertheless non-uniformly convergent.

Example. Let $$s_n(x) = nx/(1+n^2x^2).$$

Whatever the value of x, we have

$$s(x) = \lim_{n\to\infty} s_n(x) = 0,$$

and so $s(x)$ is continuous for all values of x. Also we see that $s_n(x)$ is, for each value of n, a continuous function of x.

Now $$|s_n(x) - s(x)| = \frac{n|x|}{1+n^2x^2} < \epsilon$$

if $$E \equiv x^2\epsilon n^2 - |x|\, n + \epsilon > 0.$$

If $\epsilon < \frac{1}{2}$ the expression on the left-hand side has the same sign as the coefficient of n^2, save when n lies between the two real roots of the quadratic in n,

$$x^2\epsilon n^2 - |x|\, n + \epsilon = 0.$$

The expression E is therefore positive if $n|x| > \lambda$, where

$$\lambda \equiv \{1 + \sqrt{(1-4\epsilon^2)}\}/2\epsilon.$$

Thus $$m(\epsilon, x) = [\lambda |x|^{-1}] \text{ if } |x| > 0,$$

while $$m(\epsilon, 0) = 0.$$

Hence $m(\epsilon, x)$ is not bounded, and there is non-uniform convergence in any interval which includes the point $x = 0$*.

THEOREM 2. *A* UNIFORMLY *convergent series of continuous functions of a real variable x may be integrated term-by-term.*

Suppose that we have the series

$$u_1(x) + u_2(x) + \ldots + u_n(x) + \ldots \quad \ldots\ldots\ldots\ldots(1),$$

of which each term is a continuous function of x in $a \leqslant x \leqslant b$; and suppose also that the series (1) is uniformly convergent to sum $s(x)$ in $a \leqslant x \leqslant b$.

Write $$s(x) = s_n(x) + R_n(x) \quad \ldots\ldots\ldots\ldots\ldots\ldots\ldots(2),$$

* A very interesting graphical method, due to Osgood, is described in Bromwich's *Infinite Series* (1926), § 43. The method illustrates the non-uniformity of convergence of the above sequence in a striking way, but it is not proposed to discuss the graphical illustration here.

then, since $s(x)$ is continuous by Theorem 1, the equation (2) is integrable in (a, b), and so

$$\int_a^b s(x)\,dx = \int_a^b s_n(x)\,dx + \int_a^b R_n(x)\,dx.$$

Now let $n \to \infty$, and, since the series (1) is uniformly convergent, $|R_n(x)| < \epsilon$ when $n \geqslant m(\epsilon)$, independent of x; and so, when $n \geqslant m$,

$$\left|\int_a^b R_n(x)\,dx\right| \leqslant \int_a^b |R_n(x)|\,dx < \epsilon(b-a).$$

Hence

$$\begin{aligned}\int_a^b s(x)\,dx &= \lim_{n\to\infty}\int_a^b s_n(x)\,dx \\ &= \int_a^b u_1(x)\,dx + \int_a^b u_2(x)\,dx + \ldots + \int_a^b u_n(x)\,dx + \ldots.\end{aligned}$$

COROLLARY. *If the integrations are extended only to the portion (a, x) of the interval (a, b), the integrated series is also uniformly convergent in (a, x).*

For $\left|\int_a^x R_n(x)\,dx\right| \leqslant \int_a^x |R_n(x)|\,dx \leqslant \int_a^b |R_n(x)|\,dx < \epsilon(b-a)$

for $n \geqslant m(\epsilon)$, which proves the result.

THEOREM 3*. *Let the integrable function $s(x)$ be the sum of the series of integrable functions $u_1(x), u_2(x), \ldots, u_n(x), \ldots$, and suppose that this series converges uniformly to $s(x)$ in $a \leqslant x \leqslant b$ except for a finite number of sub-intervals, the sum of whose lengths can be made less than ϵ: then, if there exists a positive number K, such that $|s_n(x)| < K$ for every value of x in (a, b) and for every positive integer n,*

$$\int_a^b s(x)\,dx = \lim \int_a^b s_n(x)\,dx = \lim \sum_{r=1}^{n} \int_a^b u_r(x)\,dx.$$

Let the sub-intervals in which the uniform convergence fails be $(a_1, b_1), (a_2, b_2), \ldots, (a_k, b_k)$, where

$$\sum_1^k (b_r - a_r) < \epsilon.$$

* This theorem is given by Carslaw in the *Mathematical Gazette*, XIII (1927), p. 438. It gives a set of sufficient conditions for term-by-term integration other than the standard condition of uniform convergence given in Theorem 2.

Then we have

$$\int_a^b s(x)\,dx - \int_a^b s_n(x)\,dx$$

$$= \left[\int_a^b \{s(x) - s_n(x)\}\,dx - \sum_1^k \int_{a_r}^{b_r} \{s(x) - s_n(x)\}\,dx\right]$$

$$+ \sum_1^k \int_{a_r}^{b_r} \{s(x) - s_n(x)\}\,dx \;\ldots\ldots(1).$$

Now $\quad |s_n(x)| < K$ and so $|s(x)| = |\lim s_n(x)| \leqslant K$,

whence
$$|s(x) - s_n(x)| < 2K.$$

The series converges uniformly in (a, b) except in the sub-intervals $(a_1, b_1), \ldots, (a_k, b_k)$, and so there is a positive integer m* such that

$$|s(x) - s_n(x)| < \epsilon$$

when $n \geqslant m(\epsilon)$, the same m serving for all values of x in the part of (a, b) which remains when the above set of sub-intervals is removed.

It follows from (1) that, when $n \geqslant m$,

$$\left|\int_a^b s(x)\,dx - \int_a^b s_n(x)\,dx\right| < (b-a)\,\epsilon + 2K\epsilon = M\epsilon,$$

where M is a constant. Hence

$$\int_a^b s(x)\,dx = \lim \int_a^b s_n(x)\,dx = \lim \sum_1^n \int_a^b u_r(x)\,dx.$$

An example of a series which is integrable term-by-term but which is non-uniformly convergent is provided by considering the series for which the sum to n terms, $s_n(x)$, is given by

$$\frac{nx}{1+n^2x^2}.$$

We have already seen that this sequence is non-uniformly convergent in any interval which includes the point $x=0$.

The reader may easily shew, however, that

$$\int_0^1 \lim_{n\to\infty} s_n(x)\,dx = 0 = \lim_{n\to\infty} \int_0^1 s_n(x)\,dx.$$

* Of course the number m must be independent of x. When the reader is quite familiar with the concept of uniform convergence it is not so necessary to *emphasise* the fact that m must be independent of x, although it is of course the fundamental and the most important fact of all.

THEOREM 4. *Let the series*

$$u_1(x)+u_2(x)+\ldots+u_n(x)+\ldots \qquad \ldots\ldots\ldots\ldots(1)$$

be a convergent series of functions of the real variable x, the sum-function of which is $s(x)$: then, if the series of derivatives

$$u_1'(x)+u_2'(x)+\ldots+u_n'(x)+\ldots \qquad \ldots\ldots\ldots\ldots(2)$$

is a UNIFORMLY *convergent series in (a, b) with sum $\sigma(x)$,*

$$s'(x)=\sigma(x).$$

By Theorem 2,

$$\int_a^x \sigma(x)\,dx=\int_a^x u_1'(x)\,dx+\int_a^x u_2'(x)\,dx+\ldots$$

$$=\sum_{r=1}^{\infty}\{u_r(x)-u_r(a)\}$$

$$=\sum_1^{\infty} u_r(x)-\sum_1^{\infty} u_r(a),$$

since the series (1) is convergent. Hence

$$\int_a^x \sigma(x)\,dx=s(x)-s(a),$$

and, on taking the derivative with respect to x, we get

$$\sigma(x)=s'(x),$$

since $\sigma(x)$ is continuous in (a, b).

The reader should observe that for the validity of term-by-term derivation, it is the *derived series* which must be *uniformly* convergent. It can be shewn that the theorem holds if the series (1) converges for at least *one* value of x in (a, b).

13·31. Weierstrass's test for uniform convergence of series.

A series $\Sigma u_r(x)$ is uniformly convergent in a given range of values of x, provided that (i) $|u_n(x)|<M_n$ *for all values of n greater than a fixed number m, where M_n is a function of n only, and* (ii) ΣM_n *is a convergent series.*

The proof is simple, for with the notation explained above,

$$|R_{m,p}(x)|<\sum_{r=m}^{m+p} M_r<\epsilon$$

for all positive integral values of p, the number m depending only on ϵ, and not on x. Hence it follows that $\Sigma u_r(x)$ is uniformly convergent in the given range.

13·32. Another test for uniform convergence.

If $s_n(x) \to s(x)$ for every x of a certain range and M_n is the upper bound of $T_n(x) \equiv |s_n(x) - s(x)|$ for fixed n and varying x, then the necessary and sufficient condition for uniform convergence is that $M_n \to 0$ as $n \to \infty$.

(i) Let $M_n \to 0$; then, given ϵ, $M_n < \epsilon$ for $n \geqslant m(\epsilon)$ and so

$$T_n(x) \leqslant M_n < \epsilon$$

for $n \geqslant m(\epsilon)$ and all x. The convergence is therefore uniform.

(ii) Let the convergence be uniform. Then $T_n(x) < \epsilon$ for $n > \mu(\epsilon)$ and all x. Hence $M_n \leqslant \epsilon$ for $n > \mu(\epsilon)$. This being true for every ϵ, $M_n \to 0$.

We deduce at once that, if $M_n \not\to 0$, $\{s_n(x)\}$ is non-uniformly convergent.

13·4. Power series.

Let us consider the power series $\Sigma a_n z^n$, where, in the general case, both z and the coefficients a_n may be complex numbers.

(i) If $|a_n|^{\frac{1}{n}} \to 0$ as $n \to \infty$ the power series is absolutely convergent for all values of z, for

$$|a_n z^n|^{\frac{1}{n}} = |a_n|^{\frac{1}{n}} |z| \to 0 \text{ as } n \to \infty,$$

and so $\Sigma |a_n| |z|^n$ converges by Cauchy's test, § 5·2 (IV), hence $\Sigma a_n z^n$ is absolutely convergent for all values of z.

(ii) If $\overline{\lim} |a_n|^{\frac{1}{n}} = \infty$ the power series never converges, for

$$\overline{\lim} |a_n z^n|^{\frac{1}{n}} = |z| \overline{\lim} |a_n|^{\frac{1}{n}} = \infty \quad (z \neq 0),$$

and so $|a_n z^n|$ does not tend to zero.

(iii) If neither of the above cases holds, then $\overline{\lim} |a_n|^{\frac{1}{n}}$ must be finite and not zero. It is usual to write

$$\overline{\lim} |a_n|^{\frac{1}{n}} = 1/R,$$

where R is a positive constant. In this case we have

$$\overline{\lim} |a_n z^n|^{\frac{1}{n}} = |z|/R,$$

and, by Cauchy's test, the series $\Sigma a_n z^n$ is absolutely convergent if $|z| < R$. If $|z| > R$ the nth term of the series does not tend to zero, and $\Sigma a_n z_n$ is divergent (or oscillates infinitely).

Cauchy's test gives no criterion if $|z|=R$, and, as we shall see later, this case needs separate discussion.

The number R is called the *radius of convergence* of the power series, and it is so called because if we consider the circle $|z|=R$ in the Argand diagram, then for values of z inside this circle the series $\Sigma a_n z^n$ is absolutely convergent, and for values of z outside this circle the series does not converge at all. For values of z on the circumference each case must be considered on its own merits, and for such values of z the series may converge, diverge or oscillate. The circle $|z|=R$ is called the *circle of convergence* of the power series.

By writing $z=R\zeta$ the series $\Sigma a_n z^n$ becomes $\Sigma b_n \zeta^n$, where $b_n = R^n a_n$, and the radius of convergence of the new series is clearly unity. *Without loss of generality therefore we can take power series whose radius of convergence is unity as typical series.*

When the variable is real the circle of convergence $|z|=1$ reduces to the end-points of the interval $-1 \leqslant x \leqslant 1$.

The theorems which follow are proved only for *real* power series $\Sigma a_n x^n$, and these will be considered as power series WHOSE RADIUS OF CONVERGENCE IS UNITY.

13·5. Properties of real power series.

THEOREM 1. *The series $\Sigma a_n x^n$ is absolutely convergent in the open interval $-1 < x < 1$.*

This is immediate, for we are assuming that the radius of convergence of our typical power series is unity.

THEOREM 2. *The series $\Sigma a_n x^n$ is uniformly convergent in the interval $-(1-\delta) \leqslant x \leqslant 1-\delta$, where $\delta > 0$, and as small as we please.*

This follows at once from Weierstrass's test, for

$$\overline{\lim}\, |a_n x^n|^{\frac{1}{n}} = |x| \leqslant 1-\delta,$$

and so $\qquad |a_n x^n|^{\frac{1}{n}} < 1 - \tfrac{1}{2}\delta$ for values of $n \geqslant \nu(\delta)$;

hence $\qquad |a_n x^n| < (1-\tfrac{1}{2}\delta)^n = M_n$ for $n \geqslant \nu(\delta)$,

and ΣM_n is clearly convergent, since it is a geometrical progression with a common ratio less than unity.

The above theorem is very important, for it shews that for power series a knowledge of the range of absolute convergence is all that is necessary to give the requisite information about uniform convergence.

Abel's Lemma. *Let* $s_n = \sum_1^n a_\nu$ *and let* $\{v_n\}$ *be a positive non-increasing sequence, then, if* $K \equiv \max |s_p|$, $1 \leqslant p \leqslant n$,

$$|\sum_1^n a_\nu v_\nu| < K v_1 \quad \text{.........................(A).}$$

Now
$$\sum_1^n a_\nu v_\nu = s_1 v_1 + (s_2 - s_1) v_2 + \dots + (s_n - s_{n-1}) v_n$$
$$= s_1 (v_1 - v_2) + s_2 (v_2 - v_3) + \dots + s_{n-1}(v_{n-1} - v_n) + s_n v_n,$$

and since $v_p - v_{p+1} \geqslant 0$, $v_p > 0$, inequality (A) follows.

Theorem 3. Abel's theorem on power series.

If Σa_n *converges, then* $f(x) \equiv \sum_0^\infty a_n x^n$ *is uniformly convergent in* $0 \leqslant x \leqslant 1$*, *and*

$$f(x) \to f(1) \text{ as } x \to 1 - 0.$$

Let $v_n = x^n$, $0 \leqslant x \leqslant 1$, then, by Abel's lemma,

$$|\sum_m^n a_\nu x^\nu| \leqslant K x^m,$$

where $K = \max |a_m + a_{m+1} + \dots + a_p|$, $m \leqslant p \leqslant n$.

Since Σa_n converges, $K < \epsilon$ for $m \geqslant \mu(\epsilon)$; and so

$$|\sum_m^n a_\nu x^\nu| \leqslant \epsilon x^m \leqslant \epsilon,$$

and since μ is *independent of* x, $\Sigma a_n x^n$ is uniformly convergent in $0 \leqslant x \leqslant 1$.

Further, by Theorem 1 of § 13·3, $f(x)$ is continuous in $0 \leqslant x \leqslant 1$, and so $f(x) \to f(1)$ as $x \to 1 - 0$.

It follows similarly that *if* $\Sigma (-)^n a_n$ *is convergent, then*

$$f(x) \to f(-1) \text{ as } x \to -1 + 0.$$

Example. Prove that $\log 2 = 1 - \frac{1}{2} + \frac{1}{3} - \frac{1}{4} + \dots$.

We have already seen that

$$\log(1+x) = x - \frac{x^2}{2} + \frac{x^3}{3} - \frac{x^4}{4} + \dots \quad \text{.....................(1)}$$

if $-1 < x < 1$, this range being the range of absolute convergence of the series on the right-hand side of (1). By Theorem 2 we deduce that the above series

* Since we are only concerned with the point $x = 1$, the interval $0 \leqslant x \leqslant 1$ has been chosen, but the theorem is equally true in $-(1-\delta) \leqslant x \leqslant 1$, if $\delta > 0$.

is uniformly convergent in $-(1-\delta) \leqslant x \leqslant 1-\delta$. We now consider each of the extreme points $x=1$ and $x=-1$ separately.

If we put $x=1$ in the series (1), it becomes

$$1-\tfrac{1}{2}+\tfrac{1}{3}-\tfrac{1}{4}+\dots \qquad \text{.............................(2)},$$

which can easily be shewn to be (conditionally) convergent by the alternating series test. Hence, by Theorem 3, the series (1) is uniformly convergent in $0 \leqslant x \leqslant 1$.

Again, because the series (2) is convergent, we deduce from Abel's theorem that the sum-function of the series (2) is the limit of the sum-function of the series (1) as $x \to 1-0$; and since

$$\lim_{x \to 1-0} \log(1+x) = \log 2,$$

the required result is established.

If we put $x=-1$ in the series (1) it becomes

$$-(1+\tfrac{1}{2}+\tfrac{1}{3}+\tfrac{1}{4}+\dots),$$

which is known to be divergent, and so Theorem 3 is not applicable in this case.

13·6. Operations on power series.

LEMMA. *The series obtained by the derivation or integration of a power series term-by-term has the same radius of convergence as the original series.*

Let the given series be $\Sigma a_n x^n$, then the reciprocal of the radius of convergence is $\overline{\lim}\,|a_n|^{\frac{1}{n}}$.

For the integrated series we have

$$\overline{\lim}\,|a_n/(n+1)|^{\frac{1}{n}} = \overline{\lim}\,\{1/(n+1)\}^{\frac{1}{n}}\,|a_n|^{\frac{1}{n}}$$

$$= \overline{\lim}\,|a_n|^{\frac{1}{n}} = 1/R.$$

Similarly for the derived series

$$\overline{\lim}\,|na_n|^{\frac{1}{n}} = \overline{\lim}\,n^{\frac{1}{n}}\,|a_n|^{\frac{1}{n}}$$

$$= \overline{\lim}\,|a_n|^{\frac{1}{n}} = 1/R,$$

since $n^{\frac{1}{n}} \to 1$ (see Examples V, 10).

THEOREM 1. *If the power series $\Sigma a_n x^n$ has radius of convergence unity, its sum-function is continuous in $|x| < 1-\delta$, $\delta > 0$.*

This follows at once from Theorem 1, § 13·3.

THEOREM 2. *If* $f(x)=\Sigma a_n x^n$ *for values of* x *such that* $|x|<1$, *then, for the same range of values of* x,

$$f'(x)=\Sigma n a_n x^{n-1}.$$

All that is needed is to shew that the conditions of Theorem 4 of § 13·3 are satisfied. By the lemma the derived series has radius of convergence unity, and is therefore uniformly convergent in

$$|x|\leqslant 1-\delta, \quad \delta>0.$$

The original series is certainly convergent, and each term possesses a derivative in $|x|<1$.

THEOREM 3. *If* $|x|<1$, $|x_0|<1$, *then*

$$\int_{x_0}^{x} f(t)\,dt=\Sigma \frac{a_n}{n+1}(x^{n+1}-x_0^{\,n+1}).$$

This follows at once from Theorem 2 of § 13·3.

The above theorem can be extended to give the following important result, which for convenience may be stated as

THEOREM 4. *A power series may be integrated term-by-term right up to the radius of convergence, provided only that the resulting series converges at that point.*

For, if $\Sigma a_n/(n+1)$ converges to sum A, then, by Abel's theorem,

$$\Sigma a_n x^{n+1}/(n+1)\to A \text{ as } x\to 1-0.$$

Hence, by taking the lower limit of integration x_0 to be zero, we deduce that

$$\lim_{x\to 1-0}\int_0^x f(t)\,dt$$

exists and is equal to A.

The following examples illustrate the application of the above theorems.

Example 1. By the binomial theorem, if $0\leqslant x<1$, we have

$$(1+x)^{-1}=1-x+x^2-x^3+\dots,$$

and so

$$\int_0^1 \frac{dx}{1+x}=1-\tfrac{1}{2}+\tfrac{1}{3}-\tfrac{1}{4}+\dots,$$

in other words

$$\log 2=1-\tfrac{1}{2}+\tfrac{1}{3}-\tfrac{1}{4}+\dots,$$

since the latter series is convergent.

Example 2. Again, if $x^2 < 1$,

$$(1-x^2)^{-\frac{1}{2}} = 1 + \frac{1}{2}x^2 + \frac{1.3}{2.4}x^4 + \frac{1.3.5}{2.4.6}x^6 + \dots;$$

hence
$$\int_0^1 \frac{dx}{(1-x^2)^{\frac{1}{2}}} = 1 + \frac{1}{2}\cdot\frac{1}{3} + \frac{1.3}{2.4}\cdot\frac{1}{5} + \frac{1.3.5}{2.4.6}\cdot\frac{1}{7} + \dots \qquad (1),$$

provided that the series on the right of (1) is convergent. This is easily proved by Gauss's test, for

$$\frac{u_n}{u_{n+1}} = \frac{2n(2n+1)}{(2n-1)^2} = \left(1+\frac{1}{2n}\right)\left(1-\frac{1}{2n}\right)^{-2} = 1 + \frac{\frac{3}{2}}{n} + O\left(\frac{1}{n^2}\right),$$

and $\mu = \frac{3}{2} > 1$.

In this case the integral on the left-hand side of (1) is not an ordinary Riemann integral, since the integrand is not bounded when $x = 1$. The integral, however, exists as an infinite integral for

$$\lim_{\epsilon\to 0}\int_0^{1-\epsilon} \frac{dx}{\sqrt{(1-x^2)}} = \lim_{\epsilon\to 0}\{\text{arc sin}(1-\epsilon)\} = \tfrac{1}{2}\pi,$$

and, even in this case, it is true that the sum of the series on the right of (1) is $\frac{1}{2}\pi$.

13·7. The general binomial theorem.

Consider the series

$$f(x) = 1 + mx + m(m-1)\frac{x^2}{2!} + m(m-1)(m-2)\frac{x^3}{3!} + \dots;$$

the convergence of the series for real values of m and x has already been mentioned, and the investigation of the convergence has been set as an example (Examples V, 8). On recalling the results we find that the series is absolutely convergent if $|x| < 1$; if $x = 1$ the series is (conditionally) convergent if $m + 1 > 0$; and if $x = -1$ (the series being then one whose terms are ultimately all of the same sign) there is convergence if $m > 0$.

If $|x| < 1$, it can be shewn by Maclaurin's theorem that the sum-function $f(x) = (1+x)^m$, but the extreme cases can only be rigorously investigated by an appeal to Abel's theorem, and so the discussion has been postponed until this point.

Since the above series is absolutely convergent when $|x| < 1$, it is uniformly convergent in any interval $(-b, b)$, where $0 < b < 1$.

Now

$$\begin{aligned} f'(x) &= m\left\{1 + (m-1)x + (m-1)(m-2)\frac{x^2}{2!} + \dots\right\} \\ &= m\phi(x), \end{aligned}$$

where $\phi(x)$ only differs from $f(x)$ by having $m-1$ in the place of m.

Now

$$(1+x)\phi(x) = 1 + mx + m(m-1)\frac{x^2}{2!} + m(m-1)(m-2)\frac{x^3}{3!} + \dots$$
$$= f'(x),$$

and so $$(1+x)f'(x) = mf(x),$$

$$\frac{d}{dx}\left\{\frac{f(x)}{(1+x)^m}\right\} = 0,$$

in other words $$f(x) = A(1+x)^m,$$

where A is an arbitrary constant. Since $f(0) = 1$ and the positive value of $(1+x)^m$ is to be chosen, it follows that $A = 1$, and so

$$f(x) = (1+x)^m.$$

The above argument only applies when x is restricted to lie in the interval $-b \leqslant x \leqslant b$, where $0 < b < 1$.

By Abel's theorem we deduce that the sum of the series when $x = -1$ is zero provided that $m > 0$; and if $m + 1 > 0$ the sum of the series when $x = 1$ is 2^m.

The discussion in the case when both m and x may be complex numbers is more difficult and is not given here*.

13·8. The circular functions.

(1) We are now in a position to give rigorous definitions of the circular functions. Consider the two series

$$C(x) = 1 - \frac{x^2}{2!} + \frac{x^4}{4!} - \dots + (-)^n \frac{x^{2n}}{(2n)!} + \dots \quad \text{......(1)},$$

$$S(x) = x - \frac{x^3}{3!} + \frac{x^5}{5!} - \dots + (-)^n \frac{x^{2n+1}}{(2n+1)!} + \dots \quad \text{...(2)},$$

both of which are absolutely convergent for all values of x†. It is clear that each of these series has an infinite radius of convergence, and so they represent functions which are everywhere continuous, and each may be differentiated term-by-term any number of times in succession. It is easily seen that, for every value of x,

$$C'(x) = -S(x),\quad C''(x) = -C(x),\quad C'''(x) = S(x),\quad C^{\text{iv}}(x) = C(x);$$
$$S'(x) = C(x),\quad S''(x) = -S(x),\quad S'''(x) = -C(x),\quad S^{\text{iv}}(x) = S(x).$$

* The reader may refer to Bromwich's *Infinite Series* (1926), § 96.

† These results are easily established by means of the ratio test.

Since the fourth derived functions are equal to the original functions, the same series of values is repeated in the same order in the successive differentiations.

We see also that $C(x)$ is an odd function and $S(x)$ is an even function, that is to say

$$C(-x) = C(x), \quad S(-x) = -S(x) \quad \ldots\ldots\ldots\ldots(3).$$

(2) *The addition theorems.*

There are several ways of proving the addition theorems, but it will suffice to indicate one of these methods. Consider the function

$$F(x) = \{C(x+x_1) - C(x)C(x_1) + S(x)S(x_1)\}^2 + \{S(x+x_1) - S(x)C(x_1) - S(x_1)C(x)\}^2.$$

It is easy to see that $F'(x) \equiv 0$, and so $F(x)$ must be a constant: but since $F(0) = 0$ it follows that $F(x) \equiv F(0) = 0$ and consequently

$$C(x+x_1) = C(x)C(x_1) - S(x)S(x_1) \ldots\ldots\ldots\ldots(4),$$

$$S(x+x_1) = S(x)C(x_1) + C(x)S(x_1) \ldots\ldots\ldots\ldots(5).$$

From these addition theorems, by writing $x_1 = x$, we get

$$C(2x) = C^2(x) - S^2(x) \quad \ldots\ldots\ldots\ldots\ldots\ldots(6),$$

$$S(2x) = 2S(x)C(x) \quad \ldots\ldots\ldots\ldots\ldots\ldots\ldots(7);$$

and by writing $x_1 = -x$ in (4) we get

$$C^2(x) + S^2(x) = 1 \quad \ldots\ldots\ldots\ldots\ldots\ldots\ldots(8).$$

(3) *Periodicity.*

It is rather more difficult to establish the properties of periodicity directly from the series: this may be done in the following way.

We have $$C(0) = 1 > 0;$$

but $$C(2) < 0,$$

for $$C(2) = 1 - \frac{2^2}{2!} + \frac{2^4}{4!} - \left(\frac{2^6}{6!} - \frac{2^8}{8!}\right) - \left(\frac{2^{10}}{10!} - \frac{2^{12}}{12!}\right) - \ldots,$$

where the expressions in brackets are all positive, since for $n \geqslant 2$,

$$\frac{2^n}{n!} - \frac{2^{n+2}}{(n+2)!} > 0,$$

and so $$C(2) < 1 - \frac{4}{2} + \frac{16}{24} = -\frac{1}{3};$$

hence $C(2)$ is certainly negative. By Theorem 7, § 3·45, $C(x)$ vanishes at least *once* between $x = 0$ and $x = 2$.

By writing

$$S(x) = x\left(1 - \frac{x^2}{2.3}\right) + \frac{x^5}{5!}\left(1 - \frac{x^2}{6.7}\right) + \ldots,$$

we easily see that between $x = 0$ and $x = 2$, $S(x) > 0$, and so $C'(x) = -S(x)$ is constantly negative in this range. It follows that $C(x)$ is a strictly monotonic decreasing function in this interval and can therefore only vanish at one point ξ in the interval.

The number ξ, which is the least positive zero of $C(x)$, is therefore a well-defined real number which, for a reason to be seen in a moment, will be denoted by $\frac{1}{2}\varpi$.

Since $C(\frac{1}{2}\varpi) = 0$, it follows from (8) that

$$S^2(\tfrac{1}{2}\varpi) = 1,$$

and, since $S(x)$ is positive between 0 and 2, it follows that

$$S(\tfrac{1}{2}\varpi) = 1.$$

From (6) and (7) we deduce that

$$C(\varpi) = -1, \quad S(\varpi) = 0, \quad C(2\varpi) = 1, \quad S(2\varpi) = 0,$$

and from the addition theorems we deduce that

$$C(x + 2\varpi) = C(x), \quad S(x + 2\varpi) = S(x),$$

and the functions $C(x)$ and $S(x)$ possess the period 2ϖ.

(4) *It now remains to shew that the number $\varpi = \pi$, the length of half of the perimeter of a circle of radius unity.* When this has been done the complete identity of the functions $C(x)$ and $S(x)$, with the functions $\cos x$ and $\sin x$, will have been established.

Consider the curve defined by the equation

$$x = C(t), \quad y = S(t);$$

then the point P whose coordinates are (x, y) is always at a distance from the origin given by

$$|OP| = \sqrt{(x^2 + y^2)} = \sqrt{\{C^2(t) + S^2(t)\}} = 1$$

by equation (8) above. The locus of P is accordingly a circle whose centre is the origin and whose radius is unity. In particular, if t increases from 0 to 2ϖ, the point P starts from the point A, $(1, 0)$, on the axis of x and describes the perimeter of the circle exactly once in the anticlockwise sense. As t increases from 0 to ϖ the function $C(t)$ is monotonic decreasing from 1 to -1, and so the abscissa of P assumes each of the values between 1 and -1 exactly once, while $S(t)$ remains positive. The point P therefore describes the upper half of the circle steadily and passes through each of its points once only. In exactly the same way, by using the appropriate relations deducible from the addition formulae, we can shew that when t increases from ϖ to 2ϖ the point P describes the lower half of the circle.

We therefore see that if x and y are any two real numbers for which $x^2+y^2=1$, then there exists one and only one number t, such that $0 \leqslant t < 2\varpi$, for which the equations $x=C(t)$, $y=S(t)$ are simultaneously satisfied.

The length of the curve described by the point P when t increases from 0 to a value t_0 is given by

$$\int_0^{t_0} \sqrt{\{C'^2(t)+S'^2(t)\}}\, dt = \int_0^{t_0} dt = t_0,$$

and so the complete perimeter of the circle is obtained by putting $t_0 = 2\varpi$. Since the perimeter of a unit circle is of length 2π we have proved that the numbers ϖ and π are identical.

We have now established the result that the abscissa $C(t)$ of the point P for which the arc $AP=t$ coincides with the cosine of the angle subtended at the centre of the circle by the arc AP, and $S(t)$, the ordinate of P, coincides with the sine of that angle. Hence we may write $\cos t$ for $C(t)$ and $\sin t$ for $S(t)$.

13·81. The other trigonometrical functions.

The two functions $\sin x$ and $\cos x$ are fundamental, and from them the remaining four may be deduced. Thus

$$\operatorname{cosec} x = 1/\sin x, \qquad \sec x = 1/\cos x,$$
$$\tan x = \sin x/\cos x, \quad \cot x = \cos x/\sin x.$$

The expansions in power series for these functions are not so simple. We consider first the expansion

$$\frac{x}{e^x-1} = B_0 + B_1 x + B_2\frac{x^2}{2!} + B_3\frac{x^3}{3!} + \ldots + B_n\frac{x^n}{n!} + \ldots \quad \ldots(1),$$

where the numbers B_n, the so-called *Bernoulli's numbers*, are not explicitly known but are easily obtainable from recurrence formulae which can be found as follows. On using the expansion of e^x in powers of x we have the identity

$$\left(1+\frac{x}{2!}+\frac{x^2}{3!}+\ldots\right)\left(B_0+\frac{B_1 x}{1}+\frac{B_2 x^2}{2!}+\ldots\right)\equiv 1,$$

hence $B_0=1$, $B_1=-\frac{1}{2}$, and for $n\geqslant 2$,

$$\frac{1}{n!}B_0+\frac{1}{(n-1)!}\frac{B_1}{1}+\frac{1}{(n-2)!}\frac{B_2}{2!}+\ldots+\frac{1}{1}\frac{B_{n-1}}{(n-1)!}=0 \qquad \ldots\ldots(2).$$

The first few of Bernoulli's numbers (which do not conform to any apparent law of formation) have the numerical values

$$B_2=\tfrac{1}{6},\quad B_3=B_5=B_7=\ldots=B_{2n+1}=0,\quad B_4=-\tfrac{1}{30},$$
$$B_6=\tfrac{1}{42},\quad B_8=-\tfrac{1}{30},\quad B_{10}=\tfrac{5}{66},\quad B_{12}=-\tfrac{691}{2730},\quad B_{14}=\tfrac{7}{6}.$$

Now
$$\frac{x}{e^x-1}+\frac{x}{2}=\frac{x}{2}\,\frac{e^{x/2}+e^{-x/2}}{e^{x/2}-e^{-x/2}}$$
$$=z\coth z$$

by writing $x=2z$. It follows that

$$z\cot z=1-\frac{B_2}{2!}(2z)^2+\frac{B_4}{4!}(2z)^4-\ldots \qquad \ldots\ldots\ldots(3).$$

Since $\tan x=\cot x-2\cot 2x$, we deduce that

$$\tan z=\sum_{r=1}^{\infty}(-)^{r-1}\frac{2^{2r}(2^{2r}-1)}{(2r)!}B_{2r}\,z^{2r-1} \qquad \ldots\ldots\ldots(4);$$

and with the help of the formula

$$\cot x+\tan\tfrac{1}{2}x=1/\sin x$$

we get
$$\frac{z}{\sin z}=\sum_{r=0}^{\infty}(-)^{r-1}\frac{(2^{2r}-2)}{(2r)!}B_{2r}\,z^{2r} \qquad \ldots\ldots\ldots\ldots\ldots(5).$$

The remaining useful series for $\sec z$ is usually written

$$\sec z=\sum_{r=0}^{\infty}(-)^r\frac{E_{2r}}{(2r)!}z^{2r} \qquad \ldots\ldots\ldots\ldots\ldots\ldots(6),$$

where the coefficients E_n, usually called *Euler's numbers*, can be found from the identity

$$\left(1-\frac{x^2}{2!}+\frac{x^4}{4!}-\ldots\right)\left(E_0-\frac{E_2}{2!}x^2+\frac{E_4}{4!}x^4-\ldots\right)\equiv 1.$$

The first few of Euler's numbers are

$$E_0=1,\quad E_2=-1,\quad E_4=5,\quad E_6=-61,\quad E_8=1385$$

and
$$E_1=E_3=E_5=\ldots=E_{2n+1}=\ldots=0.$$

13·9. Illustrative example.

In order to illustrate the procedure which is adopted to obtain power series from other known power series by means of the operations of term-by-term derivation and integration, the following example is given here.

Prove that

$$\tfrac{1}{2}\{\log(1-x)\}^2 = \tfrac{1}{2}x^2 + \tfrac{1}{3}(1+\tfrac{1}{2})x^3 + \tfrac{1}{4}(1+\tfrac{1}{2}+\tfrac{1}{3})x^4 + \ldots,$$

and that the result remains true when $x = -1$.

If $|x| < 1$ the two series

$$(1-x)^{-1} = 1 + x + x^2 + x^3 + \ldots$$

$$-\log(1-x) = x + \frac{x^2}{2} + \frac{x^3}{3} + \ldots$$

are absolutely convergent, and so, by § 5·71, their formal product is absolutely convergent in the same range and its sum is $-\dfrac{\log(1-x)}{1-x}$.

Thus, when $|x| < 1$,

$$-\frac{\log(1-x)}{1-x} = x + x^2(1+\tfrac{1}{2}) + x^3(1+\tfrac{1}{2}+\tfrac{1}{3}) + \ldots,$$

and on integrating from 0 to x, where $|x| < 1$, we get

$$\tfrac{1}{2}\{\log(1-x)\}^2 = \frac{x^2}{2} + \frac{x^3}{3}(1+\tfrac{1}{2}) + \frac{x^4}{4}(1+\tfrac{1}{2}+\tfrac{1}{3}) + \ldots \quad \ldots(1),$$

integration term by term being valid over any range strictly within the range of absolute convergence.

The result can only remain true when $x = -1$ if the series

$$\tfrac{1}{2} - \tfrac{1}{3}(1+\tfrac{1}{2}) + \tfrac{1}{4}(1+\tfrac{1}{2}+\tfrac{1}{3}) - \tfrac{1}{5}(1+\tfrac{1}{2}+\tfrac{1}{3}+\tfrac{1}{4}) + \ldots \quad \ldots(2)$$

is convergent. The series clearly cannot converge absolutely, for when all the terms of the series (2) are taken positively

$$u_n = \frac{1}{n+1}\left(1 + \frac{1}{2} + \frac{1}{3} + \ldots + \frac{1}{n}\right) > \frac{1}{n+1},$$

and so each term of the series exceeds the corresponding term of the series $\Sigma\, 1/(n+1)$ which is known to be divergent.

Now $u_n > u_{n+1}$

if $$\frac{1}{n+1}\left(1+\frac{1}{2}+\dots+\frac{1}{n}\right) > \frac{1}{n+2}\left(1+\frac{1}{2}+\dots+\frac{1}{n+1}\right),$$

that is if

$$\left(\frac{1}{n+1}-\frac{1}{n+2}\right)\left(1+\frac{1}{2}+\dots+\frac{1}{n}\right) > \frac{1}{(n+1)(n+2)},$$

which is true, because if $n > 1$,

$$1+\frac{1}{2}+\dots+\frac{1}{n} > 1$$

Further $u_n \to 0$ as $n \to \infty$, for

$$u_n = \frac{1+\frac{1}{2}+\dots+\frac{1}{n}}{n} \cdot \frac{n}{n+1} = \frac{1+\frac{1}{2}+\dots+\frac{1}{n}}{n} \cdot \frac{1}{1+\frac{1}{n}},$$

hence $u_n \to 0$ for $\dfrac{1+\frac{1}{2}+\dots+\frac{1}{n}}{n}$ tends to the same limit as $\frac{1}{n}$ (§ 2·6).

The alternating series test therefore applies, and so the series (2) is conditionally convergent, and its sum is $\frac{1}{2}(\log 2)^2$.

EXAMPLES XIII.

1. Discuss the uniformity of convergence of the sequences

(i) $s_n(x) = \text{arc}\tan nx$, (ii) $s_n(x) = x^n/(1+x^{2n})$.

2. Prove that the series

$$x^4 + \frac{x^4}{1+x^4} + \frac{x^4}{(1+x^4)^2} + \dots$$

converges in the interval $(0, k)$, where k is a positive constant, for all values of x, but that the series is not uniformly convergent in this interval.

3. Prove that the series

$$\frac{x}{x+1} + \frac{x}{(x+1)(2x+1)} + \frac{x}{(2x+1)(3x+1)} + \dots$$

is non-uniformly convergent near the point $x=0$.

4. Prove that the series

$$\frac{1}{1+x^2} - \frac{1}{2+x^2} + \frac{1}{3+x^2} - \dots$$

is uniformly convergent in the interval $x \geqslant 0$, but that it is not absolutely convergent.

5. Prove that the series

$$(1-x)+x(1-x)+x^2(1-x)+\dots$$

is absolutely convergent in the interval $-c \leqslant x \leqslant 1$, where $0<c<1$, but that it is not uniformly convergent in this interval.

6. (i) Prove that the series $\sum_1^\infty \frac{1}{n^3+n^4x^2}$ is uniformly convergent for all values of x, and that it may be differentiated term by term.

(ii) Examine for uniformity of convergence the series

$$\sum_1^\infty \frac{x}{n(1+nx^2)}.$$

7. Justify the equation

$$\frac{1}{m}-\frac{1}{m+n}+\frac{1}{m+2n}-\frac{1}{m+3n}+\dots=\int_0^1 \frac{t^{m-1}\,dt}{1+t^n}$$

when $m>0$, $n>0$.

Deduce the values of the sums of the series

$$\tfrac{1}{2}-\tfrac{1}{5}+\tfrac{1}{8}-\tfrac{1}{11}+\dots,$$
$$1-\tfrac{1}{5}+\tfrac{1}{9}-\tfrac{1}{13}+\dots.$$

8. (i) The series $\Sigma a_n x^n$ has radius of convergence R and the series $\Sigma b_n x^n$ has radius of convergence R': determine the radii of convergence of the series

$$\Sigma(a_n \pm b_n)x^n, \quad \Sigma a_n b_n x^n, \quad \Sigma(a_n/b_n)x^n.$$

(ii) Find the radii of convergence of the series

$$\Sigma\frac{a_n x^n}{n^p}, \quad \Sigma\frac{a_n x^n}{n!}, \quad \Sigma a_n n!\,x^n,$$

if the radius of convergence of $\Sigma a_n x^n$ is $R>0$.

9. Find the expansions in power series for arc sin x and arc tan x, specifying the range of values of x for which they are valid.

10. Prove that, if $|x|<1$,

$$\frac{\text{arc sin } x}{\sqrt{(1-x^2)}}=x+\frac{2}{3}x^3+\frac{2\,.\,4}{3\,.\,5}x^5+\frac{2\,.\,4\,.\,6}{3\,.\,5\,.\,7}x^7+\dots;$$

and deduce that $\quad \tfrac{1}{2}(\text{arc sin } x)^2=\frac{x^2}{2}+\frac{2}{3}\frac{x^4}{4}+\frac{2\,.\,4}{3\,.\,5}\frac{x^6}{6}+\dots.$

Examine whether the last equation remains true when $x=1$.

11. Prove that, if $|x|<1$,

$$\frac{(\text{arc tan } x)^2}{2!}=\frac{x^2}{2}-\left(1+\frac{1}{3}\right)\frac{x^4}{4}+\left(1+\frac{1}{3}+\frac{1}{5}\right)\frac{x^6}{6}-\left(1+\frac{1}{3}+\frac{1}{5}+\frac{1}{7}\right)\frac{x^8}{8}+\dots.$$

12. If $|x|<1$, prove that

$$\int_0^x \text{arc tan } x\,dx=\frac{x^2}{1\,.\,2}-\frac{x^4}{3\,.\,4}+\frac{x^6}{5\,.\,6}-\dots,$$

and deduce that the sum of the series

$$1-\tfrac{1}{2}-\tfrac{1}{3}+\tfrac{1}{4}+\tfrac{1}{5}-\dots$$

is 0·43882 approximately.

13. Prove that

$$\frac{-x}{\log(1-x)} = \int_0^1 (1-x)^t\,dt = 1 - \tfrac{1}{2}x - \tfrac{1}{12}x^2 - \tfrac{1}{24}x^3 - \dots.$$

14. Shew that, if $|x| < 1$,

$$\sqrt{(1+x^2)}\log\{x+\sqrt{(1+x^2)}\} = x + \frac{x^3}{3} - \frac{2}{3}\frac{x^5}{5} + \frac{2 \,.\, 4}{3 \,.\, 5}\frac{x^7}{7} - \dots,$$

and that the result remains true when $x = \pm 1$.

15. Prove that, if $|x| < 1$,

$$\log[\tfrac{1}{2}\{1+\sqrt{(1+x)}\}] = \frac{1}{2}\frac{x}{2} - \frac{1 \,.\, 3}{2 \,.\, 4}\frac{x^2}{4} + \frac{1 \,.\, 3 \,.\, 5}{2 \,.\, 4 \,.\, 6}\frac{x^3}{6} - \dots.$$

Is it allowable to put $x = 1$ or $x = -1$?

16. Examine the convergence, whether it is uniform, and the validity of term-by-term derivation of the series

$$e^x \sin x + e^{2x}\sin 2x + \dots + e^{nx}\sin nx + \dots.$$

17. If $s_n(x) = 2n^2 x e^{-n^2x^2}$, shew that

$$\int_0^x \lim_{n\to\infty} s_n(x)\,dx \neq \lim_{n\to\infty}\int_0^x s_n(x)\,dx,$$

and give reasons why this is so. [See § 13·32.]

18. Find the sums of the following series in terms of elementary functions:

(i) $\dfrac{1}{2} + \dfrac{x}{5} + \dfrac{x^2}{8} + \dfrac{x^3}{11} + \dots,$

(ii) $\dfrac{x^3}{1 \,.\, 3} - \dfrac{x^5}{3 \,.\, 5} + \dfrac{x^7}{5 \,.\, 7} - \dfrac{x^9}{7 \,.\, 9} + \dots,$

(iii) $x + \dfrac{x^3}{3} - \dfrac{x^5}{5} - \dfrac{x^7}{7} + \dots$

[To prove (i), observe that if $f(x)$ is the required function

$$\frac{d}{dx}\{x^2 f(x^3)\} = \frac{x}{1-x^3},$$

whence $f(x)$ may be found.]

19. Find the expansions in power series of the functions

$$\text{(i)}\ \log\frac{\sin x}{x}, \quad \text{(ii)}\ \frac{x}{e^x+1}.$$

20. Prove that the converse of Abel's theorem on power series, (which is not *in general* true), holds if the coefficients a_n are all non-negative: that is, if $a_n \geqslant 0$ for all n, and if

$$\lim_{x\to 1-0} \Sigma a_n x^n$$

exists, then Σa_n converges, and its sum is equal to that limit.

MISCELLANEOUS EXAMPLES

1. If $\{t_n\}$ is a sequence of positive numbers such that, when $a>0$,

$$2at_n = t_{n-1}^2 + a^2,$$

prove that the sequence is non-decreasing and that $t_n \to \infty$ or $t_n \to a$ according as $t_1 > a$ or $t_1 \leqslant a$.

2. If a sequence $\{s_n\}$ be defined by $s_0 = 1$, $s_1 = s_2 = 0$ and

$$2(s_{n+3} - s_{n+1}) = s_n - s_{n+2} \quad \text{for all} \quad n \geqslant 0,$$

find the upper and lower bounds and the upper and lower limits for $\{s_n\}$.

3. If the sequence $\{s_n\}$ satisfies the relation

$$s_n \cdot s_{n+1} - h(s_n - s_{n+1}) = k^2,$$

prove that $s_n \to \pm k$ according as $h/k > 0$ or $h/k < 0$.

4. Prove that, if $k>0$,

$$\lim_{n\to\infty} \frac{1^k + 2^k + \ldots + n^k}{n^{k+1}} = \frac{1}{k+1}.$$

Hence evaluate
$$\lim_{n\to\infty} \frac{F(n)}{(n+1)^{2k+2}},$$

where $F(n)$ is the coefficient of x^{n-2} in $(x-1^k)(x-2^k)\ldots(x-n^k)$.

5. Evaluate the limits as $n \to \infty$ of

(i) $$\frac{1}{\sqrt{(n^2+1)}} + \frac{1}{\sqrt{(n^2+2)}} + \ldots + \frac{1}{\sqrt{(n^2+n)}};$$

(ii) $$\frac{1}{n}\{(n+1)(n+2)\ldots(n+n)\}^{1/n}.$$

6. If $f(x, y) = y\sqrt{(x^2+y^2)}$;

$$\phi(x, y) = \frac{x^2+y^2}{x-y} \quad \text{if} \quad x \neq y, \qquad \phi(x, y) = 0, \quad \text{if} \quad x = y;$$

discuss whether at $(0, 0)$ these functions are (*a*) continuous, (*b*) differentiable, and (*c*) possess first-order partial derivatives.

7. Discuss the continuity at the origin of the functions:

(i) $$f(x, y) = \frac{x^2y}{x^2+y^2} \quad \text{if} \quad (x, y) \neq (0, 0); \qquad f(0, 0) = 0;$$

(ii) $$\phi(x, y) = \frac{x^2y}{x^4+y^2} \quad \text{if} \quad (x, y) \neq (0, 0); \qquad \phi(0, 0) = 0.$$

8. Discuss the continuity at $x=1$ of the functions:

(i) $f(x)=\lim\limits_{n\to\infty}\dfrac{x+x^n \sin x}{1+x^n}$;

(ii) $\phi(x)=\lim\limits_{n\to\infty}\dfrac{x+x^n}{1+ax^n}$.

Illustrate graphically.

9. Shew that $f'(x)$ vanishes only once in the interval $0<x<\frac{1}{2}$, where

$$f(x)=\pi x(1-x)\cos \pi x-(1-2x)\sin \pi x.$$

Deduce that $\dfrac{\sin \pi x}{x(1-x)}$ steadily increases as x increases from 0 to $\frac{1}{2}$.

10. If $f(a)=f(b)=0$ and $f(x)$ is twice differentiable, show that the function

$$\phi(t)=(t-a)(t-b)f(x)-(x-a)(x-b)f(t)$$

is such that $\phi''(t)$ vanishes at least once in $a<x<b$.

Deduce that, if $f''(x)\neq 0$ in $a<x<b$, then $f(x)\neq 0$ in that interval.

11. Determine constants α, β, γ so that the function

$$\phi(x)=\alpha+\beta(x-1)+\gamma(x-1)(x-2)$$

equals $f(0), f(1), f(2)$ when $x=0, 1, 2$ respectively.

Deduce that

$$f(x)=\phi(x)+x(x-1)(x-2)f'''(c)/3!,$$

where c lies between the greatest and the least of $0, 1, 2, x$.

12. Find the radius of convergence of each of the series for $\alpha>0$,

(i) $\sum\limits_{0}^{\infty}(\cosh n\alpha)x^n$; (ii) $\sum\limits_{2}^{\infty}\left(\dfrac{1}{\sqrt{(n-1)}}-\dfrac{1}{\sqrt{n}}\right)n^\alpha x^n$.

Shew that the second series is absolutely convergent for $x=1$ when $\alpha<\frac{1}{2}$, and conditionally convergent for $x=-1$ when $\frac{1}{2}\leqslant\alpha<\frac{3}{2}$.

13. If $u_n=(-1)^n/\sqrt{n}$ and $v_n=(-1)^n/\{\sqrt{n}+(-1)^n\log n\}$, prove that $u_n/v_n\to 1$ as $n\to\infty$, but that Σu_n is convergent and Σv_n is divergent.

14. Prove that $2\sqrt{n}-\sum\limits_{r=1}^{n}\dfrac{1}{\sqrt{r}}\to$ finite limit s as $n\to\infty$.

Hence prove that the series

$$1+\frac{1}{\sqrt{2}}-\frac{2}{\sqrt{3}}+\frac{1}{\sqrt{5}}-\frac{2}{\sqrt{6}}+\dots$$

has sum $(\sqrt{3}-1)s$.

15. If α be real, test the convergence of the series:

(i) $\Sigma\dfrac{(2n-1)!!}{(2n)!!}n^\alpha x^n$; (ii) $\Sigma\left\{\dfrac{(2n-1)!!}{(2n)!!}\right\}^\alpha x^n$.

16. If $f(x)>0$, $f''(x)\geqslant 0$ and $\phi(x)=e^x f(e^{-2x})$, verify that $\phi''(x)>0$ and hence obtain the inequality

$$2f(e^{-2x})\leqslant e^{-x}f(1)+e^x f(e^{-4x}).$$

17. Express in terms of $\Gamma(\frac{1}{3})$:

(i) $\displaystyle\int_0^1 \frac{dx}{(1-x^3)^{\frac{1}{3}}}$; (ii) $\displaystyle\int_0^{\frac{1}{2}\pi} \sin^{\frac{1}{3}}\phi\, d\phi$; (iii) $\displaystyle\int_0^\infty \frac{dx}{(9x+x^3)^{\frac{2}{3}}}$.

18. Prove that
$$\int_0^1 \left(\frac{1+x^a}{1-x^a}\right)^{\frac{1}{2}} dx=\frac{\pi^{\frac{1}{2}}}{2a}\left(G+\frac{2a}{G}\right),$$
where $a>0$ and $G=\Gamma\left(\frac{1}{2a}\right)\Big/\Gamma\left(\frac{1}{2a}+\frac{1}{2}\right)$.

19. If
$$F(n)=\int_{-1}^1 x^2 P_n(x)\,dx,$$
prove that $(2n+1)F(n)=nF(n-1)$.

Evaluate
$$\int_{-1}^1 (1-x^2)\,P_n{}^2(x)\,dx.$$

20. If
$$L_n(x)=e^x\frac{d^n(x^n e^{-x})}{dx^n}$$
for $n\geqslant 0$, shew that
$$\int_0^\infty L_m(x).L_n(x)\,e^{-x}dx=0 \quad \text{if} \quad n>m\geqslant 0,$$
$$=(n!)^2 \quad \text{if} \quad m=n.$$

21. Sketch the curve $r^4=a^4\sin 4\theta$ and shew that the length of the arc of one loop is $a\{\Gamma(\frac{1}{8})\}^2/2^{11/4}\,\Gamma(\frac{1}{4})$.

22. If u, v, w are differentiable functions of t such that $u^2+v^2+w^2=0$ and t is a differentiable function of the independent variables x, y, z satisfying $t=xu+yv+zw$, prove that
$$\frac{\partial^2 t}{\partial x^2}+\frac{\partial^2 t}{\partial y^2}+\frac{\partial^2 t}{\partial z^2}=0.$$

23. If $f(x, y, z)=0$ is a homogeneous function of degree n with z as the dependent variable, prove that

(i) $\displaystyle x^2\frac{\partial^2 z}{\partial x^2}=-xy\frac{\partial^2 z}{\partial x\partial y}=y^2\frac{\partial^2 z}{\partial y^2}$,

(ii) $\displaystyle x^3\frac{\partial^3 z}{\partial x^3}-y^3\frac{\partial^3 z}{\partial y^3}=xy\left(y\frac{\partial^3 z}{\partial x\partial y^2}-x\frac{\partial^3 z}{\partial^2 x\partial y}\right)$.

24. If $u=f(x, y, z)$ and $v=g(x, y, z)$ where z is defined by $h(x, y, z)=0$, prove that
$$\frac{\partial h}{\partial z}\cdot\frac{\partial(u, v)}{\partial(x, y)}=\frac{\partial(f, g, h)}{\partial(x, y, z)}.$$

If $u=x/z^2$, $v=y/z^2$ and $x^4+y^4+z^4=1$, prove that
$$\frac{\partial(u, v)}{\partial(x, y)}=\frac{2}{z^8}-\frac{1}{z^4}.$$

25. Shew that $u=xyz$ has fourteen stationary values on the sphere

$$x^2+y^2+z^2=a^2$$

of which one is $u=a^3/3\sqrt{3}$. Determine whether this stationary value is a maximum or minimum.

26. If $l, m, n, p>0$, prove that the extreme value of $u=lx+my+nz$ subject to the condition

$$x^p+y^p+z^p=c^p$$

is $c\,(l^q+m^q+n^q)^{1/q}$ where $q=p/(p-1)$. Shew that this value is a maximum if $p>1$ and a minimum if $p<1$.

27. If $0<\alpha<\frac{1}{2}\pi$ and the domain D is the interior of the circle $x^2+y^2=a^2$, prove that

$$\iint_D \frac{dx\,dy}{(a\sec\alpha-x)^2}=2\pi\,(\operatorname{cosec}\alpha-1).$$

28. Prove that the volume common to the two cylinders $x^2+z^2=a^2$ and $y^2+z^2=a^2$ is $16a^3/3$. Find also the volume common to these cylinders and the cylinder $x^2+y^2=a^2$.

29. A vessel is in the shape of the part of the hyperboloid $x^2+y^2-3z^2=b^2$ which lies between the planes $z=0$ and $z=b$. Calculate the volume and the curved surface area of the vessel.

30. If C consists of the two straight lines LM and MN where $L\,(0, 1)$, $M\,(3, 1)$ and $N\,(3, 4)$, prove that

$$\int_C \left\{\left(1+\frac{1}{x+y^2}\right)dx+\left(2y-\frac{2x}{y\,(x+y^2)}\right)dy\right\}=18+\log\tfrac{19}{16}.$$

Shew that this curvilinear integral has the same value if C is any simple curve joining L to N.

If the integrand is $d\phi$, determine the function ϕ and verify the above result.

31. Evaluate by Green's Theorem, or otherwise,

$$\iint\left\{\frac{x}{a^2}(x^2+yz)\,dy\,dz+\frac{y}{b^2}(y^2+xz)\,dx\,dz+\frac{z}{c^2}(z^2+xy)\,dx\,dy\right\}$$

over the outside of the ellipsoid

$$\frac{x^2}{a^2}+\frac{y^2}{b^2}+\frac{z^2}{c^2}=1.$$

32. If $0<\beta<\frac{1}{4}\pi$, prove that the volume common to the sphere

$$x^2+y^2+z^2-\sqrt{2}\,ay-\sqrt{2}\,az=0$$

and the cone $z^2\tan^2\beta=x^2+y^2$ is

$$\frac{\sqrt{2}}{6}\pi a^3\sin^2\beta\,(5-\cos^2\beta).$$

33. Shew that

$$\iiint\frac{1}{z^3}\left(\frac{x^2}{a^2}+\frac{y^2}{b^2}+\frac{z^2}{c^2}\right)^{\frac{1}{2}}dx\,dy\,dz,$$

taken over the region common to the ellipsoid $x^2/a^2+y^2/b^2+z^2/c^2=1$ and the part of the cone $\alpha^2x^2+\beta^2y^2=z^2$ for which $z>0$, has the value $\pi/\alpha\beta$.

34. If $a^2 > b^2 > c^2$, shew that the volume common to the ellipsoids

$$x^2/a^2 + y^2/b^2 + z^2/c^2 = 1, \quad x^2/c^2 + y^2/b^2 + z^2/a^2 = 1$$

is $\frac{16}{3}abc \tan^{-1} c/a$.

35. Discuss the uniform convergence in $x \geqslant 0$ of the series

$$\sum_{n=1}^{\infty} \frac{x}{(1+x)^{n-1}}, \quad \sum_{n=1}^{\infty} \frac{x^2}{(1+x^2)^{n-1}}, \quad \sum_{n=1}^{\infty} \frac{x^3}{(1+x^2)^{n-1}}.$$

36. If $s(x)$ is the sum of the power series

$$x + \frac{x^3}{3} - \frac{x^4}{4} - \frac{x^6}{6} + \frac{x^7}{7} + \frac{x^9}{9} - \frac{x^{10}}{10} - \frac{x^{12}}{12} + \dots$$

when it is convergent, find $s(x)$ in terms of elementary functions.

37. Shew that the function

$$f(x) = \sum_{n=1}^{\infty} \frac{1}{n^2 + n^{\alpha} x^2}$$

is uniformly convergent for all values of x and that term-by-term differentiation is valid when $\alpha < 4$.

If $\alpha = 4$, show that

$$\frac{f(0) - f(x)}{x} > \int_1^{\infty} \frac{x\,dt}{1 + x^2 t^2},$$

when $x > 0$. Hence deduce that $f'(0)$ does not exist.

38. Shew that, if $p > 0$, the series

$$(1-x)^p + 2x(1-x)^p + \dots + nx^{n-1}(1-x)^p + \dots$$

is convergent when $0 \leqslant x \leqslant 1$, but that it is not uniformly convergent in this range unless $p > 2$.

39. If

$$\phi(x) = \sum_{n=2}^{\infty} \frac{x^n}{n(n-1)},$$

prove that, when $|x| < 1$, $\phi(x) = (1-x)\log(1-x) + x$. Does the result still hold when $x = -1$?

40. Shew that

$$f(x) = \sum_{n=1}^{\infty} \frac{1}{n^2} \log(1 + nx^2)$$

can be differentiated term-by-term for all values of x.

ANSWERS

EXAMPLES I

5. Parabola $x^2=4y$. 7. (i) Appolonius' circle for points α and β. (ii) Circle on line α and β as chord. 8. Perpendicular bisector of line α to $-\beta/\gamma$.

9. $\Pi\left\{z-\rho^{1/5}\exp\left(\dfrac{i\phi+2n\pi i}{5}\right)\right\}$ $(n=0, 1, 2, 3, 4)$ where $a+ib=\rho e^{i\phi}$.

EXAMPLES II

1. (a) $m=\frac{1}{2}$, $M=1$, $\lambda=\Lambda=1$; (b) $m=1$, $M=2$, $\lambda=\Lambda=l=\frac{3}{2}$; (c) $m=0=\lambda$, $M=3=\Lambda$; (d) $m=1$, $M=\sqrt[3]{3}$, $\lambda=\Lambda=1$; (e) $m=-2$, $M=7$, $\lambda=1$, $\Lambda=4$.

2. $s=b$.

EXAMPLES III

3. (i) $3\frac{1}{2}$; (ii) 4; (iii) $\frac{1}{2}$; (iv) 0; (v) 1. 4. Yes. 6. Yes. 7. $1-1/(nx+1)$. 9. Continuous. 10. (i) Discontinuous; (ii) continuous. 14. $f(x)$ not continuous at $x=1$. 16. (i) Removable discontinuity at $x=0$; (ii) discontinuity of second kind at $x=a$; (iii) discontinuity of first kind at $x=0$; (iv) discontinuities of first kind at $x=n+\frac{1}{2}$. 24. Upper bound $1+2/\epsilon$.

EXAMPLES IV

4. (i) R.H. $=\frac{1}{2}\pi$, L.H. $=-\frac{1}{2}\pi$; (ii) R.H. $=0$, L.H. $=1$. 7. Discontinuity at $x=1$. 12. (i) 1; (ii) 1; (iii) 1; (iv) $2a/b$. 13. (i) ∞ if $n>1$; (ii) 1; (iii) 1; (iv) $e^{-\frac{1}{6}}$; (v) $-1/a$; (vi) $e^{-\frac{1}{2}\theta^2}$; (vii) $-\frac{1}{3}$; (viii) $2a/3$; (ix) $(abc)^{\frac{1}{3}}$. 15. $\left(\dfrac{\pi^2}{4}+\dfrac{1}{n^2}\right)\Big/4n^2$.

18. $1/b$. 21. Minimum (0, 0), maximum $\left(\dfrac{2}{5}, \dfrac{2^2.3^3}{5^5}\right)$. 22. (ii) No extreme for $\phi(x)$, minimum for $\psi(x)$.

EXAMPLES V

(C. = convergent, D. = divergent, A.C. = absolutely convergent, C.C. = conditionally convergent.)

1. (i) D.; (ii) C.; (iii) D. $s\leqslant 1$, C. $s>1$; (iv) D.; (v) D.; (vi) D. $(a\neq 1)$; (vii) D. $t\leqslant 1$, C. $t>1$; (viii) C. $s>1$, D. $s<1$, $s=1$, same as (vii); (ix) C.; (x) C. $x<1$, D. $x>1$, $x=1$ D. with +, C. with − sign. 3. (i) D.; (ii) C. 5. (i) A.C.; (ii) C.C. 6. (ii) $R=1$, C. for $x=1$ if $b-a>c$; C.C. for $x=-1$ if $0<b-a\leqslant c$. 7. (i) $R=1$, A.C. for $x=1$ if $s>1$, C.C. for $x=-1$ if $0<s\leqslant 1$.

9. $R=1$, A.C. for $x=1$ if $c>a+b$; C.C. for $x=-1$ if $c\leqslant a+b<c+1$.

12. $1+\sum_{1}^{\infty}(-1)^m\{\pi^n(x-\frac{1}{4})^n/n!\}$ where $m=[n/2]$. 13. $\dfrac{\pi}{4}-\sum_{0}^{\infty}\dfrac{(-1)^n}{(2n+1}\left(\dfrac{x}{a}\right)^{2n+1}$.

14. $x+\sum_{1}^{\infty}4^r(r!)^2x^{2r+1}/(2r+1)!$.

Examples VI

2. (i) Concave for $t>0$; (ii) concave for $t>0$; (iii) convex for $t>0$; (iv) convex for $t>0$; (v) convex for all t; (vi) convex except at $t=0$; (vii) convex.

9. (i) Converges if $p>1$, diverges $p\leqslant 1$; (ii) converges for all α if $p>1$, converges for $\alpha>1$ if $p=1$, diverges otherwise.

Examples VII

11. (i) $\frac{1}{6}\log\{(x-1)/(x+1)\}+\dfrac{\sqrt{2}}{3}\tan^{-1}x/\sqrt{2}$;

(ii) $\frac{1}{12}\log\dfrac{(1+x)^3(1-x^3)}{(1-x)^3(1+x^3)}+\dfrac{1}{2\sqrt{3}}\tan^{-1}\dfrac{x\sqrt{3}}{1-x^2}$;

(iii) $\frac{1}{3}\log\{x^3/(x^3+1)\}+(2x^3+3)/6(x^3+1)^2$; (iv) $6t^4\left(\dfrac{t^5}{9}+\dfrac{t^4}{8}+\dfrac{t^3}{7}+\dfrac{t^2}{6}+\dfrac{t}{5}+\dfrac{1}{4}\right)$,

where $t^6=1+x$; (v) $-(2x^5+x^4+8x^2-2x-9)/18(x^3+1)^3-\dfrac{2}{9\sqrt{3}}\tan^{-1}\dfrac{2x-1}{\sqrt{3}}$; (vi) $\frac{3}{56}(4x^2-3)(1+x^2)^{4/3}$.

12. $u/3(u^3-1)-\frac{1}{18}\log\{(u-1)^2/(u^2+u+1)\}+\dfrac{1}{3\sqrt{3}}\tan^{-1}\dfrac{2u+1}{\sqrt{3}}$, where $u^3-1=1/x^3$.

13. (ii) $-\frac{1}{2}\pi\log 2$.

Examples VIII

1. No, yes, no, yes.

3. (i) $2n(n+1)/(2n+1)$;
(ii) $2/(2n+1)-\{2(n+1)^2/(2n+3)+2n^2/(2n-1)\}/(2n+1)^2$.

11. (i) $c\sinh 1$. 13. (i) $92a^3/15$; (ii) $\frac{3}{2}\log\left(\dfrac{a^3}{2}+1\right)$.

14. $\frac{1}{2}\log(x^2+y^2)+\tan^{-1}x/y$; yes.

Examples IX

7. $2\left(\dfrac{\partial^2 z}{\partial u^2}+\dfrac{\partial^2 z}{\partial v^2}\right)=0$. 14. $\theta=\frac{11}{18}$.

Examples X

1. $(1-u^2v^{u-1})/\{(1+v^2u^{v-1})(1-u^2v^{u-1})+u^v v^u(1+v\log u)(1+u\log v)\}$.

5. (i) $\left(\dfrac{\partial u}{\partial x}\right)_y=\cos x.\cosh y$; $\left(\dfrac{\partial u}{\partial x}\right)_z=\cos x\cosh y-\sin x.\sinh y/(x+5y)$;

(ii) $\left(\dfrac{\partial x}{\partial z}\right)_u=1/2u^3v$; $\left(\dfrac{\partial x}{\partial z}\right)_v=-2/3uv^2$; $\left(\dfrac{\partial x}{\partial z}\right)_y=(4uv+1)/2u^2v(u-3v^2)$.

$\left(\dfrac{\partial y}{\partial z}\right)_u=\dfrac{1}{u^3}$; $\left(\dfrac{\partial y}{\partial z}\right)_v=\dfrac{1}{3u^2v^2}$; $\left(\dfrac{\partial y}{\partial z}\right)_x=\dfrac{1+4uv}{u^2v(3v+4u^2)}$.

12. (i) $\frac{\partial^2 z}{\partial u^2}+\frac{\partial^2 z}{\partial v^2}+\left(\frac{\partial z}{\partial u}\right)^2\cdot\frac{\partial^2 z}{\partial v^2}+\left(\frac{\partial z}{\partial v}\right)^2\cdot\frac{\partial^2 z}{\partial u^2}-2\frac{\partial z}{\partial u}\cdot\frac{\partial z}{\partial v}\cdot\frac{\partial^2 z}{\partial u\,\partial v}$;

(ii) $a^2\left(\frac{\partial z}{\partial u}\right)^2+\{(a^2-u^2)/u^2\}\left(\frac{\partial z}{\partial v}\right)^2$.

17. (i) Maximum at (1, 0), minimum at (7, 0); (ii) maximum at $x=\pm\sqrt{2}$; (iii) maximum at (0, 0).
18. Stationary at $(0, 0, \pm\frac{1}{2})$. 20. $\{2a-\sqrt{(a^2+b^2+c^2)}\}/\sqrt{3}$ if $a>b>c$.
22. $u=0$ and roots of $l^2/(u-a^2)+m^2/(u-b^2)+n^2/(u-c^2)=0$, $(u=r^2)$.
24. $abc+2fgh-af^2-bg^2-ch^2=0$. 28. (i) $u^2+v^2=2w$.

EXAMPLES XI

3. $(51\sqrt{2}/4-\frac{1}{8}\cosh^{-1}3)/a^3$. 7. (i) $\frac{2}{3}a^3$; (ii) $\frac{1}{3}a^2b$; (iii) $-\frac{1}{4}\pi a^4$.
8. $\frac{4}{3}ab$. 11. $\frac{1}{12}B(\frac{7}{3},\frac{3}{2})$. 12. $\frac{2}{7}abB(\frac{1}{10},\frac{1}{4})$.
14. $\frac{3}{8}\pi a^3$. 15. $2\log 2-\frac{1}{2}$. 17. $\frac{1}{4}\pi ab\,(pa^2+qb^2)$.

19. (i) $\left(\frac{4a}{3\pi},\frac{4b}{3\pi}\right)$; (ii) $\left(\frac{8a}{15},\frac{8b}{15}\right)$; $M\frac{a^2}{4}$, $M\left(\frac{a^2+b^2}{4}\right)$; $M\frac{a^2}{3}$, $M\left(\frac{a^2+b^2}{3}\right)$, where M is mass of lamina.

EXAMPLES XII

2. $\frac{2}{3}\sqrt{2}$.
8. (i) $2\pi\left(a^2+\frac{ac}{\epsilon}\sin^{-1}\epsilon\right)$, if $a<c$, where $c\epsilon=\sqrt{(c^2-a^2)}$;

(ii) $2\pi\left\{a^2+(ac^2/\sqrt{[a^2-c^2]})\log\left(\frac{a}{c}+\frac{\sqrt{[a^2-c^2]}}{c}\right)\right\}$, if $a>c$.

11. (i) $\frac{3}{8}a\,(1+\cos\alpha)$; (ii) $\frac{2}{5}a\,(1+\cos\alpha)$. 12. $\frac{2}{5}Ma^2$.
14. $(8ab+3a^2)/(6\pi b+8a)$.

EXAMPLES XIII

1. (i) Non-U.C. at $x=0$; (ii) non-U.C. at $x=1$. 6. (ii) U.C. for all x.
7. (i) $\frac{1}{3}(\pi/\sqrt{3}-\log 2)$; (ii) $\{\pi+2\log(\sqrt{2}+1)\}/4\sqrt{2}$.
8. (i) (a) $\geqslant\min(R, R')$, (b) $\geqslant RR'$, (c) $\leqslant R/R'$; (ii) R; ∞; 0.

9. (i) $x+\frac{1}{2}\left(\frac{x^3}{3}\right)+\frac{1.3}{2.4}\left(\frac{x^5}{5}\right)+\ldots$ for $|x|\leqslant 1$; (ii) $x-\frac{1}{3}x^3+\frac{1}{5}x^5-\frac{1}{7}x^7+\ldots$ for $|x|\leqslant 1$.

15. Yes. 16. Converges if $x\leqslant 0$; U.C. for $x<0$.

18. (i) $x^{-2}\left\{\pi/6\sqrt{3}-\frac{1}{\sqrt{3}}\tan^{-1}\frac{2x+1}{\sqrt{3}}+\frac{1}{6}\log(x^2+x+1)/(1-x^2)\right\}$;

(ii) $\frac{1}{2}(x^2+1)\tan^{-1}x-\frac{1}{2}x$; (iii) $\frac{1}{\sqrt{2}}\tan^{-1}\{\sqrt{2}x/(1-x^2)\}$.

19. (i) $\sum_{n=1}^{\infty}\frac{(-1)^n(2x)^{2n}B_{2n}}{2n\,(2n)!}$; (ii) $\frac{x}{e^x+1}=\frac{x}{e^x-1}-\frac{2x}{e^{2x}-1}$ and use (1), p. 351.

Answers to the Miscellaneous Examples

2. $\lambda = -\frac{1}{3}$, $\Lambda = \frac{2}{3}$, $m = -\frac{1}{3}$, $M = 1$. 4. $\frac{1}{2}(k+1)^{-2}$.
5. (i) 1; (ii) $4/e$. 6. (i) Yes, no, yes; (ii) no, no, yes.
7. (i) Continuous; (ii) discontinuous.
8. (i) Discontinuous; (ii) continuous only if $a = 1$.
11. $\alpha = f(1)$, $\beta = f(2) - f(1)$, $\gamma = \frac{1}{2}\{f(0) - 2f(1) + f(2)\}$.
12. (i) $e^{-\alpha}$; (ii) 1.
15. (i) $R = 1$, A.C. if $|x| = 1$ when $\alpha < -\frac{1}{2}$, C.C. if $x = -1$ for $-\frac{1}{2} \leqslant \alpha < \frac{1}{2}$; (ii) $R = 1$, A.C. if $|x| = 1$ when $\alpha > 2$, C.C. if $x = -1$ for $0 < \alpha \leqslant 2$.
17. (i) $\{\Gamma(\frac{1}{3})\}^3/\pi . 3^{\frac{1}{2}} . 2^{\frac{4}{3}}$; (ii) $\pi^2 . 2^{\frac{4}{3}}/\{\Gamma(\frac{1}{3})\}^3$; (iii) $\{\Gamma(\frac{1}{3})\}^3/\pi . 2^{\frac{7}{3}}$.
19. $4(n^2 + n - 1)/(2n-1)(2n+1)(2n+3)$. 25. Maximum.
28. $8a^3(2 - \sqrt{2})$. 29. $V = 2\pi b^3$, $S = \pi b^2 \left\{\sqrt{13} + \frac{1}{2\sqrt{3}} \log(2\sqrt{3} + \sqrt{13})\right\}$.
30. $\phi = y^2 - 2\log y + \log(x + y^2) + x$. 31. $12\pi abc/5$.
36. $\frac{1}{6}\log(x+1)^3(x^3+1) + \frac{1}{\sqrt{3}}\tan^{-1}\frac{\sqrt{3}x}{2-x}$.

LIST OF AUTHORS QUOTED

[*The numbers refer to the pages*]

GENERAL INDEX

[*The numbers refer to the pages. Numbers in brackets refer to the examples*]

Printed in the United States
By Bookmasters